86 Topics in Current Chemistry

Fortschritte der Chemischen Forschung

Spectroscopy

Springer-Verlag
Berlin Heidelberg GmbH 1979

This series presents critical reviews of the present position and future trends in modern chemical research. It is addressed to all research and industrial chemists who wish to keep abreast of advances in their subject.

As a rule, contributions are specially commissioned. The editors and publishers will, however, always be pleased to receive suggestions and supplementary information. Papers are accepted for "Topics in Current Chemistry" in English.

ISBN 978-3-662-15718-3   ISBN 978-3-540-35219-8 (eBook)
DOI 10.1007/978-3-540-35219-8

Library of Congress Cataloging in Publication Data. Main entry under title: Spectroscopy. (Topics in current chemistry ; 86) Bibliography: p. Includes index. 1. Spectrum analysis – Addresses, essays, lectures. I. Series. QD1.F58. vol. 86 [QD95] 540'.8s. [543'.085]. 79-16331

© by Springer-Verlag Berlin Heidelberg 1979
Originally published by Springer-Verlag Berlin Heidelberg New York in 1979
Softcover reprint of the hardcover 1st edition 1979

2152/3140 – 543210

Contents

Photochemistry and Spectroscopy of Simple Polyatomic Molecules in the Vacuum Ultraviolet

M. N. R. Ashfold, M. T. Macpherson, and J. P. Simons

Department of Chemistry, The University, Birmingham B15 2TT, England

Table of Contents

1 Introduction

The distinction between the ultraviolet and the vacuum ultraviolet regions is essentially terrestrial, arising from the unique planetary atmosphere we still (thankfully) enjoy. This distinction has, however, a spectroscopic quality as well, since excitation in the vacuum ultraviolet explores the energy range from intravalency shell transitions to photoionization thresholds and beyond. Any attempt to describe the nature of the primary photochemical process must begin with the characterization of the excited state initially populated by light absorption, and in consequence spectroscopic considerations will be of central importance in this review. Transitions which lie at energies below the first ionization limit may be intravalence and/or Rydberg in nature, and this aspect of the characterization is a problem which exercises theoreticians, spectroscopists and photochemists alike. In considering photochemistry, the utility (and limitations) of electronic state correlation diagrams in rationalising experimentally observed branching ratios as a function of the exciting wavelength will be emphasised, but the more important information concerning the details of potential energy surfaces over which photodissociation proceeds is only just beginning to appear. The discussion will stop short of photoionization processes, except where these compete against dissociation to neutral products or help in the understanding of electronic transitions in the vacuum u.-v.

Apart from its intrinsic interest, the investigation of photochemical processes in the vacuum u.-v. is of much value in relation to other studies, for example, the determination of bond dissociation energies, the molecular dynamics of photodissociation, the kinetics of electronically excited atomic and molecular species, the photochemistry of planetary atmospheres, and the development of chemical laser and laser amplifier systems based on metastable products of vacuum u.-v. photodissociation. There is a great deal of current interest among plasma physicists in the use of rare gas-metastable atom exciplexes, derived from $O\,(^1S)$, $S\,(^1S)$ and $Se\,(^1S)$, as energy storage media in high-power gas laser amplifier systems. The variety of applications will become apparent as the review progresses. It is not intended to be a comprehensive catalogue, however, but rather an extended essay discussing in depth the spectroscopy and photochemistry of a representative sample of simple polyatomic molecules. We feel this is a more valuable exercise than the listing of a large number of systems at a relatively superficial level.

The review was completed in June 1978 and covers literature published up to March 1978; some earlier reviews and monographs are listed[1-25] to provide leading references.

2 The Nature of Molecular Electronic States Populated in the Vacuum Ultraviolet

The vacuum u.-v. region begins at photon energies around 600 kJ mol^{-1} (6 eV), sufficient to excite electronic transitions in all but the very simplest of molecules. The initial semantic problem to be overcome concerns the language used to define the "nature" of the photoexcited state. This may be based on an operational, experimental description which follows an analysis of the absorption spectrum, or it may be a theoretically based description which follows an analysis of the results of molecular orbital/configuration interaction calculations. In developing a proper understanding of the character of the photoexcited molecule, each approach borrows from the results of the other until the two types of description converge.

The most important operational description in the vacuum u.-v. region identifies "Rydberg" transitions on the basis of two criteria.

(i) Their membership of a well-characterised Rydberg series with term values

$$T_n = \text{I.P.} - \nu_n = \frac{R}{(n - \delta)^2} \tag{1}$$

where I.P. is a molecular ionization potential (normally obtained from the photoelectron spectrum), ν_n is the frequency of the absorbed photon, R is the Rydberg constant, n is an integer and δ is the quantum defect. The magnitude of δ reflects the degree to which the upper orbital penetrates the molecular core: increasing penetration increases the electron binding energy and hence the term value T_n. It also reflects departure from simple H-atom behaviour and varies with the "quantum number l". Typical values are δ $(ns) \sim 1$, δ $(np) \sim 0.6$, δ $(nd) \sim 0.1$ for Rydberg states when $l = 0,1,2$, respectively (but see later discussion).

(ii) Rydberg excited states are highly sensitive to external perturbation, e.g. collisional broadening under the influence of high pressures of inert diluent gases, because the large volume of the Rydberg orbitals allows substantial electron density in regions remote from the molecular core. Robin in particular has exploited the onset of pressure broadening as a diagnostic experimental method[1].

Theoretical descriptions measure the degree of Rydberg and/or intravalency shell character in the upper electronic state by

(i) assessing the effect of adding long-range Rydberg functions to the basis set,

(ii) considering the LCAO expansion of the excited molecular orbital and calculating its Coulomb integral (as a measure of diffuseness), and

(iii) comparing SCF/CI predictions for the electronic state energy with that observed experimentally[26–28].

However, as later discussions will indicate, the molecular orbital character may change dramatically with changes in the molecular geometry, particularly as the molecule distorts along the path to dissociation, so that an electronic state initially of Rydberg character may subsequently become largely intravalence. Such changes will affect the contours of the potential surface over which the molecule moves as it evolves from its initial photoexcited state into the separating molecular fragments. Unfortunately, detailed knowledge of the geography of the potential surface, which is so important in any discussion of the molecular dynamics of photodissociation, is almost entirely unknown, (except to the molecule, which enjoys the privilege of exploring that unfamiliar terrain).

When the upper states possess Rydberg character, the electronic transitions populate molecular orbitals which are so large and diffuse that to a first approximation they are centred on a molecular core which is "seen" as atom-like, so that the details of the molecular geometry exert only a minor perturbation on the orbital character. From the photochemist's point of view, the larger the volume of the Rydberg orbital, the less influence it will have on the molecular bonding; thus the photoexcited molecule will tend to an equilibrium geometry characteristic of the molecular ion. It might then be supposed that, unless the ion is unstable with respect to dissociation, the population of states with a high degree of Rydberg character would not be expected to lead to direct dissociation. In conformity with this, it is often found that Rydberg transitions exhibit sharp spectral features, especially when the originating orbital is non-bonding in character (e.g. the transition $5pe \rightarrow 6sa_1$ ($\widetilde{B}^1 E \leftarrow \widetilde{X}^1 A_1$) in CH_3I, largely localised on the I atom). In some instances, the sharp structure may even include resolvable rotational features, even though neighbouring transitions are wholly diffuse or at best have ill-defined vibrational band structure. A particularly dramatic illustration of this type of behaviour is provided by H_2O (and D_2O) (see Section 3.1); the first two electronic transitions each generate broad continua which reflect intravalence, antibonding character in the upper state, or the introduction of such character as the molecule moves on the upper potential surface, but another transition which is superimposed on the second continuum has a rotationally structured contour. The structure is sharp enough to display a reduction in the spectral line-width upon deuteration, associated, no doubt, with the change in the rate of radiationless transfer out of the bound Rydberg state into the underlying continuum (see Sect. 3.1). If there are large differences in the equilibrium geometry and force constants for the two states, Franck-Condon considerations may restrict the rate of predissociation from the Rydberg state.

A high degree of intravalence character can be expected when the electronic transition has a term value larger than that of the least energetic Rydberg transition in the atoms on which the molecular orbitals are centred. For example, it might be expected that the first continuum in the u.-v. absorption of H_2O, which has a term of 41800 cm^{-1}, would be largely intravalence in character, since the largest term value in the O atom is 33100 cm^{-1}, corresponding to the transition $2p \rightarrow 3s$. The broad continuum is believed to reflect the strong O–H_2 antibonding character of the orbital populated in the upper state (see Sect. 3.1.1). In contrast to a Rydberg state, the nature of the upper orbital has a major influence on the character of the photoexcited molecule, a feature that was first recognised and rationalised in Walsh's Rules[29].

Mixed Rydberg/intravalency character may be expected when the intravalency orbital has an energy close to that of an atomic Rydberg orbital, and the same symmetry. The mixing may depend on the molecular geometry and it will, in principle, generate a conjugate pair of excited states, one of which will emphasise Rydberg and the other, intravalence character. Transitions into the latter may not be readily identified if they produce broad continuous absorption underlying sharp, primarily Rydberg, transitions. These descriptions have been discussed at length by Robin[1] who has argued the case for the separate existence of these conjugate Rydberg/intravalency shell transitions in the "no-man's land" where the two extremes overlap. Higher Rydberg orbitals retain their purity since they lie too far above the valence energy levels for mixing to be significant.

2.1 Influence of the Molecular Core

In the lower lying Rydberg states, orbitals with $l > 0$ will be split by the asymmetry of the molecular core. The splittings decrease as n increases and the orbitals become increasingly diffuse; in a linear system, for example, the separate components $\lambda = l, l - 1 \ldots 0$ become nearly degenerate. Under these conditions, the Rydberg states form "complexes", e.g. the "p-complex" comprising $p\sigma$ and $p\pi$ states, or the "d-complex" comprising $d\sigma, d\pi$ and $d\delta$ states (see Ref.[30]). In practice, there are several other interactions which may affect the energies and the character of Rydberg states and complicate the task of spectral analysis. These include

(i) Interactions between states differing in both n and l; for example, an nd state for which the quantum defect $\delta \sim 0$ will be nearly iso-energetic with the $(n + 1)s$ state, for which $\delta \sim 1$, since they will have very similar term values. If the two states have the same symmetry under the point group of the molecular core, they will interact strongly.

(ii) The electron spin coupling scheme may change as n increases, from the Russell-Saunders scheme when n is low, to separate core (Ω_c) – optical electron (ω), (Ω_c, ω) coupling when n is large. In the latter case the distinction between pairs of singlet and triplet states, whose energies are separated by the electron exchange integral

$$2K = 2 < \phi_C \phi_R \left| \frac{e^2}{r_{12}} \right| \phi_C \phi_R >$$

disappears, since the overlap between the core orbitals ϕ_C and the Rydberg orbital ϕ_R becomes very small. A $\pi_C \rightarrow \pi_R$ transition in a linear molecule such as ICN, for example, would generate an ionic core for which $\Omega_c = \frac{3}{2}$ or $\frac{1}{2}$ and a Rydberg orbital for which $\omega = \frac{3}{2}$ or $\frac{1}{2}$. Under (Ω_c, ω) coupling these will lead to two pairs of states $\left\{ \left(\frac{3}{2}, \frac{3}{2} \right), \left(\frac{3}{2}; \frac{1}{2} \right) \right\}$ and $\left\{ \left(\frac{1}{2}, \frac{3}{2} \right), \left(\frac{1}{2}, \frac{1}{2} \right) \right\}$ separated by an energy close to the spin-orbit coupling in the molecular ion, while each pair is split further to generate ten states in all[31]. The correlations for this situation are represented in Fig. 2.1.1. When n is low and Russell-Saunders coupling prevails, the singlet transitions will be far more intense, but as n increases the "singlet" and "triplet" components will approach equal intensity, although the reducing overlap between the core and the Rydberg orbitals will lead to a decline in their nett oscillator strength.

(iii) The degree of penetration into the molecular core, and hence the magnitude of the quantum defect δ, is dependent on the size and symmetry of the Rydberg orbital and the size and architecture of the molecular core. The more the Rydberg state is aware of its molecular parentage, the less will it conform to the simple, atom-like Rydberg formula. Strong mixing of intravalency and Rydberg character will lead to large departures from the latter behaviour; since this mixing may be prevalent in (but not exclusive to) the lower lying states, the first member of the Rydberg progression may often be out of step with the succeeding ones, though in view of the interactions

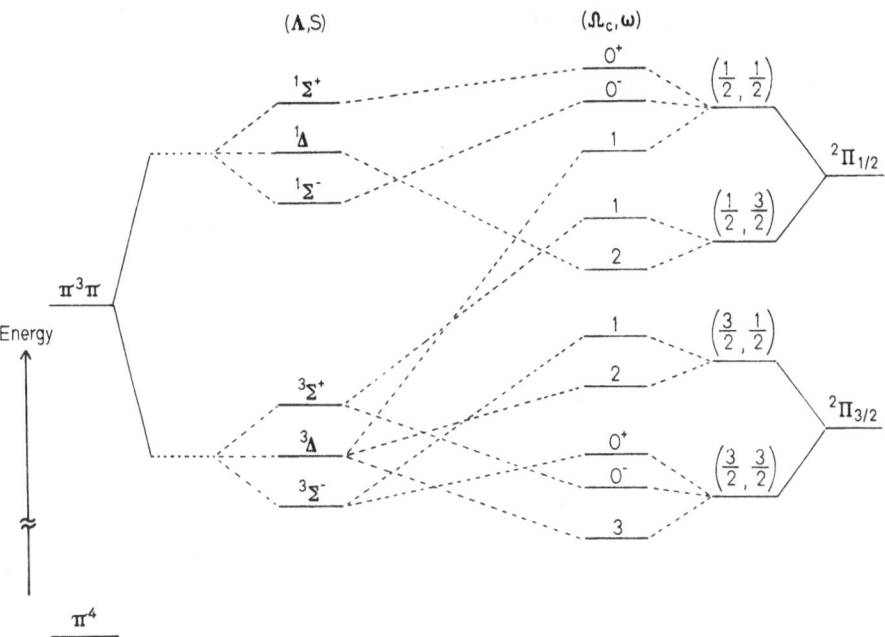

Fig. 2.1.1. Schematic correlation between (Λ, S) and (Ω_c, ω) coupling for the $\pi^3\pi$ configuration in a linear molecule

which can arise from other sources, generalisations should be treated with caution. Robin[1] has presented a thorough discussion of the consequences of orbital penetration into the molecular core and of the influence of substituents on the values of δ in the ns Rydberg progressions (where the orbital penetration is greatest) in homologous series of organic molecules.

To summarise, it should be clear that the characterization of the lower-lying states of polyatomic molecules populated by absorption in the vacuum ultraviolet region is no easy task, either theoretically or experimentally. The transitions may generate broad continua which can only be assigned on the basis of theoretical considerations: furthermore, the orbital composition in the upper state may be strongly dependent on the interatomic distances. Where the transitions are structured, it is relatively rare for the rates of radiationless transfer to be slow enough to allow resolution of the rotational structure and hence allow the assignment of the orbital symmetry of the excited state. Assignments are generally achieved through comparisons with isovalent molecules and "united atoms", by appeal to theoretical calculations, and by comparison of observed and anticipated term values. A fourth route which is developing involves the analysis of photofragment excitation spectra, the polarization of fragment fluorescence, the predictions of correlation diagrams linking the excited parent molecular states with the observed photofragment states and the energy disposal amongst them. The remainder of the review is devoted to a survey of the inter-relation between the molecular spectroscopy of the parent molecule and its photochemistry in the vacuum ultraviolet. The discussion is limited to simple polyatomic systems.

7

3 Spectroscopy and Photochemistry

3.1 H_2O and D_2O

3.1.1 Spectroscopy

In its ground state the electronic configuration of the water molecule may be written

$$(1a_1)^2 \, (2a_1)^2 \, (1b_2)^2 \, (3a_1)^2 \, (1b_1)^2; \tilde{X}^1A_1$$

Fig. 3.1.1 summarises the nodal properties of the valency shell orbitals, including the two virtual antibonding orbitals $4a_1$ and $2b_2$. These might be expected to contribute to the character of the lower-lying excited states, though the electronic absorption spectrum of water lies well in the vacuum u.-v. region, and the term values of some of the lower-lying transitions are comparable with those for the $2p \rightarrow 3s$ and $3p$ transitions of the O atom. Much of the debate regarding the assignment of electronic transitions in H_2O (and D_2O) has centred on the relative proportion of Rydberg/intravalency character in the upper states[1]. There is less argument over the description of the vacated orbitals since the "lone pair" electrons in $3a_1$ and $1b_1$ are much less tightly bound than the rest.

The absorption spectrum of H_2O is shown in Fig. 3.1.2: that of D_2O is broadly similar[32]. The first continuum, with $\lambda_{max} \sim 167$ nm, was originally assigned by Mulliken[33] to the leading member of the Rydberg progression $1b_1 \rightarrow nsa_1$ $(\tilde{A}^1B_1 \leftarrow \tilde{X})$,

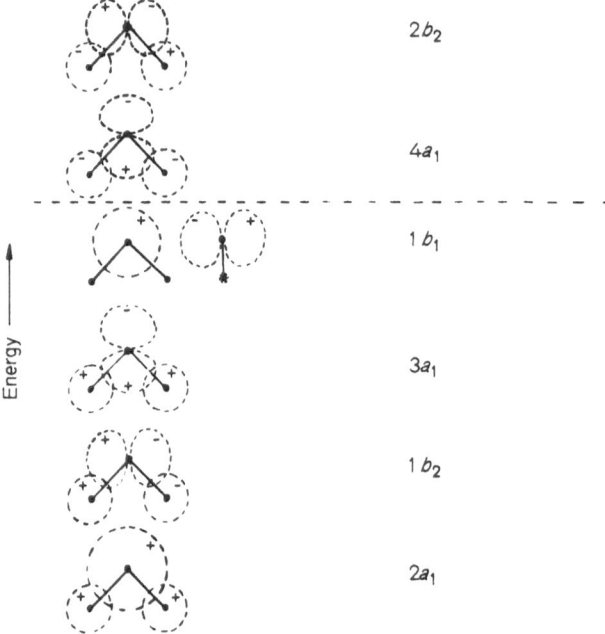

$2b_2$

$4a_1$

$1b_1$

$3a_1$

$1b_2$

$2a_1$

Energy ⟶

Fig. 3.1.1. Molecular orbitals of the ground electronic state of H_2O in terms of LCAO's. Orbitals are symmetrical with respect to the molecular plane except for the $1b_1$ orbital, where two views are given (after Herzberg, Ref.[3])

$n = 3,4,5 \ldots$; however, the excited states associated with higher members of the series[32, 34] and the corresponding molecular ion[35-38] (in the $^2B_1$ state) are all bound rather than dissociative, with geometries and frequencies similar to those of the neutral molecule in its ground state[3]. McGlynn and co-workers, who have presented a thorough review of the vacuum u.-v. absorption spectroscopy of water, discerned a very diffuse progression in the bending vibrational frequency ν_2', superimposed on the broad continuum[32], with a value $\sim 20\%$ greater than the frequency in the ground state, ν_2''. This behaviour is consistent with excitation of an electron from the non-bonding orbital $(1b_1)$ into an upper orbital which encourages H-H bonding, i.e. one of symmetry a_1, since this must allow the possibility of in-phase overlap between the $1s_H$ orbital functions. By definition, the Rydberg orbital $3sa_1$ is less likely to achieve this result than the intravalency shell $4a_1$ m.o., and McGlynn et al.[32] conclude that the continuum $\lambda_{max} \sim 167$ nm is associated with a predominantly intravalency shell transition into a strongly bent upper state, $1b_1 \to 4a_1$ $(\widetilde{A}\ ^1B_1 \leftarrow \widetilde{X})$. In support of this, they note that the photo-electron band associated with the ionization from the $1b_1$ orbital has a very different structure[35-38] (see Fig. 3.1.3) implying that in the optical transition the geometrical changes are promoted by the character of the newly populated m.o., rather than by that of the vacated orbital. The inclusion of intravalency shell character in states which involve the $3sa_1$ or $3pb_2$ Rydberg orbitals can also be justified on theoretical grounds[39]; the two O-H antibonding orbitals are of symmetry a_1 and b_2 (see Fig. 3.1.1).

In a linear configuration the $\widetilde{A}\ ^1B_1$ state correlates with one component of $^1\Pi_u$; the other component, $^1A_1$, has been assigned to the second broad continuum centred around 128 nm, upon which is superimposed much more pronounced vibrational structure (see Fig. 3.1.4). Its appearance is similar to the $\widetilde{X}\ ^1A_1 \to\ ^2A_1$ photoelectron band[35-38] (see Fig. 3.1.3) suggesting excitation of an electron from the H-H bonding orbital, $3a_1$, into the non-bonding Rydberg orbital, $3sa_1$. This view is reinforced by the long progression in the bending frequency with $\nu_2' < \nu_2''$ and by the comparable values for the quantum defect δ in equivalent transitions in the te-

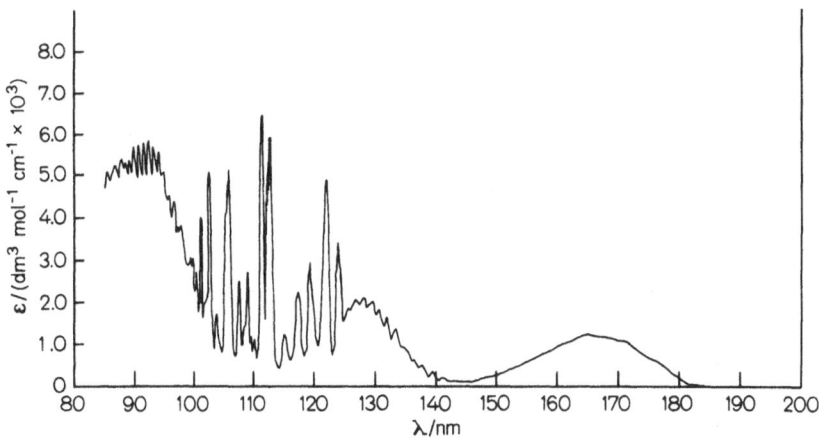

Fig. 3.1.2. Vacuum ultraviolet absorption spectrum of H_2O (after Watanabe and Zelikoff, Ref.[86])

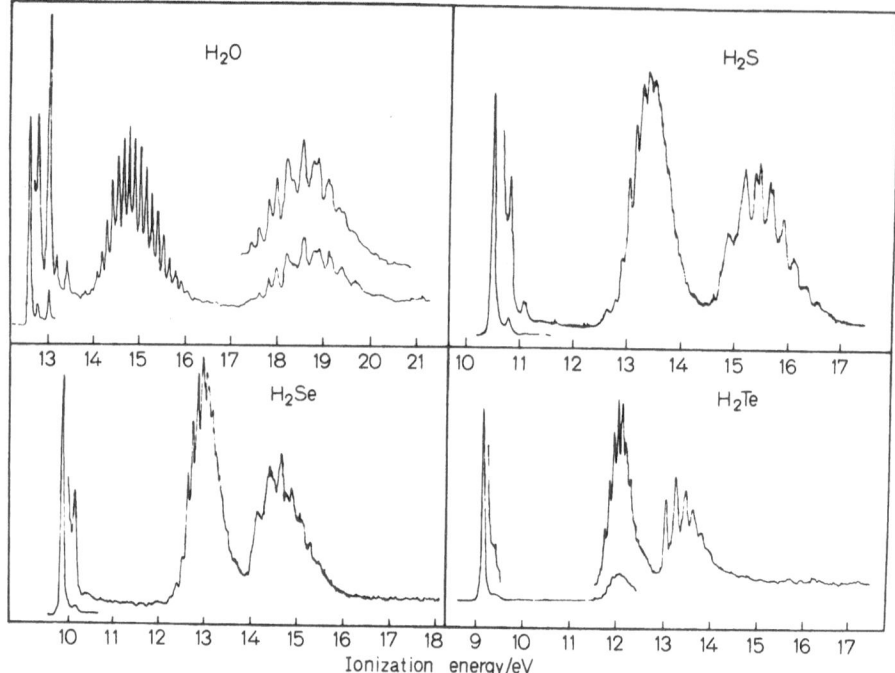

Fig. 3.1.3. He I photoelectron spectra of H_2O, H_2S, H_2Se and H_2Te (after Potts and Price, Ref.[36])

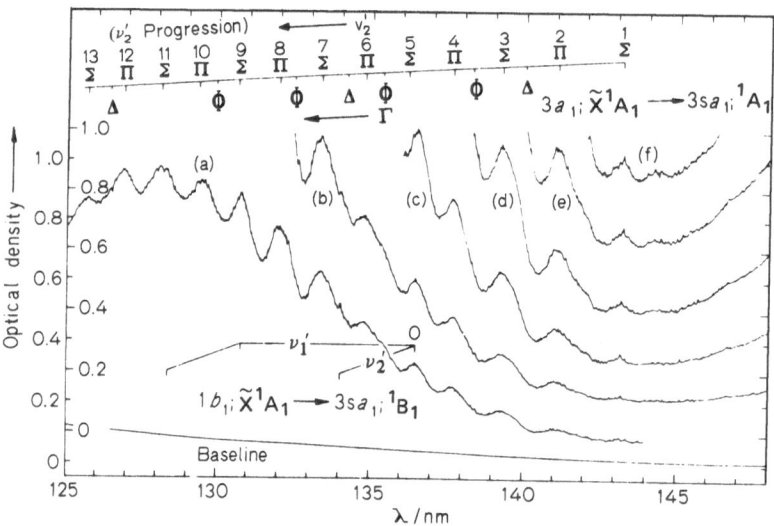

Fig. 3.1.4. Absorption spectrum of H_2O showing the beginning of the second continuum. Spectrum (a), recorded at 0.75 Torr, and the baseline, are referred to the lower "optical density" scale. Spectra (b) – (f) were recorded at increasing pressures and are referred to the upper "optical density" scale. The figure shows the proposed electronic and vibrational assignments of McGlynn et al. (after Ref.[32])

tratomic hydrides NH_3, PH_3 and AsH_3[32]. As expected for an electronic state correlating with a degenerate, linear configuration, the vibrational structure displays the
effects of Renner-Teller interaction[32] and McGlynn and his co-workers have summarised arguments which establish that the upper state is the linear, $3sa_1/4a_1$;
$\widetilde{B}\,^1A_1$ component of $^1\Pi_u$. They suggest that as the molecule bends, the lower component ($\lambda_{max} \sim 167$ nm) becomes predominantly intravalence in character
$1b_1 \rightarrow 4a_1$ ($\widetilde{A}\,^1B_1 \leftarrow \widetilde{X}$), while the upper component (~ 128 nm) remains predominantly Rydberg $3a_1$ (π_u) $\rightarrow 3\,sa_1$ (σ_g) ($\widetilde{B}\,^1A_1$ ($^1\Pi_u$) $\leftarrow \widetilde{X}$)[32].

The assignment of the continuum centred at 167 nm to a mainly intravalency
$^1B_1$ state implies the existence of a conjugate $^1B_1$ Rydberg state, associated with
the transition $1b_1 \rightarrow 3\,sa_1$. Two independent experimental measurements have
indicated the population of an additional electronic state at $\lambda \lesssim 136.5$ nm[32, 40].
The first follows the identification of irregularities in the vibrational band structure superimposed on the second continuum (see Fig. 3.1.4) which led to the
postulate of a bent Rydberg state at $\lambda \sim 136$ nm, with an effective quantum number ($n - \delta$) = 1.96. This was readily associated with the electron promotion
$1b_1 \rightarrow 3sa_1$, and subsequent members of the series could be identified at shorter
wavelengths, with ($n - \delta$) = 2.95 (4s) and 3.96 (5s)[32]. The second study involved
measurement of the polarization of the OH ($A \rightarrow X$) (and OD ($A \rightarrow X$)) fluorescence
excited by photodissociation at wavelengths $\lambda \leqslant 137$ nm[40]. This confirmed that the ma-
jority of the OH($A^2\,\Sigma^+$) was produced via excitation of the $\widetilde{B}\,^1A_1$ state (leading to
negative polarization of the OH ($A \rightarrow X$) fluorescence) but it also revealed contributions from an underlying excited state of symmetry B_1 (which produces positive
fluorescence polarization). Furthermore, Fig. 3.1.5 reveals that the minima in the
polarization spectra of H_2O (and D_2O) coincide with band maxima identified by

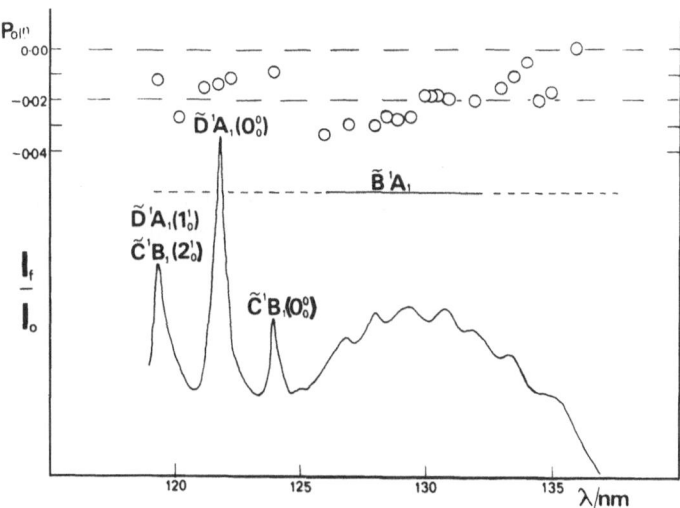

Fig. 3.1.5. OH($A^2\Sigma^+ \rightarrow X^2\Pi$) photofragment fluorescence excitation spectrum from H_2O. Open
circles indicate measured polarizations (after Macpherson and Simons, Ref.[40])

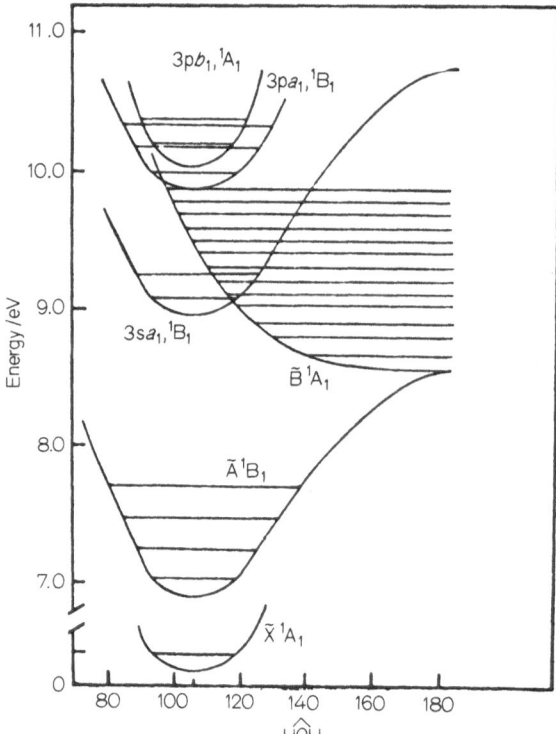

Fig. 3.1.6. Potential energy diagram (schematic) for the lower lying singlet electronic states of H_2O as a function of bond angle. The figure shows the proposed electronic assignments of McGlynn et al. (after Ref.[32])

McGlynn et al.[32] and assigned by them to the $1b_1 \rightarrow 3sa_1$ ($^1B_1 \leftarrow \tilde{X}$) electronic transition. In the polarization experiments it was suggested that the upper state was vibronically B_1[40], derived from the forbidden electronic transition $1b_1 \rightarrow 3pb_2$ ($^1A_2 \rightarrow \tilde{X}$). SCF/CI calculations[39, 41, 42] predict a $^1A_2$ state in this spectral region (correlating with the lower component of $^1\Pi_g$). This assignment would also be consistent with the original analysis of the K-shell excitation spectrum by Wight and Brion[43]. However, McGlynn et al.[32] note that the assignment to $3pb_2$; $^1A_2$ would be inconsistent with a quantum defect $\delta \simeq 1$ and revise the earlier analysis in favour of $3sa_1$; $^1B_1$ despite the absence of any predicted $^1B_1$ state at the required energy[39, 41, 42]. Thus while there is strong experimental evidence for the existence of an additional electronically excited state associated with weak absorption at $\lambda \sim 136$ nm, its parentage remains problematic.

The sharp series of bands beginning at 124 nm in H_2O, and extending to shorter wavelengths, have long been associated with Rydberg transitions populating linear, bound excited states[44-46]. The first sharp band in the absorption spectrum of H_2O, which occurs at 124.1 nm[44, 45] has been assigned to the transition $1b_1 \rightarrow 3pa_1$ ($\tilde{C}$ $^1B_1 \leftarrow \tilde{X}$)[32, 34, 44-46]. The band has a clearly defined rotational structure, which is sharper in D_2O, although diffuseness sets in at higher rotational levels. It is likely that the excited state is undergoing heterogeneous predissociation[3], with rotation coupling the bound $\tilde{C}$ $^1B_1$ and diffuse $\tilde{B}$ $^1A_1$ states. The next major feature in H_2O occurs at 121.9 nm[44, 46] and has been assigned to the transition $1b_1 \rightarrow 3pb_1$ ($\tilde{D}$ $^1A_1 \leftarrow \tilde{X}$)[32, 34, 44, 46]. The spectral lines associated with it are wide and diffuse[46] and as the transition involved is essentially an electron promotion from a nonbonding orbital to a Rydberg orbital, it is clear that the upper state is predissociated, probably via the $\tilde{B}\,^1A_1$ state. A detailed survey of the molecular Rydberg states in H_2O and D_2O, extending from the original work of Price[44] to the more recent studies of Katayama et al.[47] and McGlynn et al.[32] may be found in the latter's review. The Rydberg structure has also been analysed by Gürtler et al.[34] and by Ishiguro et al.[48] who have proposed some alternative assignments; Fig. 3.1.6 summarises the suggested assignments[32] of the lower-lying singlet excited states in a schematic potential energy diagram.

Singlet-triplet transitions in H_2O have been studied by electron impact spectroscopic techniques[49, 50] and reviewed by Robin[1] and Tsurubuchi[51]. A number of theoretical calculations[39, 41, 42] agree in assigning a strong band centred at 7.2 eV (equivalent to absorption at $\lambda \sim 190$ nm) to the electronic transition $1b_1 \rightarrow 3sa_1;\,^3B_1$. McGlynn et al.[32] have commented that an intravalence description for the optical transition $1b_1 \rightarrow 4a_1;\,^1B_1$ at $\lambda \sim 167$ nm, would enhance the predicted singlet-triplet separation and lower the energy of the conjugate triplet state, $4a_1;\,^3B_1$. Robin[1] suggests the latter might be associated with a weak, broad electron impact band lying between 4 eV and 5 eV.

A survey of the current literature reveals that despite the molecule's deceptive simplicity, the description of the nature of the lower-lying electronic states which contribute to the vacuum u.-v. absorption remains controversial. Particular difficulties arise over the number and identity of the excited states which are associated with the broad-band continua[1].

3.1.2 Photochemistry

Many photochemical studies of water in the gas phase have been concerned with identification of the alternative primary products of photodissociation, assessment of their relative contributions and the estimation of the energy disposal and distribution among the separating fragments[10, 51−53]. Much theoretical work has been devoted to calculating potential energy functions for photo-excited states (particularly the $\tilde{B}\,^1A_1$ state) and interpreting the observed energy disposal in terms of the motion on the upper potential surface.

Photochemistry in the First Absorption Continuum, $\tilde{A}\,^1B_1 \leftarrow \tilde{X}$. At wavelengths corresponding to absorption in the first continuum ($\lambda_{max} \sim 167$ nm) the following dissociation channels are energetically accessible

$$H_2O \rightarrow H_2(X^1\Sigma_g^+) + O(^3P) \qquad\qquad \lambda < 246 \text{ nm} \qquad\qquad \text{(i)}$$

$$\rightarrow H(^2S) + OH(X^2\Pi) \qquad\qquad \lambda < 242 \text{ nm} \qquad\qquad \text{(ii)}$$

$$\rightarrow H_2(X^1\Sigma_g^+) + O(^1D) \qquad\qquad \lambda < 177 \text{ nm} \qquad\qquad \text{(iii)}$$

Black and Porter[54] detected OH (OD) as a primary product of the flash photolysis of H_2O (D_2O) in the range $150 < \lambda < 190$ nm and suggested that $O(^1D)$ might also be a primary product (though their arguments were later queried by McNesby and Okabe[10]). Welge and Stuhl[55] reported an increased yield of OH following flash pho-

13

tolysis of H_2O in the presence of H_2 at $\lambda \geqslant 130$ nm and inferred that $O(^1D)$ was produced (though this could well have been associated with excitation in the second continuum at shorter wavelengths). In general, the primary production of $O(^1D)$ or of
$O(^3P)$ is not thought to be important at $\lambda \gtrsim 145$ nm. For example, a number of static experiments at 147 nm employing isotopic labelling and scavenging techniques
indicate a quantum efficiency for the molecular dissociation channels (i) and (iii),
$\phi \lesssim 0.03^{56-61)}$, while a flash photolysis study of H_2O over the range
145 nm $<\lambda <$ 185 nm employing resonance fluorescence detection of O and H[56)]
found no evidence for O atoms and concluded $\phi[O] \leqslant 0.01$, $\phi[OH] \geqslant 0.99$. Static
experiments at 174 nm[61)] (which is nearly coincident with the threshold for $O(^1D)$)
and 185 nm[57, 58, 62)] (which lies below the threshold) also found no evidence for
the primary production of atomic oxygen. It seems, therefore, that despite being
energetically accessible and spin allowed, the dissociation (iii) is unable to compete
against the alternative dissociation (ii) in this region, even though both channels can
correlate adiabatically with the $\tilde{A}\,^1B_1$ state[51)]. The explanation must lie in the nature
of the potential energy surface for the excited molecule; potential surface calculations suggest a low barrier ($\sim$ 1 eV) towards dissociation (iii)[63)] but instability with
respect to dissociation (ii)[64)]. The latter calculation also indicated little dependence
on the bond angle, which was consistent with Welge and Stuhl's experimental finding
that very little of the excess available energy appeared in internal excitation of the
$OH^{65)}$. The disparity in the masses of the separating fragments must result in very
high translational excitation of the H atom[53, 65)].

Photochemistry at Wavelengths $\lambda <145$ nm. The second absorption continuum
begins at $\lambda \sim$ 143 nm and extends to wavelengths underlying the banded region of
the absorption spectrum (see Fig. 3.1.2); unlike the first continuum, it is associated
with at least two electronic states (see earlier discussion). Thresholds for the additional dissociation channels which become accessible at the shorter wavelengths are
as listed:

$$H_2O \rightarrow H(^2S) + OH(A^2\Sigma^+) \qquad\qquad \lambda < 137 \text{ nm} \qquad\qquad \text{(iv)}$$

$$H_2\,(X^1\Sigma_g^+) + O(^1S) \qquad\qquad \lambda < 134 \text{ nm} \qquad\qquad \text{(v)}$$

$$2H(^2S) + O(^3P), O(^1D), O(^1S) \qquad\qquad \lambda < 129 \text{ nm, } 108 \text{ nm, } 90 \text{ nm (vi)}$$

$$H(^2P) + OH(X^2\Pi) \qquad\qquad \lambda < 81 \text{ nm} \qquad\qquad \text{(vii)}$$

There have been relatively few quantitative estimates of the branching ratios, though
it is clear that production of OH, either in the $X^2\Pi$ ground state or the fluorescent
$A^2\Sigma^+$ state constitutes the main primary process, with the ground state providing
the more important channel. Stief et al.[56)] were able to excite fluorescence from
both H and $O(^3P)$ following flash photolysis of H_2O in the presence of Ar between
105 nm and 145 nm. When He was substituted for Ar, the induced fluorescence
from $O(^3P)$ disappeared and it was inferred that the original signal followed primary
production of $O(^1D)$ which was subsequently quenched by Ar to $O(^3P)$; they estimated values for the quantum yields $\phi[OH(A^2\Sigma^+ + X^2\Pi)] = 0.89$ and $\phi[O(^1D)] =$
$= 0.11$. Primary production of atomic oxygen following photolysis of $H_2O^{57, 66)}$
and $D_2O^{58)}$ at 123.6 nm has also been proposed, on the basis of radical scavenging
experiments, though the earlier results were probably distorted by the reaction of

"hot" H atoms with the parent molecule. Branching ratios between ground state $(X^2\Pi)$ and excited $(A^2\Sigma^+)$ OH radicals have been estimated from measurements of the OH$(A \rightarrow X)$ fluorescence. The most recent work indicates a threshold wavelength at $\lambda \approx 137$ nm, above which the quantum yield rises quite sharply[40, 67, 68], but to a maximum, at ~ 120 nm, which only corresponds to $\phi[\mathrm{OH}(A)] \approx 0.05$[68].

A rationale for the observed branching ratios following excitation into the second continuum can be developed by considering correlation diagrams and the contours of calculated potential energy surfaces. Figure 3.1.7 summarises the important correlations and potential energy contours for the lower-lying electronic states. The second continuum is associated principally with the $\tilde{B}\,^1A_1$ state[32, 40], which is expected to be linear[3, 29, 32, 51]. In the linear configuration, the symmetry becomes $^1\Pi_u$ which correlates with OH$(X^2\Pi)$ + H(^2S); the bent $\tilde{B}\,^1A_1$ state correlates with excited OH$(A^2\Sigma^+)$ + H(^2S) (see Fig. 3.1.7). Thus the main feature of the calculated potential energy surface for the $\tilde{B}\,^1A_1$ state (represented schematically in Fig. 3.1.8) is a conical intersection with the ground state surface at the linear configuration[69, 70]. In this region radiationless transfer to the ground state surface, leading to OH$(X^2\Pi)$ + H(^2S), is probable[3, 71]. It has also been noted that rotationally induced Renner-Teller coupling may mix the bent $\tilde{A}$ and $\tilde{B}$ states[69] and it is apparent that a very detailed knowledge of the multi-dimensional potential surfaces would be required before any quantitative prediction of the OH(A):OH(X) branching ratios could be made.

The $\tilde{B}\,^1A_1$ state also correlates with O(^1D) + H$_2(X^1\Sigma_g^+)$ under C_{2v} point group symmetry. Flouquet and Horsley[69] calculated the contours of sections through the potential surface leading to this exit channel but found an intervening potential barrier with an energy ~ 5 eV, associated with an avoided crossing, and concluded (in agreement with experiment) that O(^1D) was unlikely to be an important primary

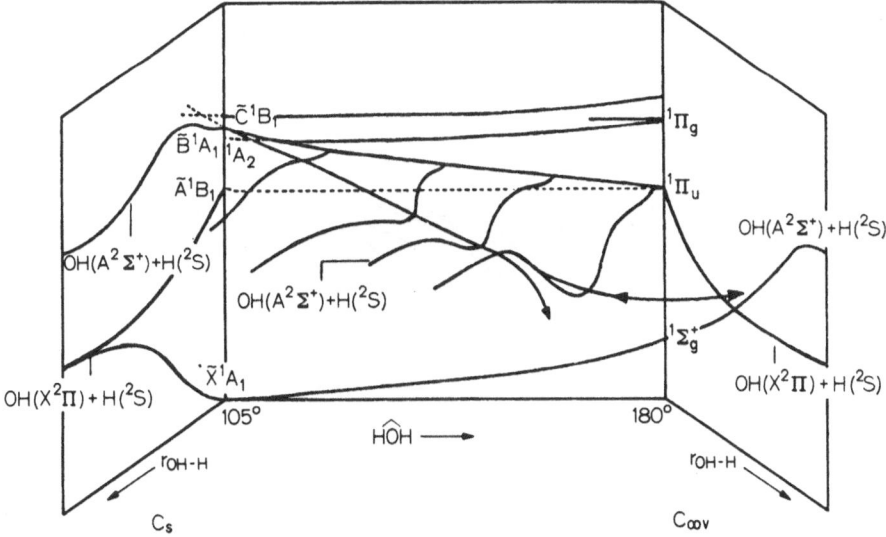

Fig. 3.1.7. Correlations and potential energy contours for lower lying singlet electronic states of H$_2$O, showing dependence upon bond angle (after Tsurubuchi, Ref.[51])

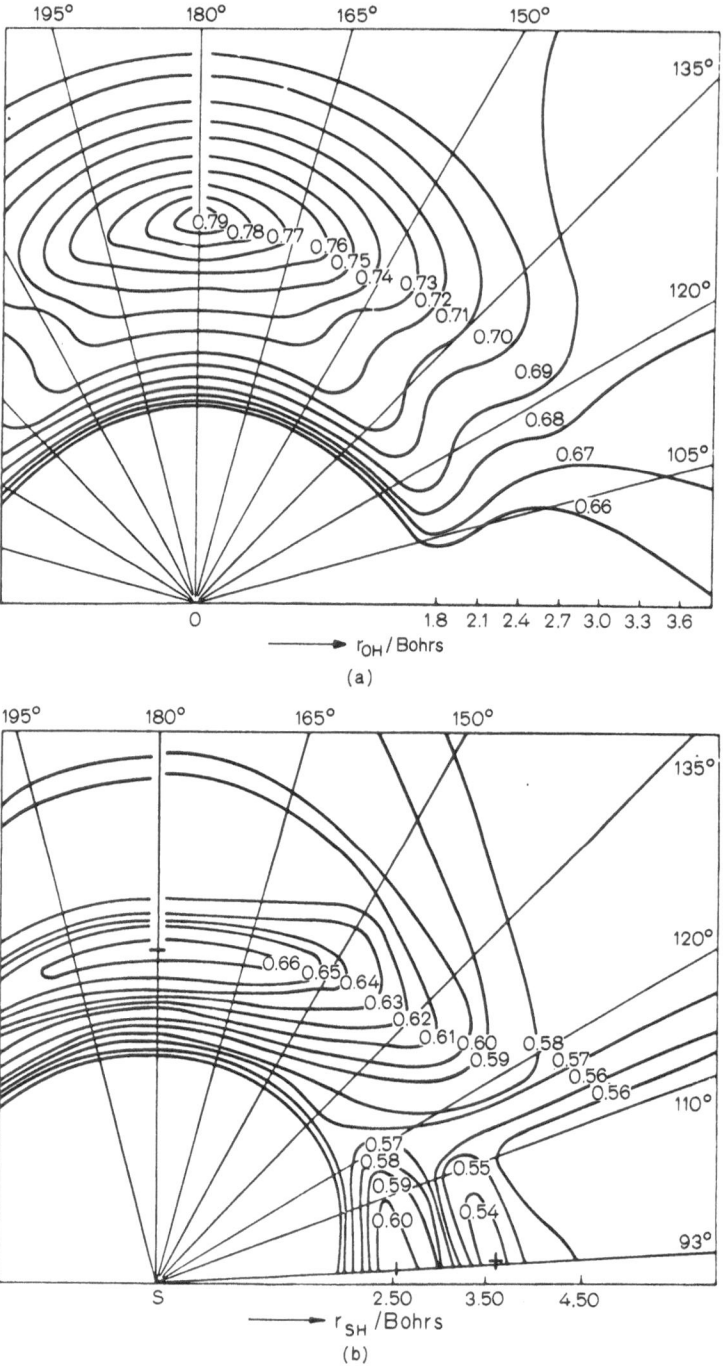

Fig. 3.1.8. (a) Potential energy surface for the $\tilde{B}\,^1A_1$ excited state of H_2O for dissociation to H + OH (after Flouquet and Horsley, Ref.[69])
(b) Potential energy surface for the lowest excited state of $^1A_1$ symmetry in H_2S for dissociation to H + SH (after Flouquet, Ref.[70])

product. The other low-lying energetically accessible channels leading to $O(^1S) + H_2$ $(^1\Sigma_g^+)$ or $O(^3P) + 2H(^2S)$ do not correlate with the $\widetilde{B}\,^1A_1$ state.

In a classic experiment, Carrington[72] analysed the $OH(A \to X)$ fluorescence excited by photodissociation of H_2O (and D_2O) at 130.2 nm, 123.6 nm, 121.6 nm and 116.5 nm and found that much of the excess available energy was channelled into rotation. Some of the absorbed wavelengths excite transitions in the banded, as well as the continuum, region of the absorption spectrum (e.g. 121.6 nm) but neither the energy partitioning nor the fluorescence quantum yields[68] were affected by changes in the identity of the upper state initially populated. The high levels of rotational excitation of the $OH(A\,^2\Sigma^+)$ were confirmed in later studies[73-75], usually conducted with Lyman α radiation at 121.6 nm. At this wavelength, the rotational distribution in $OH(A)_{v=0}$ peaked at the highest energetically accessible level, $K' \simeq 22$[72-75] and exhibited a "plateau" between $K' \approx 3$ and 10, though not all experiments agreed as to its precise structure. The differences were probably associated with partial rotational relaxation during the lifetime of the fluorescent state, $\tau \sim 1\ \mu s$[72]. In an effort to assess the importance of relaxation processes in the "static" experiments, Yamashita[75] monitored the fluorescence excited under near collision-free conditions by crossing the Lyman-α beam with a molecular beam. The resulting rotational distribution at the high values of K' was very similar to that observed previously[72-74], but the populations in levels $K' \lesssim 11$ were negligible. This agreed with the results of Carrington[72] and Kley[73] but not with those of Tanaka et al.[76]. By comparing the beam results with those of further experiments conducted under static conditions, Yamashita[75] concluded that the rotational distribution could be distorted by deactivation at the surface or in the gas phase and possibly by secondary production of $OH(A)$ from $OH(X)$. In view of these results the existence of two distinct rotational population distributions must be regarded as doubtful. The next rotational level above the highest accessible level $K' = 23$, corresponds to the predissociation threshold for $OH(A)_{v=0}$[77], and in the small population of $OH(A)$ fragments excited into the vibrational state $v' = 1$, the observed rotational populations appear to peak not at the highest accessible level $K' = 18$ but at the lower level $K' = 14$, which matches the onset of the predissociation threshold in the excited vibrational state[77].

There seems little doubt that the abnormal rotational excitation of $OH(A)$ arises from the motion on the $\widetilde{B}\,^1A_1$ potential surface excited directly via the $\widetilde{B} \leftarrow \widetilde{X}$ continuum, or indirectly following radiationless transition from the bound series of Rydberg states (e.g. $\widetilde{C}\,^1B_1$, $\widetilde{D}\,^1A_1$). Thus theoretical discussions of the energy disposal have focussed on the molecular dynamics of the excited molecule on the $\widetilde{B}\,^1A_1$ potential surface[51, 69, 70, 78, 79]. Carrington[72] drew the analogy between the dissociation and an orbiting collision, and trajectory calculations on an *ab initio* potential surface[69, 70] show the configuration point for the bent molecule descending through the conical intersection at the linear configuration; the bond angle would subsequently widen beyond 180° as the departing H atom spiralled away, to give an exit impact parameter ~ 4 a. u. and $\sim 75\%$ of the available energy channelled into rotation. Similar conclusions have been reached by Akamatsu and O-ohata[79] using the "time dependent wave packet" approach of Heller[80] and a potential surface calculated by O-ohata and Liu[81] to determine the motion on the upper potential surface. None

17

of the theoretical discussions would "predict" a double-humped rotational distribution and there is now strong experimental evidence against such a population spread. [Double peaked distributions have been observed following electron-impact dissociation[82] but it is likely that with this method of excitation the final distributions represent the nett result of alternative excitation pathways]. The vibrational energy disposal, $[v = 1]:[v = 0] = 0.3:1.0$ at 121.6 nm[72], has been reproduced solely on the basis of Franck-Condon considerations[78] with no contribution from "final state" interactions between the separating fragments. In view of their relative masses, and the very low energy released into translation, this is entirely to be expected[83], and it is confirmed by Lee and Judge[84], who have photolysed H_2O at wavelengths as low as 76 nm. They also found evidence of strong rotational excitation of $OH(A)$ fragments, as did Masanet and Vermeil[85] following excitation of H_2O and D_2O at the argon resonance lines (104.8 nm and 106.7 nm). The latter workers found that their results could be fitted to Boltzmann distributions over rotational states of the OH and OD fragments ($T_{rot} \sim 9000$ K and ~ 7000 K, respectively), in contrast to the results obtained at longer photolysis wavelengths, where the rotational distributions peaked towards the highest energetically accessible levels. Masanet and Vermeil[85] rationalized their observations by suggesting that dissociation was occurring specifically from a (Rydberg) state with similar bond angle to the ground state. Three Rydberg states in addition to the underlying continuum have been observed in this region however[32, 34], where the quantum yield for $OH(A^2\Sigma^+)$ is ~ 0.02, having gradually declined below ~ 115 nm[68]. As at the longer wavelengths, these rotational distributions may be explained in terms of dissociation over the $\tilde{B}\,^1A_1$ potential surface, although here a faster process would be expected because of the greater energy available. Since the motion would begin in a steeply repulsive region, the departing H atom would separate more rapidly from the OH fragment and follow a more open spiral trajectory corresponding to a smaller exit impact parameter and hence less rotational excitation of the diatomic fragment.

3.2 H_2S, H_2Se, H_2Te

3.2.1 Spectroscopy

The outer orbitals in H_2S, H_2Se and H_2Te necessarily have the same symmetry and ordering as those in H_2O, so that their ground electronic configurations may be written as

$$(a_1)^2\,(b_2)^2\,(a_1)^2\,(b_1)^2; \tilde{X}\,^1A_1$$

The outermost a_1 m.o. will be largely composed of the ns orbital centred on the $S(n = 3)$, $Se(n = 4)$ or $Te(n = 5)$ atom, since the bond angles in the ground state lie close to $90°$[3, 29].

H_2S^*. The absorption spectrum of H_2S, shown in Fig. 3.2.1, begins with a very long continuum, extending from ~ 270 nm to ~ 165 nm, $\lambda_{max} \sim 196$ nm, although the absorption is very weak at wavelengths $\lambda \geqslant 230$ nm[87]. Five diffuse vibrational bands

---

* *Note added in proof:* the vacuum ultra-violet absorption spectrum of H_2S has recently been recorded and analysed in great detail by Masuko et al.[366].

are superimposed on the continuum near its maximum, with a spacing
$\sim (1100-1150)$ cm^{-1} [88, 89] and a similar, more pronounced progression with a spacing ~ 820 cm^{-1} can be distinguished in the corresponding region in D_2S [89]. The frequencies are close to those of the bending mode in the ground electronic state
$(v_2''$ $(H_2S) = 1183$ cm^{-1}, v_2'' $(D_2S) = 934$ cm$^{-1})$ [90] and the progressions have been assigned to the bending frequency in the upper state [88, 89]. The small changes in frequency and the relatively short progressions indicate only a small change in bond angle on excitation, consistent with a transition originating from the non-bonding np (b_1) m.o.

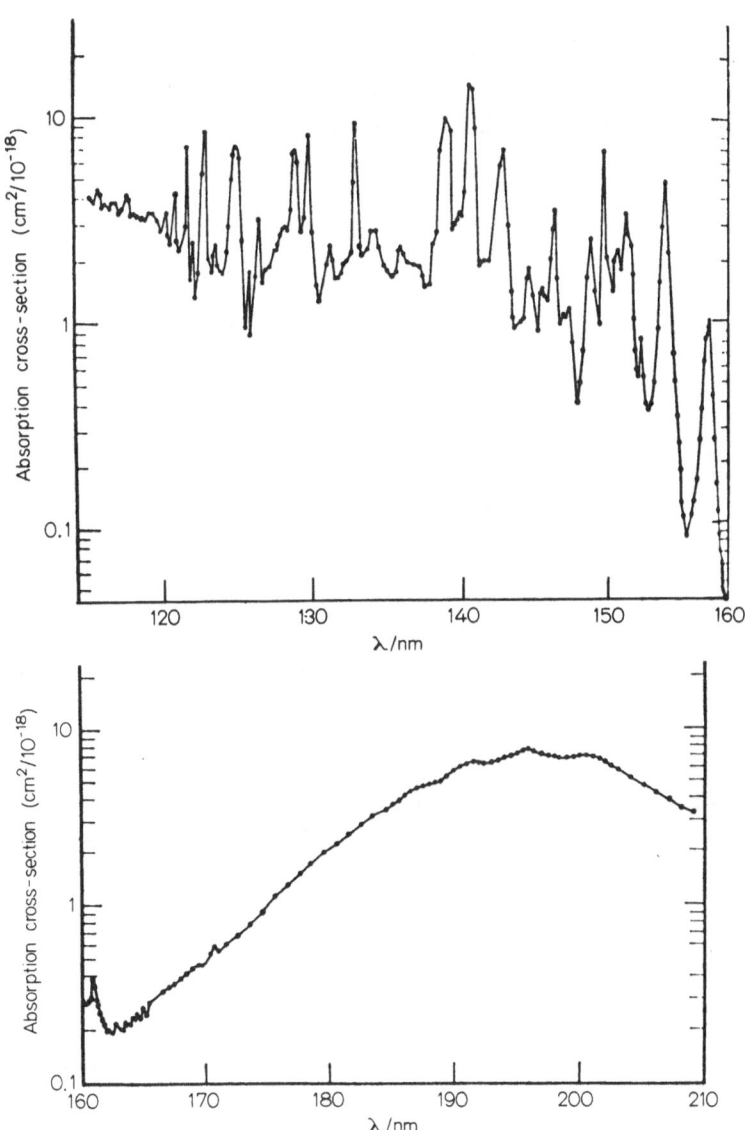

Fig. 3.2.1. Vacuum ultraviolet absorption spectrum of H_2S (after Watanabe and Jursa, Ref.[88])

The bending frequency in the $^2B_1$ state of the molecular ion is almost identical to that in the neutral molecule[91]. However, the diffuseness and breadth of the band in the optical spectrum contrasts with the very sharp transition in the photoelectron band (see Fig. 3.1.3), and it is evident that the terminating orbital cannot be purely non-bonding, Rydberg.

Robin[1] has assembled convincing arguments which imply that more than one transition contributes to the very broad continuum, and that the original assignment[33] to the first member of the Rydberg series $np \rightarrow (n + 1)s$ (cf. H_2O/D_2O) was only a part of the story. The $3p \rightarrow 4s$ term value in the S atom is $\sim 28{,}000$ cm^{-1} (which compares with a value for the continuum $\sim 33{,}000$ cm^{-1} (vert.) in H_2S) and theoretical calculations[92, 93] place the $2b_1$ $(3p) \rightarrow 4sa_1$ Rydberg band in the observed spectral region. Extrapolation from the term values for higher members of the series places the leading member at $\lambda \sim 179$ nm. None the less, its diffuse nature implies significant intravalence character, which would be contributed from the first antibonding $6a_1$ m.o., by analogy with H_2O. Calculations[92–95] also indicate the presence of at least one other electronic transition in the same spectral region and comparisons with the spectra of alkyl sulphides and thiols[1] reinforce this view. In the latter molecules, two distinct continua can be resolved, the first of which is very weak[96]; it has been assigned to the intravalence transition $2b_1 (3p) \rightarrow 3b_2$ ($^1A_2 \leftarrow \widetilde{X}$), forbidden under C_{2v} symmetry[1, 96, 97] and its analogue in H_2S could contribute to the broad continuum. Various m.o. calculations[92, 94, 95] place a $^1A_2$ state in the appropriate region; the most recent SCF-CI calculations of Buenker and Peyerimhoff[93] describe the state as $2b_1 \rightarrow 3d_{xz} (b_2)$, though not purely Rydberg in character.

Several Rydberg series have been identified[44, 98] in the banded absorption region of H_2S ($\lambda < 160$ nm) and the vacuum u.-v. spectroscopy has been thoroughly reviewed by Robin[1]. The first sharp member of the E series of Price[44] $(2b_1 (3p) \rightarrow nsa_1)$ $n = 5$, at 131.9 nm, has been studied by Gallo and Innes[99], who confirmed the assignment to this $^1B_1$ Rydberg state, but found that the corresponding band of D_2S was nearly continuous, although Bell[100] had seen little shift of this band on deuteration. There is some debate over the assignments for the first two diffuse peaks at 157.9 nm and 154.5 nm, which Robin[1] suggests may be due to transitions from $2b_1 \rightarrow 4p$, with the $4p$ degeneracy lifted by the symmetry of the molecular core, as in H_2O. Carroll et al.[101] had earlier assigned the bands to members of the $2b_1 \rightarrow 3d$ group, and this has since been supported by McDiarmid[102], although Robin had argued against this assignment because of their term values ($\sim 20{,}000$ cm^{-1}). In complete contrast, Clark and Simpson[96] have assigned the bands to "non-Rydberg" transitions, on the basis of comparisons with the spectra of alkyl sulphides. Figure 3.2.1 shows a substantial continuum underlying the banded region, which may be associated with transitions from lower-lying occupied orbitals to the $4s (6a_1)$ orbital, e.g. $5a_1 \rightarrow 4s (6a_1)$ ($^1A_1 \leftarrow \widetilde{X}$) analogous with the second continuum in H_2O. Calculations based on experimental ionisation potentials predict that such transitions will occur here; however, Robin has commented that, in view of its overlapping nature, it is possible to imagine continua contributing at any and all frequencies in this region of the H_2S spectrum[1].

H_2Se, H_2Te. The absorption spectra of H_2Se and H_2Te are well known[98], and again regions of continuous absorption can be distinguished at the longer wavelengths, with maxima at $\lambda \sim 200$ nm and $\lambda \sim 197$ nm, respectively. These presumably correspond to the similar feature in the H_2S spectrum. Vibrational progressions have been observed in H_2Te, centred around ~ 190 nm[103], which shifted on deuteration and whose displacements were consistent with the assignment to the bending mode ν_2'. There is also a close similarity between the photoelectron spectra in the molecular series (see Fig. 3.1.3).

Rydberg series have been identified at the shorter wavelengths and the reader is referred to Robin's excellent book[1] for further details.

3.2.2 Photochemistry

H_2S. Investigations of the vacuum u.-v. photochemistry of H_2S are distinguished by their rarity. The lowest excited states are believed to be those of B_1 and A_2 symmetry (see earlier). The B_1 state correlates with $SH(X^2\Pi) + H(^2S)$ (cf. Fig. 3.1.8) and studies of the photolysis of H_2S at $\lambda \geqslant 185$ nm[104–110] indicate that the reaction

$$H_2S \rightarrow H(^2S) + SH(X^2\Pi) \tag{i}$$

represents the major primary process with most of the excess energy carried away in translation of the H atom[53, 107–110]. At $\lambda < 200$ nm, there may be some contribution from channels (ii) and (iii) leading to atomic sulphur[111].

$$H_2S \rightarrow H_2(^1\Sigma_g^+) + S(^3P), S(^1D), S(^1S)$$
$$\lambda < 406 \text{ nm}, 245 \text{ nm}, 186 \text{ nm} \tag{ii}$$

$$2H(^2S) + S(^3P), S(^1D), S(^1S)$$
$$\lambda < 165 \text{ nm}, 143 \text{ nm}, 121 \text{ nm} \tag{iii}$$

At $\lambda < 162$ nm, reaction (iv)

$$H_2S \rightarrow H + SH(A^2\Sigma^+) \tag{iv}$$

has been observed, and, in contrast to H_2O, a thermal distribution over rotational energy states has been found for the $SH(A)$ fragments[112]. The electronically excited SH radical may be considered, by analogy with H_2O dissociation, to arise via the $\tilde{B}^1A_1$ state of H_2S (which, again by analogy, may be associated with the continuum underlying the Rydberg bands at $\lambda < 160$ nm – this is consistent with the results of SCF-CI calculations[93]). Flouquet[113] has calculated the potential surface for the dissociation of such a state into H + SH, and found that, as with H_2O, there is a conical intersection with the ground state $\tilde{X}^1A_1$ surface. Unlike the H_2O surface, however, an exit valley is found which lies close to the SH bond direction at the equilibrium (bent) geometry (see Fig. 3.1.8), so whereas H_2O must dissociate by oscillating around the conical intersection, thus imparting rotational angular momentum to the $OH(A)$ fragment, H_2S may dissociate along this exit valley, with little change of geometry, and consequently little rotational excitation of the SH fragment.

H_2Se, H_2Te. The vacuum ultraviolet photochemistry of these molecules has received little attention, although ground state SeH and TeH radicals have been observed as transients in the vacuum u.-v. following photolysis of the parent molecules at $\lambda \geqslant 130$ nm[103]. Photolysis of hydrogen selenide at $\lambda > 190$ nm in the region of the broad continuum has suggested that the primary process

$$H_2Se \rightarrow H(^2S) + SeH(^2\Pi), \tag{v}$$

proceeds with unit quantum efficiency[114].

3.3 NH_3, ND_3, PH_3, PD_3

3.3.1 Spectroscopy

Considerable information is available on the electronic spectrum of the ammonia molecule (see Fig. 3.3.1) and the principal findings have been reviewed by Robin[1]. In its ground state, the electronic configuration of NH_3 may be expressed

$$(1a_1)^2 (2a_1)^2 (1e)^4 (3a_1)^2; \tilde{X}^1A_1 \qquad (C_{3v}).$$

The lower excited states have all been assigned to equilateral, planar Rydberg states populated by excitation of an electron from the "lone pair" $3a_1 (1a_2'')$ m.o. in the pyramidal ground state, and under D_{3h} symmetry the ground state orbitals correlate with

$$(1a_1')^2 (2a_1')^2 (1e')^4 (1a_2'')^2; \tilde{X}^1A_1' \qquad (D_{3h}).$$

The first excited singlet state appears in absorption as a long progression in the out-of-plane bending mode ν_2' with origin at $\lambda = 216.8$ nm and with maximum intensity

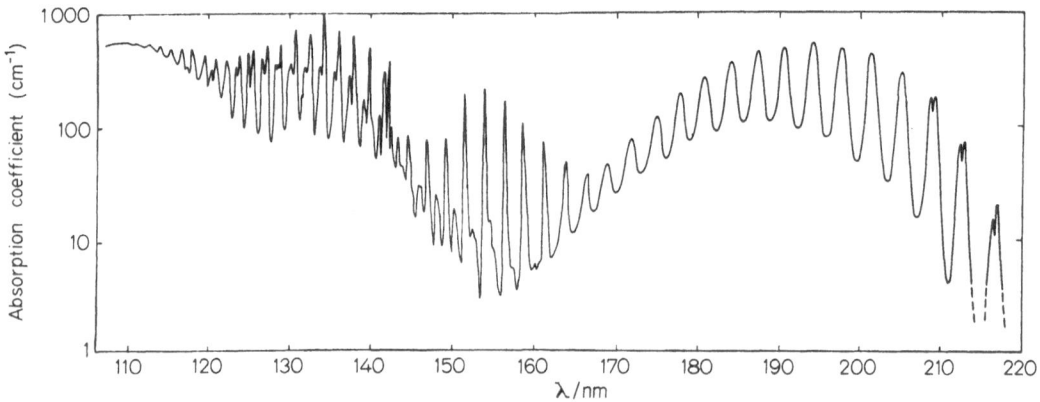

Fig. 3.3.1. Vacuum ultraviolet absorption spectrum of NH_3 (after Watanabe, Ref.[115])

at $v_2' \sim 6$[1, 115, 116]; it has been assigned to the Rydberg transition $3a_1\,(1a_2'') \rightarrow 3sa_1'\,(\tilde{A}\,^1A_2'' \leftarrow \tilde{X})$ with an appropriately large term value of $36{,}230\ \text{cm}^{-1}$ [1]. Absorption features attributable to the $n = 4$ and 5 members of the ns series have been identified at 143.5 nm and 133 nm (adiab.) respectively, with $\delta \simeq 1.02$[116]. The $(\tilde{A} \leftarrow \tilde{X})$ bands are predissociated with no resolvable rotational structure, though structure can be distinguished in those bands of ND_3 for which $v_2' \leqslant 1$[117]. The barrier to predissociation has been tentatively attributed to a "near intersection" of the bound $\tilde{A}\,^1A_2''$ state and a second repulsive $^1A_2''$ state which correlates with the ground state products $NH_2(^2B_1) + H(^2S)$[118]. Difficulties in accounting for the presence of a second low-lying $^1A_2''$ state[3, 118] could be resolved if the continuum underlying the $(\tilde{A} \leftarrow \tilde{X})$ band system were associated with the conjugate intravalency transition $3a_1(1a_2'') \rightarrow 4a_1'$ [1]. However, recent theoretical studies[28] have been able to reproduce a realistic potential barrier to dissociation in the $\tilde{A}$ state, which may be penetrated by quantum mechanical tunnelling, without recourse to the postulate of an avoided crossing. The SCF/CI calculations indicate a continuous evolution of the upper orbital character as the N–H distance increases, from a predominantly Rydberg $3s_N$ orbital through $(3s_N/\sigma^*_{N-H})$ to a pure hydrogenic $1s$ orbital at the dissociation limit.

The second excited singlet state, which generates a banded spectrum with its origin at 168.8 nm[120] [1], has been assigned to the Rydberg transition $3a_1 \rightarrow 3p_{xy}\,(e')\,(\tilde{B}\,^1E'' \leftarrow \tilde{X})$, while the third excited state is generated by the other $3a_1 \rightarrow 3p$ component $3a_1 \rightarrow 3p_z\,(a_2'')\,(\tilde{C}\,^1A_1' \leftarrow \tilde{X})$[28, 117, 119, 120, 121]*. Both transitions populate planar upper states and the out-of-plane bending mode v_2' is strongly excited in the two absorption systems (see Fig. 3.3.1). At wavelengths $\lambda < 130$ nm, the absorption becomes very diffuse and no high resolution spectra have been reported. However, the electron impact spectrum displays a broad peak centred around 109 nm[122] which has been assigned, on the basis of its term value, to the Rydberg transition $1e \rightarrow 3s\,(a_1)\,(^1E' \leftarrow \tilde{X})$. The lower orbital is $N-H_3$ σ-bonding so the broad, diffuse spectrum is easily understood.

PH₃, PD₃. The electronic spectrum of phosphine bears a close resemblance to that of its nitrogen analogue; all the absorption features observed in the vacuum ultraviolet can be attributed to Rydberg transitions involving an electronic promotion from what is essentially the central phosphorus atom lone pair orbital[1, 123, 124]. In the $4a_1\,(np) \rightarrow 4sa_1\,(\tilde{A}\,^1A_2'' \leftarrow \tilde{X})$ system of PH_3 and PD_3 (term value $\sim 29{,}800\ \text{cm}^{-1}$ [1]), the predissociation is more complete than in NH_3, and only a continuum centred around 180 nm is observed[123, 124].

To higher energies this region of continuous absorption is overlapped by a long progression of vibrational bands which, as in NH_3, have been attributed to the $4a_1\,(3p) \rightarrow 4p$ Rydberg excitation[1]. Again, the progression is composed of conse-

---

1   Recent calculations[28] imply that the band origin may actually lie one vibrational quantum to longer wavelengths and thus be hidden under the neighbouring $\tilde{A} \leftarrow \tilde{X}$ system.

*   *Note added in proof:* Nieman and Colson have discovered another excited state $\tilde{C}'$ close to the $\tilde{C}$ state using multiphoton spectroscopic techniques[367], which they assign to the transition originally associated with $\tilde{C} \leftarrow \tilde{X}$.

cutive quanta of the out of plane bending vibration, v_2', but, in contrast to the situation in NH_3, the v_2' frequency in this second excited state of phosphine (~ 490 cm^{-1})[1]) is only about one half the ground state value (992 cm^{-1})[125]. Adopting similar arguments to those used to account for the long progression in v_2' in ammonia[116, 126], Maier and Turner[127] conclude that the ground ionic state, and consequently the upper Rydberg states, of PH_3 are nearly planar, with a small barrier to inversion (~ 0.07 eV) accounting for the slight diffuseness of the lower vibrational bands in the photoelectron spectrum. A more recent theoretical study of the ground state geometry of PH_3^+ has led to a similar conclusion[128].

The absorption spectrum of PH_3 and PD_3 at $\lambda < 140$ nm consists of a "tangled mass of vibronic structure"[1] for which no definite assignments have been reported.

3.3.2 Photochemistry

The dissociation products resulting from photolysis of NH_3 and ND_3 within the diffuse, banded $\tilde{A} \leftarrow \tilde{X}$ absorption region have been extensively studied, and the general features of the primary photochemical processes are now reasonably well established[10, 12]. Reference to Table 3.3.1 shows that the following primary processes are energetically allowed at wavelengths longer than 165 nm:

Table 3.3.1 Threshold energies for primary product channels in the photodissociation of NH_3 (upper values, in brackets, are in electron volts, lower values are in nanometres)

NH ↓		$H_2(^1\Sigma^+)$ (0.00)	2H (4.48)[b]
$X^3\Sigma^-$	(0.00)	(3.92) 316.3	(8.40) 147.6
$a^1\Delta$	(1.56)[a]	(5.48) 226.2	(9.96) 124.5
$b^1\Sigma^+$	(2.62)[b]	(6.54) 189.6	(11.02) 112.5
$A^3\Pi_i$	(3.69)[b]	(7.61) 162.9	(12.09) 102.5
$c^1\Pi$	(5.44)[b]	(9.36 ± 0.05)[c] 132.5 ± 0.7	(13.84) 89.6
$d^1\Sigma^+$	(10.31)[b]		
NH_2 ↓		$H(^2S)$ (0.00)	
$\tilde{X}(^2B_1)$	(0.00)	(4.38)[d] 283.1	
$\tilde{A}(^2A_1)$	(1.27)[d]	(5.65) 219.4	

a Ref.[129].   c Ref.[131].
b Ref.[130].   d Ref.[3].

$$NH_3 \longrightarrow NH(X^3\Sigma^-) + H_2 \qquad \lambda < 316 \text{ nm} \qquad\qquad\qquad \text{(i)}$$

$$NH_2(\tilde{X}^2B_1) + H \qquad \lambda < 283 \text{ nm} \qquad\qquad\qquad \text{(ii)}$$

$$NH(a^1\Delta) + H_2 \qquad \lambda < 226 \text{ nm} \qquad\qquad\qquad \text{(iii)}$$

$$NH_2(\tilde{A}^2A_1) + H \qquad \lambda < 219 \text{ nm} \qquad\qquad\qquad \text{(iv)}$$

$$NH(b^1\Sigma^+) + H_2 \qquad \lambda < 190 \text{ nm} \qquad\qquad\qquad \text{(v)}$$

Experimentally, production of $NH_2(\tilde{X}^2B_1)$ has been observed in absorption following v.u.-v. flash photolysis over the entire range of the NH_3 parent molecular absorption above 120 nm[132]. Reaction (ii) has been shown to be the sole primary process operating at 206.2 nm[133, 134], and to account for at least 96% of the reaction products at 184.9 nm[66]. A near unit quantum yield for production of $NH_2(\tilde{X}^2B_1)$ and H atoms was measured at $\lambda > 185$ nm[133, 134]. However, weak fluorescence from electronically excited $NH_2(\tilde{A}^2A_1)$ fragments has been observed following photolysis of NH_3 in the v.u.-v.[129, 131, 135] at wavelengths as long as 193 nm[135]. The structure displayed in the $NH_2(\tilde{A} \rightarrow \tilde{X})$ fluorescence excitation spectrum closely mirrors the profile of the $NH_3(\tilde{A} \leftarrow \tilde{X})$ absorption system at the longer wavelengths, down to at least the (7 − 0) band of the progression in ν_2', and at the shorter wavelengths it faithfully reproduces the absorption profiles of the $(\tilde{B} \leftarrow \tilde{X})$ and $(\tilde{C} \leftarrow \tilde{X})$ systems as well. A previous experimental study had failed to detect $NH_2(\tilde{A} \rightarrow \tilde{X})$ fluorescence at exciting wavelengths $\lambda > 164$ nm[131] due to photomultiplier insensitivity for emission above 600 nm. Masanet et al.[129] have observed the fluorescence from the $\tilde{A}^2A_1$ state of $NH_2(ND_2)$ following the photodissociation of $NH_3(ND_3)$ at the Xe (147, 129.5 nm) and Kr (123.6, 116.5 nm) resonance-wavelengths, but not at the Ar resonance lines (104.8, 106.7 nm), possibly because the competing ionisation process:

$$NH_3 \rightarrow NH_3^+ + e \qquad\qquad\qquad\qquad\qquad\qquad\qquad \text{(vi)}$$

which is energetically allowed at $\lambda \leqslant 122.3$ nm[136], has an estimated quantum yield of 0.40 at this wavelength[137]. A value for the quantum yield $\phi[NH_2(\tilde{A})] < 0.01$ has been estimated for all wavelengths in the range 116 nm − 193 nm[131, 135]; this tends to rule out the theoretical proposals of Back and Koda[138] in which increased competition from dissociation channel (iv) was used to explain the observed *decrease* in the translational excitation of the H(D) atomic fragments from NH_3 and ND_3 as the photolysis wavelength was reduced from 213.9 nm to 184.9 nm. The radiative lifetime of the $NH_2(\tilde{A}^2A_1)$ electronic state has been measured using modulated atomic resonance lamps (147 nm and 123.6 nm) to give a value of $\sim 15 \, \mu s$[139] in good agreement with earlier determinations[140, 141].

The predissociation of NH_3 from the planar $\tilde{A}$, $\tilde{B}$ and $\tilde{C}$ states has been discussed on the basis of symmetry conservation considerations by Runau, Buenker and Peyerimhoff[28]. As the H−NH_2 distance increases the point group symmetry reduces to C_{2v}. *Electronic* symmetry conservation then allows formation of $NH_2(\tilde{X}^2B_1)$ *via* the $\tilde{A}$ state and $NH_2(\tilde{A}^2A_1)$ *via* the $\tilde{C}$ state, but in practice both products are formed from all three states; it is clear that vibronic interactions must play a significant role.

25

In this respect, there appears to be an alternation in the relative intensities of the progression in the bending mode ν_2' (a_2'') in the $\tilde{B}\,(^1E'') \leftarrow \tilde{X}$ excitation spectrum of the photofragment fluorescence from $NH_2(\tilde{A}\,^2A_1)^{131)}$. With the revised vibrational numbering of this progression[28] the more intense bands would correspond to odd values for ν_2'; under C_{2v} symmetry, these correspond to vibronic states transforming as $(A_1 \oplus B_2)$, one of which correlates with the observed product, whereas the even numbered members, vibronically $(A_2 \oplus B_1)$, correlate with unexcited $NH_2(\tilde{X}\,^2B_1)$.

There is an extensive literature relating to studies of the alternative mode of photodissociation into molecular H_2 (or $2H$) and NH; Table 3.3.1 summarises the energetic accessibility of alternative channels. Symmetry correlations have also been considered[28] but they appear to be of limited value in the absence of detailed information on the contours of the excited state potential energy surfaces.

Back and Koda[138] report a quantum yield, $\phi[D_2] = 0.009$ following photodissociation of ND_3 within the $\tilde{A} \leftarrow \tilde{X}$ band system at 184.9 nm, and a similar result was obtained for H_2 from NH_3; at this wavelength the sister product is most likely singlet $NH(ND)$ produced via the spin-allowed processes (iii) or (v), though neither $NH(a^1\Delta)$ nor $NH(b^1\Sigma^+)$ have been detected directly at $\lambda > 135$ nm[142]. The observations conflict with an earlier study which failed to detect molecular H_2 at 184.9 nm[66], but agree with those of Masanet et al.[143] who reported $\phi[H_2]/\phi[H] =$ $= \phi[D_2]/\phi[D]$ at this wavelength.

At 147 nm (in the $\tilde{B} \leftarrow \tilde{X}$ band system) $\phi[H_2]/\phi[H] > \phi[D_2]/\phi[D]^{143)}$, although the quantum yields remain small ($\phi[H_2] \sim 0.07$, $\phi[D_2] \sim 0.03)^{144)}$, but the isotopic dependence is very much less pronounced at the shorter wavelengths ($\phi[H_2(D_2)] \sim 0.25$ at 123.6 nm and ~ 0.30 at 104.8 nm)[144]. The additional spin-allowed channel

$$NH_3 \rightarrow NH(c^1\Pi) + H_2 \qquad\qquad \lambda < 133 \text{ nm} \qquad\qquad\qquad \text{(vii)}$$

which becomes energetically accessible at these shorter wavelengths, is unlikely to contribute to be observed increase in $\phi[H_2]$, since the quantum yield for production of $NH(c^1\Pi)$ at 123.6 nm is only $\sim 0.01^{131)}$). It may be pertinent to note here that under C_{2v} symmetry the products $NH(c^1\Pi) + H_2(^1\Sigma_g^+)$ do not correlate with the C^1A_1' state populated by absorption at 123.6 nm, but do correlate with the higher $^1E'$ state assigned to the continuum ~ 110 nm.

Production of triplet ground state $NH(X^3\Sigma^-)$ in the range $\lambda\,(125-160)$ nm has been demonstrated by observation of the $NH(A^3\Pi \leftarrow X)$ absorption band around 336 nm following flash photolysis[132, 145] 2. This channel may be associated with the spin-allowed process

$$NH_3 \rightarrow NH(X^3\Sigma^-) + 2H \qquad\qquad \lambda < 147 \text{ nm.} \qquad\qquad\qquad \text{(viii)}$$

Formation of $NH(a^1\Delta)$ through channel (iii) has also been proposed to account for the primary yield of molecular hydrogen at 147 nm and 123.6 nm[131] but the ex-

---

2 A recent study employed NH resonance fluorescence excitation to monitor the kinetics of the reaction of $NH(X)$ with NO, following flash photolysis of NH_3 at $\lambda > 105$ nm[146].

pected NH($c \leftarrow a$) absorption system could not be detected following flash photo-lysis[147]. However, NH(ND) ($c^1\Pi$) has been detected in emission (via the $c^1\Pi$-$a^1\Delta$ band system) during photolysis of NH_3(ND_3) at 129.5 nm[148, 149], 123.6 nm[66, 129, 145, 148–150], 121.6 nm[149] and 104.8 nm[129]. The NH($c \to a$) fluorescence yield rises gradually from threshold at (132.5 ± 0.7) nm and then more steeply in the continuum region below 120 nm to reach a value ~ 0.1 at $\lambda \sim 110$ nm[131]. Examination of the NH($c \to a$) fluorescence spectrum reveals a high level of rotational excitation in the NH(c) produced at 123.6 nm[129, 150], corresponding to around 35% of the excess available energy. This excitation is associated, no doubt, with the geometrical change from pyramidal to planar on excitation into the $\tilde{C}^1A_1'$ state (cf. the excitation of OH following excitation of H_2O into the linear $\tilde{B}^1A_1$ state, discussed in Sect. 3.1). A far lower level of excitation was recorded at 104.8 nm[129], where the rotational energy disposal in NH($c^1\Pi$) only accounted for 4% of the excess energy (cf. the rotational distribution of OH($A^2\Sigma^+$) following photolysis of H_2O at this wavelength).

In 1974, observation of the NH($b^1\Sigma^+ \to X^3\Sigma^-$) radiative intercombination around 471 nm following the photodissociation of NH_3 at 123.6 nm and 104.8, 106.7 nm provided the first experimental determination of the energy difference between the triplet $X^3\Sigma^-$ ground state and the first excited singlet state, $a^1\Delta$[129, 151], (see Table 3.3.1). A value for the radiative lifetime of the metastable NH($b^1\Sigma^+$)$_{v=0}$ state, 17.8 ± 4 ms[142], was subsequently derived from experiments in which the NH($b \to X$) emission was quenched by a variety of added molecules[142, 152] and atoms[153]. No NH($b \to X$) emission has been observed from the photolysis of ammonia at $\lambda > 135$ nm[142] or at 147 nm[129], (cf. the thermochemical threshold (v)). This observation has led to the suggestion that part, or all, of the NH($b^1\Sigma^+$) produced in the photodissociation of ammonia at $\lambda \leqslant 125$ nm is formed via NH($c^1\Pi$) and not directly via reaction (v)[142].

Photochemistry of PH_3 and PD_3. The vacuum ultraviolet photochemistry of PH_3 and PD_3 has received relatively little attention. Ground state PH_2 ($\tilde{X}^2B_1$) radicals have been observed in absorption following the flash photolysis of phosphine[154, 156], and a number of high resolution studies of the PH_2 ($\tilde{A}^2A_1 \leftarrow \tilde{X}^2B_1$) system have been reported[157]. A recent study involving H-atom resonance fluorescence has confirmed the primary production of PH_2 radicals via reaction (ix)

$$PH_3 \to PH_2(\tilde{X}^2B_1) + H \qquad \text{(ix)}$$

at wavelengths longer than 105 nm[158]. The production of electronically excited PH_2 radicals has been observed at $\lambda < 208 \pm 1$ nm[159]. Although no quantum yield measurements have been reported, the overall intensity of the PH_2 ($\tilde{A}^2A_1$) emission has been found to be about the same as that of the NH_2 ($\tilde{A}^2A_1$) emission from the photolysis of NH_3[160], from which we can infer $\phi[PH_2(\tilde{A})] \lesssim 0.01$. The fluorescence excitation spectrum of $PH_2(\tilde{A}^2A_1)$ closely resembles the absorption spectrum of the parent molecule[159] at energies up to the threshold for photoionisation (124.2 nm)[161] from which point it rapidly decreases to zero.

The results from the few studies focussed on the alternative dissociation channels involving formation of the PH radical (see Table 3.3.2) reveal certain interesting dissimilarities when compared with the equivalent dissociations in NH_3. Kley and Welge[156] have observed ground state $PH(X^3\Sigma^-)$ radicals in absorption following quartz ultraviolet flash photolysis of PH_3. Although the $(^1\phi \leftarrow a^1\Delta)$ and $(^1\Pi \leftarrow a^1\Delta)$ systems of the PH radical have been observed in absorption at $\sim$ 162.5 nm and $\sim$ 159.5 nm respectively[166], the only currently available estimates for the vertical excitation energy of the $PH(a^1\Delta)$ state have been provided by theoretical calculations[162–164].

The production of $PH(b^1\Sigma^+)$ fragments in the photolysis of PH_3 at all wavelengths shorter than 147 nm, presumably via the spin-allowed process:

$$PH_3 \longrightarrow PH(b^1\Sigma^+) + 2H \qquad \lambda < 147\ nm \qquad (x)$$

has recently been established by observation of the $PH(b^1\Sigma^+ \to X^3\Sigma^-)$ emission at 697.3 nm[160]. It has been proposed that the production of $NH(b^1\Sigma^+)$ in the photodissociation of NH_3 arises via a secondary reaction involving the well-documented $NH(c^1\Pi)$ state[161]. In contrast, there have been no reported experimental observa-

Table 3.3.2 Threshold energies for primary product channels in the photodissociation of PH_3 (upper values, in brackets, are in electron volts, lower values are in nanometres)

PH ↓		$H_2(^1\Sigma^+)$ (0.00)	2H $(4.48)^e$
$X^3\Sigma^-$	(0.00)	$(2.14)^b$ 579.3	(6.62) 187.3
$a^1\Delta$	$(0.95)^{a*}$	(3.09) 404	(7.57) 164
$b^1\Sigma^+$	$(1.78)^b$	(3.92) 316.3	$(8.40)^b$ 147.6
$A^3\Pi_i$	$(3.63)^c$	(5.77) 214.9	$(10.25)^g$ 121.0
$c^1\Pi$	$(4.72)^{d*}$	(6.86) 181	(11.34) 109
$d^1\Sigma^+$	$(7.13)^{d*}$	(9.27) 134	(13.75) 90

PH_2 ↓		$H(^2S)$ (0.00)
$\tilde{X}(^2B_1)$	(0.00)	3.69 ± 0.05^g 336 ± 4
$\tilde{A}(^2A_1)$	$(2.27)^f$	$(5.96 \pm 0.05)^g$ 208 ± 1

a Ref.162–164.   c Ref.165.   e Ref.130.   g Ref.159.
b Ref.160.   d Ref.162.   f Ref.157.

* No experimental verification.

tions of the $PH(c^1\Pi)$ electronic state, and the observation of $PH(b^1\Sigma^+ \rightarrow X^3\Sigma^-)$ fluorescence at wavelengths where the formation of $PH(c^1\Pi)$ by a spin-allowed process is energetically inaccesible has led to the suggestion that $PH(b^1\Sigma^+)$ is a primary product in the photodissociation of PH_3 [160].

$PH(A^3\Pi_i)$ fragments have been observed in emission around 342 nm [148, 159] and high resolution analyses of the PH and $PD(A^3\Pi_i \rightarrow X^3\Sigma^-)$ systems have been reported [164, 167]. A very low yield of $PH(A^3\Pi \rightarrow X^3\Sigma^-)$ fluorescence was observed for all $\lambda < 159 \pm 2$ nm [159] (which approximately coincides with the onset of the $PH_3(\tilde{B} \leftarrow \tilde{X})$ absorption system); however, the fluorescence yield increases dramatically at $\lambda < 121$ nm, the thermodynamic threshold for the spin-allowed process:

$$PH_3 \rightarrow PH(A^3\Pi_i) + 2H \qquad \lambda < 121 \text{ nm} \qquad\qquad (xi)$$

The lack of any observed fluorescence from excited vibrational levels in the $PH(A^3\Pi_i)$ electronic state [159] has been attributed to a competing predissociation from levels with $v' > 0$ [164].

It is not altogether surprising that the analogous $NH(A^3\Pi \rightarrow X^3\Sigma^-)$ emission at ~ 336 nm has not been observed in the photodissociation of NH_3, as the thermodynamic threshold for the spin-allowed proces (xii) is 102.5 nm (see Table 3.3.1)

$$NH_3 \longrightarrow NH(A^3\Pi_i) + 2H \qquad\qquad (xii)$$

3.4 HCN and DCN

3.4.1 Spectroscopy

In its ground electronic state the electronic configuration of HCN may be represented as

$$\dots (4\sigma)^2 \, (5\sigma)^2 \, (1\pi)^4; \, {}^1\Sigma^+.$$

The lower electronically excited states will be populated by the electron promotions $1\pi \rightarrow 2\pi$ (which generates states of symmetry $^{3,1}\Sigma^+$, $^{3,1}\Sigma^-$ and $^{3,1}\Delta$) and $5\sigma \rightarrow 2\pi$ (which generates $^{3,1}\Pi$ states). Some of these are expected to be bent on the basis of the Walsh diagram [168] [3]; the correspondence between the molecular orbitals under the $C_{\infty v}$ and C_s point groups is also shown in Fig. 3.4.1.

Transitions populating the $^1\Delta$ or $^1\Sigma^-$ states are forbidden in the linear configuration and so it is not surprising that the first absorption bands of HCN and DCN, which lie between 160 nm and 200 nm, are weak [170, 171]. A detailed high resolution study of the absorption in this region was made by Herzberg and Innes [171] who observed long progressions in the bending mode v_2', and established the bond angle in the upper state as $125°$. Diffuseness gradually sets in as the level of vibrational exci-

---

3 Modifications to the Walsh diagram have been proposed by Absar and McEwen [169] on the basis of fixed geometry SCF calculations. Schwenzer et al. [172] also comment that the original Walsh diagram may not be entirely appropriate for the case of HCN.

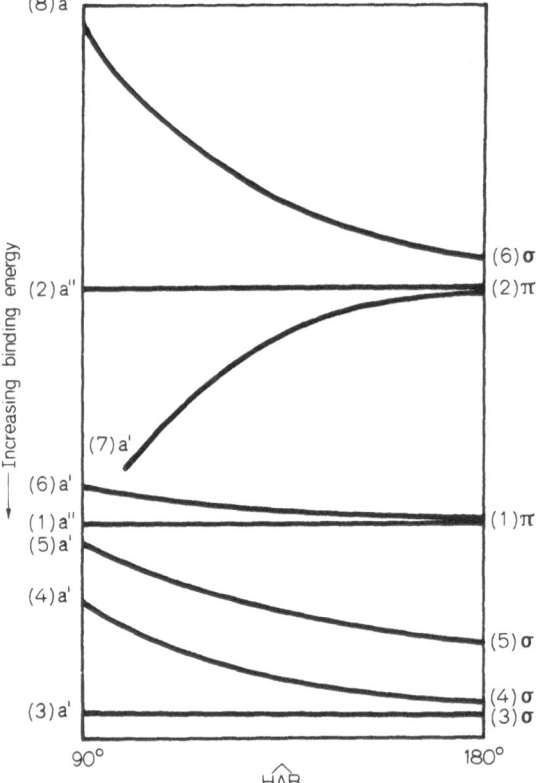

Fig. 3.4.1. Walsh diagram for molecule HAB (after Walsh, Ref.[168])

tation increases, and in the higher vibrational levels the molecules are quite strongly predissociated[171]. Analysis of the K-type doubling resolvable in the rotational structure of the lower levels indicated an upper state of symmetry A''[171]. The transition, which in HCN has its band origin at 191.4 nm, has been associated with the $1a'' \rightarrow 7a'$ ($\tilde{A}\,^1A'' \leftarrow \tilde{X}$) transition correlating with excitation into the linear $^1\Delta$ state[171]. SCF/CI calculations also find the lowest excited singlet state to be one of A'' symmetry, stabilised by bending and almost entirely associated with the $1a'' \rightarrow 7a'$ orbital promotion[172,173] (see Table 3.4.1). However, some calculations and comparisons with isoelectronic species have placed the $^1A''(^1\Sigma^-)$ state below the $^1A''(^1\Delta)$ state, and on this basis the transition has been reassigned as $\tilde{A}\,^1A''(^1\Sigma^-) \leftarrow \tilde{X}$[173].

Weaker features associated with the C–N stretching mode (ν_3') have been identified in the absorption spectra of HCN and DCN, but the spectrum of DCN in this region is further complicated by the appearance of a band system with its origin at 183.1 nm, the intensity of which is comparable with that of the $\tilde{A}\,(0, \nu_2', 0) \leftarrow \tilde{X}\,(0, 0, 0)$ progression[171]. This additional system displays a progression in a frequency similar to that of the bending mode ν_2' in the $\tilde{A}$ state[171]. Rotational analysis indicated "unambiguously" a bent upper state of A'' symmetry and the system was assigned to the transition $\tilde{B}\,^1A''(^1\Sigma^-) \leftarrow \tilde{X}$[171]. No analogous progression appears in the absorption spectrum of HCN and its absence was ascribed to rapid predissociation or to fortuitous overlap with $\tilde{A} \leftarrow \tilde{X}$ system[171]. SCF/CI calculations predict that the second excited $^1A''$ state should be near linear, which is inconsistent with the observed long progression in the bending frequency in DCN, and it should lie at a higher energy than is observed experimentally[172-174]. It was suggested, therefore, that the original assignment was incorrect, and to obtain better agreement with the calculations the $\widetilde{B}$ state was reassigned to the $2\,{}^1A'({}^1\Delta)$ state[172, 173], although this assignment conflicted with Herzberg and Innes's rotational analysis[171].

The non-linear geometry of the $\widetilde{B}$ system, the assignment to a state of symmetry A'' as originally proposed, and the absence of a second bent ${}^1A''$ state in this region, could all be reconciled, of course, if the $\widetilde{B} \leftarrow \widetilde{X}$ band system were, in fact, the $(1, v_2', 0)$ progression of the $\widetilde{A}$ state (where v_1' is the C–D stretching mode); a situation analogous to this is indeed believed to occur in the $\widetilde{C}$ state of DCN (see later discussion). This explanation has however been rejected on the basis of anharmonicity constants[171] and calculated Franck-Condon factors[173]. Péric et al.[173] further suggest that the first bands associated with the "$\widetilde{B}$" state are actually hot bands of the $\widetilde{A} \leftarrow \widetilde{X}$ system, and re-locate the $\widetilde{B}$ state band origin at 165.7 nm, close to their calculated value for the $2\,{}^1A'$ state.

HCN begins to absorb strongly at $\lambda \lesssim 150$ nm, and the absorption spectrum above 130 nm is dominated by a progression with spacing $\sim (750-850)$ cm^{-1} [3, 175-177], which has been associated with excitation of the v_2' bending mode of the $\widetilde{C}$ electronic state[3, 175, 177] (see Fig. 3.4.2).

The vibrational bands are diffuse, the diffuseness increasing with vibrational quantum number, again indicating strong predissociation[3, 175-177]. Rotational structure has been resolved on the lower members of the progression, and analysis of this has indicated that the state has A' symmetry and a bond angle of 141° [3, 175]. Consistent with a bent upper state geometry, the absorption spectrum reaches a maximum at $v_2' = 7$. Herzberg and Innes[3, 175] assigned this system to the A' component of the ${}^1\Pi$ state arising from the $(5\sigma \rightarrow 2\pi)\,(5a' \rightarrow 7a')$ orbital promotion; SCF/CI calculations place a ${}^1A'$ state in this region with bond angle 141.2°, and find that for this state the configuration $(\ldots 5a', 6a'^2, 1a''^2, 1a''^2, 7a')$ is dominant[172] (see

Table 3.4.1 Theoretical predictions for the singlet states of HCN (after Schwenzer et al., Ref.[172])

Symmetry	T_e(eV)	r_e(HC) (Å)	R_e(CN) (Å)	θ_e° (HCN)	Most important configurations	Coefficients
$\widetilde{X}\,{}^1\Sigma^+({}^1A')$	0.00	1.055	1.180	180	$5a'^2\,6a'^2\,1a''^2$	0.9684
$\widetilde{A}\,{}^1A''$	6.48	1.096	1.318	127.2	$5a'^2\,6a'^2\,1a''\,7a'$	0.9714
$2\,{}^1A'$	6.78	1.102	1.287	124.9	$5a'^2\,6a'\,1a''^2\,7a'$	0.7024
					$5a'\,6a'^2\,1a''^2\,7a'$	0.5123
					$5a'^2\,6a'^2\,1a''\,2a''$	0.3852
$2\,{}^1A''$	7.52	1.076	1.316	164.4	$5a'^2\,6a'\,1a''^2\,2a''$	0.8209
					$5a'^2\,6a'^2\,1a''\,7a'$	0.5372
$3\,{}^1A'$	7.85	1.092	1.264	141.2	$5a'\,6a'^2\,1a''^2\,7a'$	0.7442
					$5a'^2\,6a'^2\,1a''\,2a''$	0.4879
					$5a'^2\,6a'\,1a''^2\,7a'$	0.3264
$3\,{}^1A''$	8.97	1.045	1.229	180	$5a'\,6a'^2\,1a''^2\,2a''$	0.9392
$4\,{}^1A'$	9.54	1.313	1.254	180	$5a'^2\,6a'\,1a''^2\,8a'$	0.8639

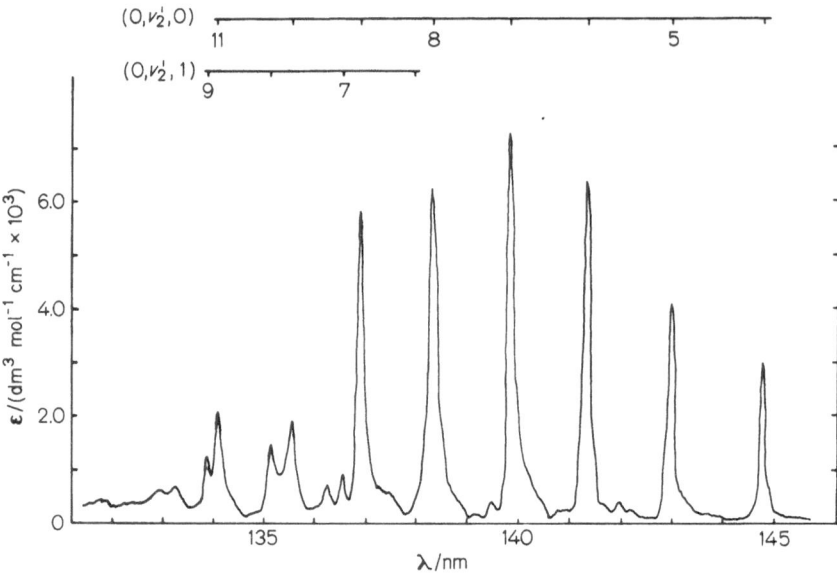

Fig. 3.4.2. Absorption spectrum of HCN in the wavelength range 130–150 nm (after Macpherson and Simons, Ref.[177])

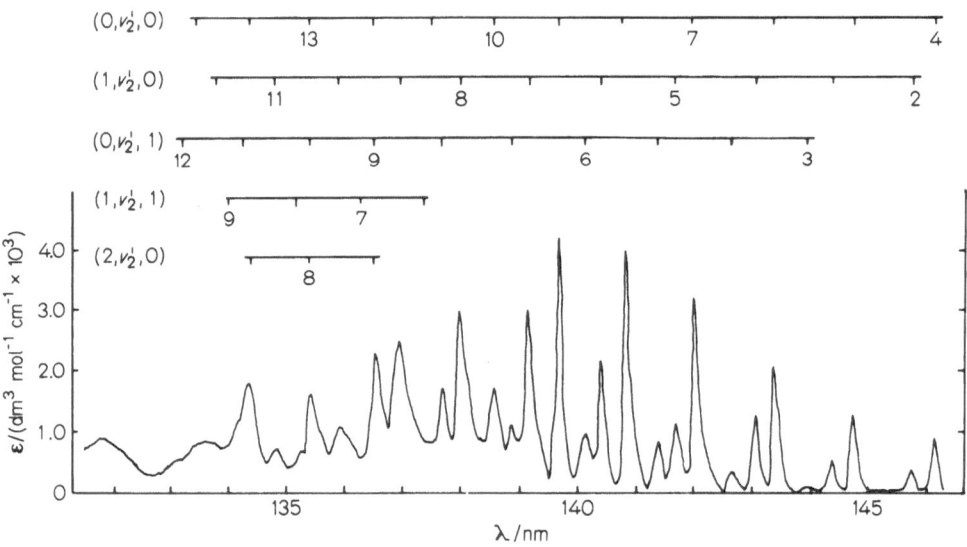

Fig. 3.4.3. Absorption spectrum of DCN in the wavelength range 130–150 nm (after Macpherson and Simons, Ref.[177])

Table 3.4.1). The transition is allowed under both $C_{\infty v}$ and C_s point group symmetry and the absorption is accordingly more intense than for the systems at longer wavelengths.

The $\tilde{C} \leftarrow \tilde{X}$ absorption region in DCN is considerably more complex than in HCN and the strong progression in the bending frequency $(0, v_2', 0)$, peaking at

$v_2' \sim 7-8$, is augmented by another system of comparable intensity, desplaced from it by a frequency appropriate to one quantum of the C–D stretching frequency v_1' [177]. The behaviour is very reminiscent of that observed in the $\tilde{A} \leftarrow \tilde{X}$ region, but in the present case both systems are believed to belong to a single electronic transition [177]; a vibrational analysis of the spectra of the two isotopic species is shown in Figs. 3.4.2 and 3.4.3. The assignments are supported by photofragment fluorescence polarization and relative quantum yield measurements for $CN(B^2 \Sigma^+)$, produced *via* predissociation at $\lambda \lesssim 149$ nm (see Sect. 3.4.2). The absence of any vibrational progressions based on excitation of v_1' (C–H stretch) in $HCN(\tilde{C} \leftarrow \tilde{X})$ has been explained in terms of an efficient tunnelling predissociation mechanism which is very sensitive to isotopic substitution and leads to line broadening [177].

Another electronic state of HCN, the $\tilde{D}$ state, has been reported to have its band origin at 139.6 nm [3], and some features in the electron impact spectra of HCN[178–181] and DCN[178] have been assigned to overlapping vibrational progressions into the $\tilde{C}$ and $\tilde{D}$ states. However, nearly every vibrational feature in the optical spectra of HCN and DCN between 147 nm and 130 nm can be assigned without recourse to the presence of an additional electronic state [177].

At shorter wavelengths still, particularly below 115 nm, strong diffuse peaks appear in the absorption spectrum of HCN (see Fig. 3.4.4)[175,182]; the most intense peak lies at 112.1 nm. In an analysis of their electron impact spectrum, Tam and Brion[178] assigned this peak to the $(1\pi \rightarrow 3p\pi)$ Rydberg transition, which could correspond to the intense $^1 \Sigma^+ \leftarrow \tilde{X}\, ^1 \Sigma^+$ system expected from the electronic configuration $(\pi)^3 (\pi)$. Other Rydberg features have been assigned to the excitation of electrons from the 1π and 5σ orbitals, and Rydberg series have been identified leading to the first ionization potential at 13.61 eV[178].

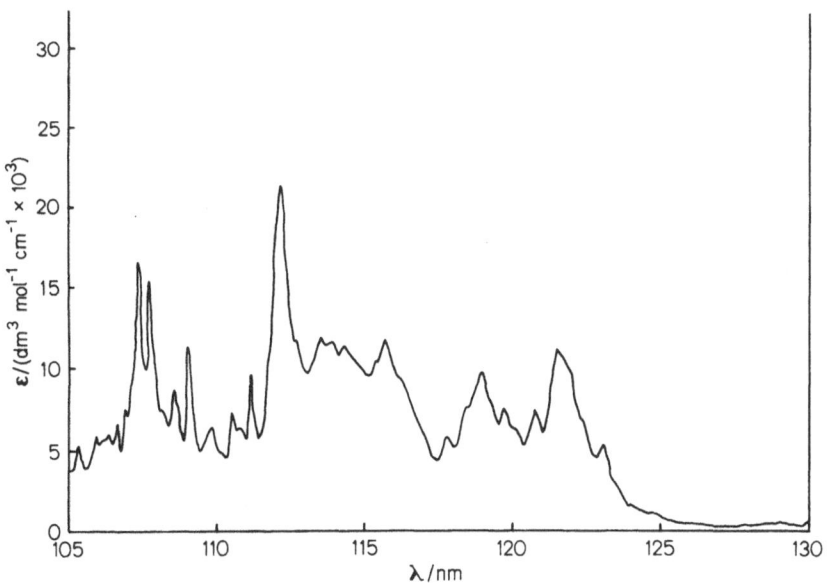

Fig. 3.4.4. Absorption spectrum of HCN in the wavelength range 105–130 nm (after West, Ref.[176])

3.4.2 Photochemistry

The first three energetically accessible photodissociation channels in HCN are

$$\text{HCN} \quad \rightarrow \text{H}(^2S) + \text{CN}(X^2\Sigma^+) \qquad \lambda < 237.5 \text{ nm} \qquad \qquad \text{(i)}$$

$$\rightarrow \text{H}(^2S) + \text{CN}(A^2\Pi) \qquad \lambda < 190.0 \text{ nm} \qquad \qquad \text{(ii)}$$

$$\rightarrow \text{H}(^2S) + \text{CN}(B^2\Sigma^+) \qquad \lambda < 149.0 \text{ nm} \qquad \qquad \text{(iii)}$$

Most photochemical studies of HCN have involved observation of fluorescence either from CN(A) in the red and near infrared (band origin at 1082 nm) or CN(B) in the violet and near u.-v. (band origin at 388 nm). HCN is transparent at the threshold wavelength for channel (i) and the threshold for channel (ii) coincides roughly with the band origin of the weak $\tilde{A} \leftarrow \tilde{X}$ system. The absorption spectrum does not become noticeably broadened until $\lambda < 180$ nm[171] and although fluorescence has been reported from HCN ($\tilde{A}\,^1A''$)* it is likely that predissociation occurs throughout the entire absorption spectrum.

West and Berry observed CN($A \rightarrow X$) fluorescence following excitation of HCN in the $\tilde{A} \leftarrow \tilde{X}$ region and reported a photodissociation threshold for CN($A^2\Pi$) at $\lambda > 180$ nm[183]. They also observed laser action on the CN($A \rightarrow X$) system and used a correlation diagram to account for the preferential formation of CN(A) rather than

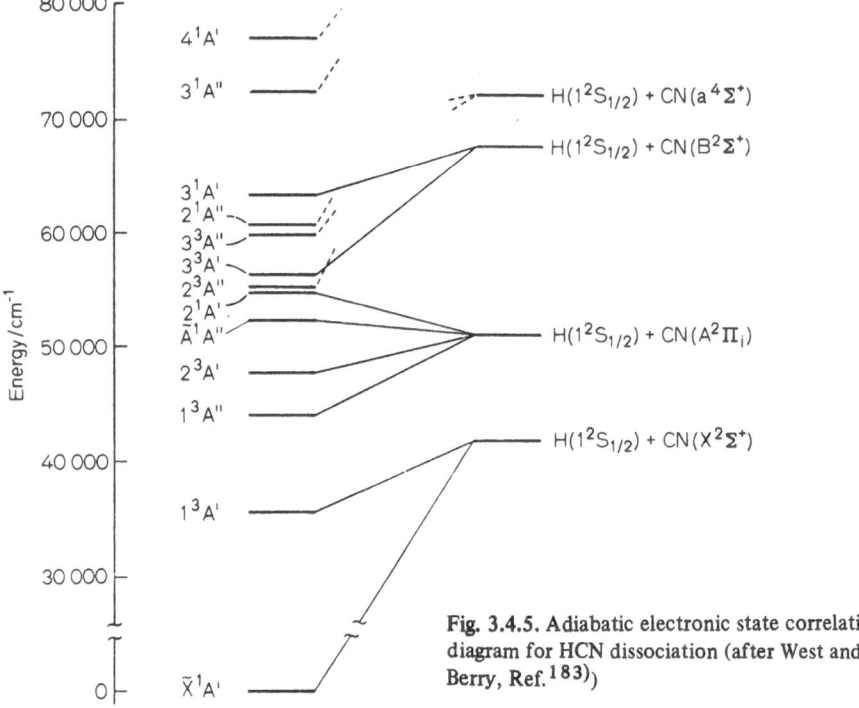

Fig. 3.4.5. Adiabatic electronic state correlation diagram for HCN dissociation (after West and Berry, Ref.[183])

---

* Added in proof: A. P. Baronavski, Chem. Phys. Lett., 61, 512 (1979).

CN(X). The diagram (see Fig. 3.4.5) shows that the $\tilde{A}\,^1A''$ (and $2\,^1A'$) states are the only singlet states which correlate with $H(^2S) + CN(A\,^2\Pi)$ and that predissociation via the ground state would lead to $H(^2S) + CN(X^2\Sigma^+)$. Since the latter channel is suppressed, West and Berry[183] suggested that channel (ii) proceeded via vibrational predissociation in the $\tilde{A}\,^1A''$ state with the C–H stretching mode identified as the reaction coordinate. The absence of progressions based on excitation of the C–H mode in HCN was consistent with this mechanism[171].

The quantum yield of CN(A) from HCN has been measured at wavelengths $\lambda < 150$ nm[176], which corresponds to excitation into the $\tilde{C}\,^1A'$ state, though at 150 nm, where $\phi[CN(A)] = 0.11$, the absorbed energy still lies below the threshold for channel (iii). Figure 3.4.5 indicates a direct correlation between HCN ($\tilde{C}\,^1A'$) and $H(^2S) + CN(B^2\Sigma^+)$, (under C_s point group symmetry though not under $C_{\infty v}$, where the $^1A'$ state becomes $^1\Pi$) and fluorescence from CN(B) has been detected once the threshold for channel (iii) is crossed[176, 184]. However, the combined quantum yields of CN(A) and CN(B) remain smaller than unity throughout the range 150 – 105 nm[176] and presumably all three channels contribute in this region.

The CN($A \rightarrow X$) photofragment fluorescence was first observed by Mele and Okabe[185] following photolysis of HCN at 147.0 nm and 123.6 nm, and West[176] has recorded the photofragment fluorescence excitation spectrum over the region occupied by the $\tilde{C}\,^1A'$ state; the spectrum closely follows the absorption spectral profile. The quantum yield $\phi[CN(A)]$ varies between 0.10 and 0.18 over the wavelength range 150 nm – 130 nm[176]. Attempts have been made to monitor vibrational

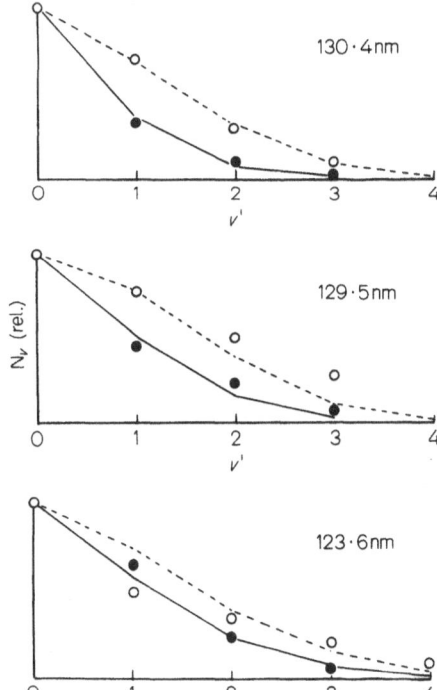

Fig. 3.4.6. Relative vibrational populations in CN(B)$_v$, from the photodissociation of HCN (•) and DCN (○) at three wavelengths. Modelled distributions for HCN (——) and DCN (–––) are included for comparison (after Ashfold, Macpherson and Simons, Ref.[189])

energy disposal in the $CN(A)$[186] but its long radiative lifetime ($\tau \sim 7\ \mu s$[187]) and the weakness of its fluorescence which lies in the red and near infrared, prevent any measurements being made under collision free conditions. The reported distributions are undoubtedly relaxed.

Energy partitioning in the photodissociation channel (iii) yielding $CN(B^2\Sigma^+)$ has been much more thoroughly investigated[185, 188, 189] despite the lower quantum yield for its production (< 0.06 at $\lambda > 130$ nm), since its radiative lifetime ($\tau \sim 60-65$ ns[190-192]) is far shorter and its fluorescence lies in a convenient spectral region. Mele and Okabe[185] measured vibrational energy disposal in $CN(B)$ following photolysis of HCN at the Kr and Xe atomic resonance lines, and the $CN(B \rightarrow X)$ fluorescence spectra produced through photodissociation of HCN and DCN at wavelengths between 147.0 nm and 123.6 nm, have been recorded under high resolution in the authors' own laboratory[189]. The vibrational distributions determined from the spectra are shown in Fig. 3.4.6. The availability of the original results of Mele and Okabe[185] has prompted a number of attempts at theoretical modelling of the vibrational energy disposal[193]. Most of these distinguish contributions to vibrational excitation arising from the changes in geometry and force constants on transfer to the final repulsive potential surface ("Franck-Condon" considerations), and from recoil between the separating fragments ("final state interactions"). With a light atom such as hydrogen, final state interactions are expected to be relatively unimportant[83], and Franck-Condon factors will largely determine the observed levels of vibrational excitation. A quasi-diatomic global model which incorporates the correct parameters for the HC–N and C–N bond lengths and frequencies in the $\tilde{C}^1A'$ and $B^2\Sigma^+$ states, respectively, has successfully reproduced the observed vibra-

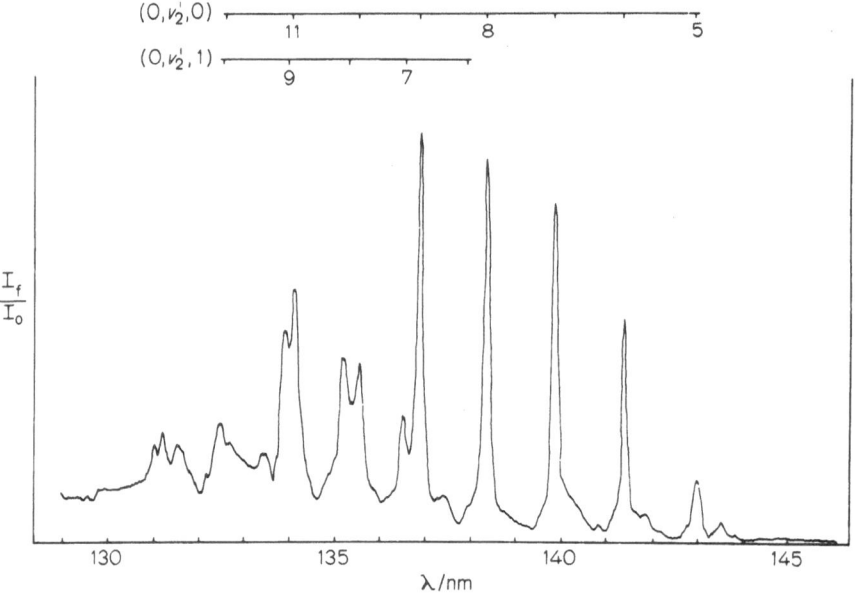

Fig. 3.4.7. Photofragment excitation spectrum of $CN(B^2\Sigma^+)$ from HCN (after Macpherson and Simons, Ref.[177])

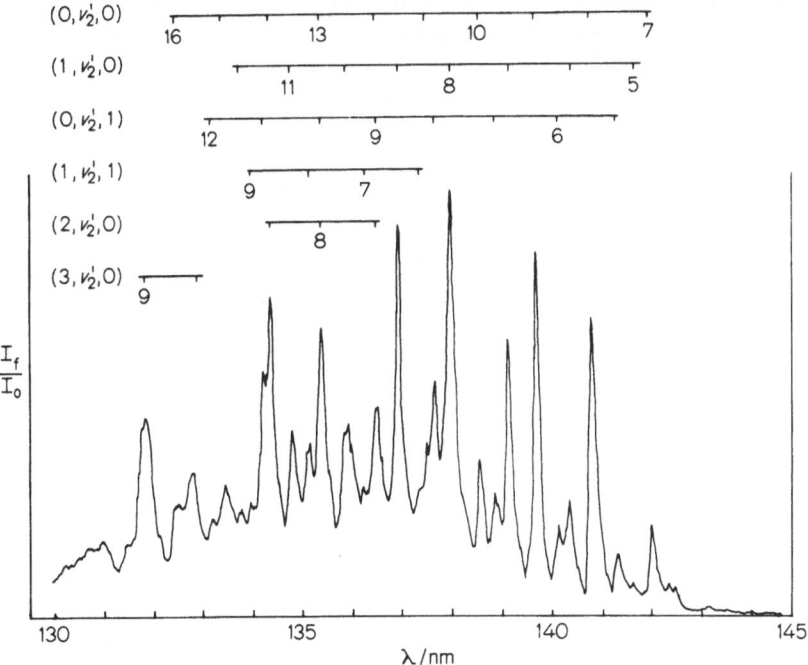

Fig. 3.4.8. Photofragment excitation spectrum of CN $(B^2\Sigma^+)$ from DCN (after Macpherson and Simons, Ref.[177])

tional distributions at energies near threshold, as well as the increased excitation introduced by substitution of the heavier isotope in DCN[189] (see Fig. 3.4.6). The rotational distributions near threshold reflect the mapping $J_{parent} \rightarrow j_{fragment}$, which satisfies the requirement of angular momentum conservation following detachment of the H or D atom when there is no appreciable recoil between the separating fragments[194].

A detailed study of the predissociation channel (iii) from the $\widetilde{C}\,^1A'$ state in HCN and DCN, which included analysis of the absorption and photofragment fluorescence excitation spectra, and measurements of the relative quantum yields and polarizations of the fluorescence following excitation into different vibronic levels, provides an example of the interrelation between photochemistry and spectroscopy[177]. This was used to help characterise the nature of the excited electronic parent molecular states and provide information on the topography of the potential surfaces over which the predissociation proceeds. The photofragment fluorescence excitation spectra from HCN and DCN are displayed in Figs. 3.4.7 and 3.4.8; all the major vibronic bands can be attributed to progressions in the $\widetilde{C}\,^1A'$ state, and the CN $(B \rightarrow X)$ fluorescence is polarized positively, confirming the A' symmetry in the upper state[177]. The absence of any progressions which involve excitation of the C–H stretching mode v_1' in HCN contrasts with the structure observed in the corresponding spectrum for DCN, where progressions with $v_1' \leqslant 2$ are quite sharp and spectral line broadening only becomes pronounced when $v_1' = 3$ (see Fig. 3.4.8). This has led to the suggestion that the rate of predissociation in HCN is so greatly accelerated by excitation of v_1' that the corresponding spectral features merge into a continuum. Indeed, comparison of the absorption and fluorescence excitation spectra of HCN reveals an enhanced contribution from an underlying continuum at $\lambda \sim (130\text{--}135)$ nm (see Fig. 3.4.7) and it is in this region that the $(1, v_2', 0)$ progression, if it were resolved, would be most intense.

The isotopic sensitivity of the rates of predissociation in HCN and DCN can be understood in terms of a tunnelling mechanism out of the $\widetilde{C}^1A'$ state, since penetration through the non-classical region would be reduced by deuteration, but enhanced by excitation of v_1'. By combining the results for the vibrational energy disposal, (which lead to an estimate of the steepness of the repulsive potential between the separating fragments), with the observed isotope effect on the rates of predissociation, (which leads to an estimate of the potential barrier through which tunnelling may proceed), it has been possible to map the contour of the potential surface along the reaction coordinate (assumed to be Q_1', the H(D)–CN distance[177]) (see Fig. 3.4.9).

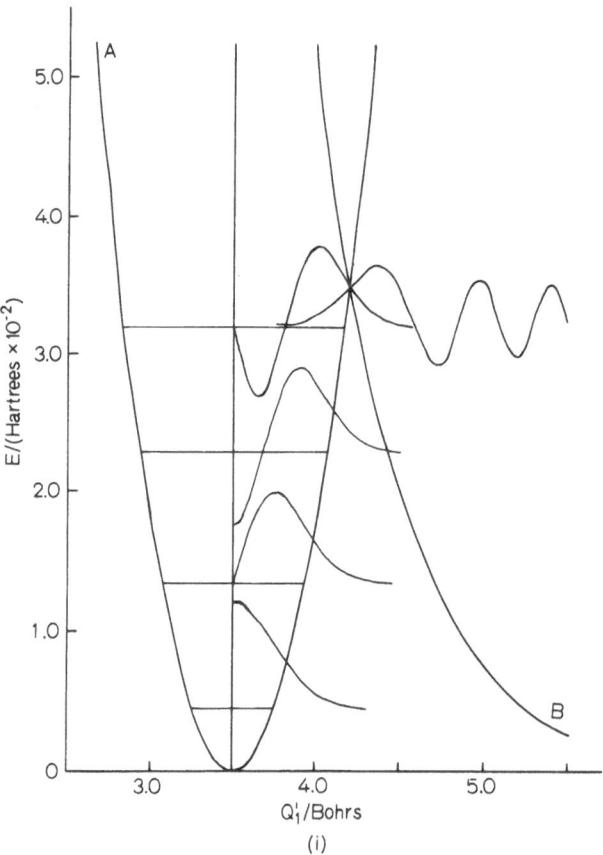

Fig. 3.4.9 (i) and (ii). Bound and continuum potential energy functions for (i) HCN and (ii) DCN showing variation in Franck-Condon overlap with isotopic substitution.
Bound simple harmonic oscillator function with (i) $v_1' = 2000$ cm^{-1} and (ii) $v_1' = 1450$ cm^{-1}.

If the tunnelling mechanism provides the only predissociative pathway, the isotopic sensitivity should be greatest when $v_1' = 0$, since this would correspond to the maximum barrier width. In fact, the rates of predissociation from levels with $v_1' = 0$ in HCN and DCN, as reflected by the observed line widths, polarization ratios and relative CN(B) quantum yields, are very similar, and an alternative pathway must still be available when the tunnelling route is too slow to compete[177], i.e. when $v_1' = 0$ in HCN and when $v_1' \leqslant 2$ in DCN.

Photofragment excitation spectra of CN(A) and CN(B) have been recorded at shorter wavelengths ($\lambda \geqslant 110$ nm) by West[176], who also measured absolute quantum yields. The excitation and absorption spectra are similar, except for the peaks at ~ 121.5 nm and 119 nm, where ϕ[CN(A)] increases, and ϕ[CN(B)] falls slightly. The peaks at ~ 121.5 nm and 119 nm have been assigned to vibrational bands in the Rydberg transition $1\pi \to 3s\sigma$[178] which would correlate with CN($A\,^2\Pi$) but not CN($B\,^2\Sigma^+$).

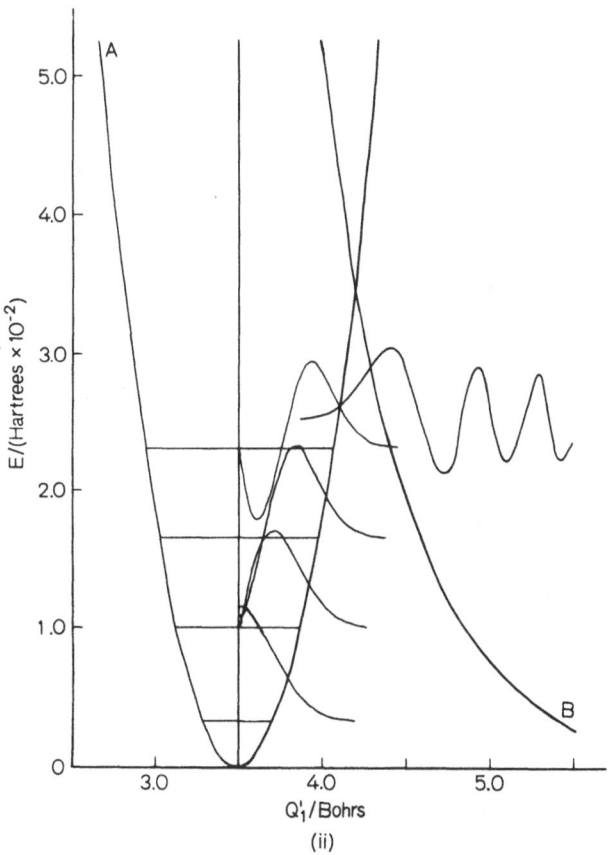

Exponentially repulsive function of the form $V = V_0 e^{-Q_1'/L}$ where $L = 0.5$ Bohr and $V_0 = 140$ Hartrees; the excess available energy is (i) E = 0.0355 Hartrees, (ii) E = 0.0302 Hartrees, corresponding to an incident photon with λ = 134 nm (after Macpherson and Simons, Ref.[177])

3.5 CO_2, OCS, OCSe, CS_2, CSe_2 and N_2O

3.5.1 Spectroscopy

Each of the molecules in this group has sixteen valence electrons, and in the linear ground state their electronic configurations can be written:

$$\dots (3\sigma_{(g)})^2 \, (1\pi_{(u)})^4 \, (4\sigma_{(g)})^2 \, (2\pi_{(g)})^4 \quad {}^1\Sigma_{(g)}^+$$

Assuming, for the present, a pure intravalency shell description, the lowest-lying electronically excited states will be populated through the orbital promotion $2\pi_{(g)} \to 3\pi_{(u)}$ which generates singlet and triplet states of symmetry $\Delta_{(u)}^{(\pm)}$, $\Sigma_{(u)}^-$, $\Sigma_{(u)}^+$. Walsh's rules[195] predict that the excited states may be bent and, in particular, Renner-Teller interaction will lift the degeneracy of the $\Delta_{(u)}^{(\pm)}$ state. The lowered symmetry may also alter the ordering of the states derived from the $\Delta_{(u)}$ and $\Sigma_{(u)}^-$ configurations. Table 3.5.1 shows the symmetry species for each state for both linear and bent geometries.

At the other extreme, the lowest-lying electronically excited states associated with pure Rydberg transitions could be generated by the orbital promotions $2\pi_{(g)} \to ns\sigma_{(g)}$ (producing ${}^{1,\,3}\Pi_{(g)}$ states), $2\pi_{(g)} \to np\sigma_{(u)}$ (producing ${}^{1,\,3}\Pi_{(u)}$ states) and $2\pi_{(g)} \to np\pi_{(u)}$ (producing ${}^{1,\,3}\Delta_{(u)}^{(\pm)}$, ${}^{1,\,3}\Sigma_{(u)}^-$ and ${}^{1,\,3}\Sigma_{(u)}^+$ states).

In practice, neither of the two descriptions will be wholly accurate since configuration interaction will introduce Rydberg character into the more highly excited electronic states. For example, the $\Sigma_{(u)}^+$ states which can be generated by intravalency shell or Rydberg excitation would be expected to form a conjugate pair[1] the lower of which would assume the greater contribution from the intravalency shell description (though still possessing a high degree of Rydberg character if it lies at high enough energy). Similarly, the ${}^1\Pi_{(g)}$ Rydberg state has the same symmetry as the intravalency shell state generated by the transition $2\pi_{(g)} \to 5\sigma_{(g)}$, but since the latter transition is expected to lie at energies above, or close to, the ${}^1\Sigma_{(u)}^+$ state, it also assumes a high degree of Rydberg character and is generally identified as the first member of the $2\pi_{(g)} \to ns\sigma_{(g)}$ Rydberg Series[6, 196-198]. When the excited states have a high degree of intravalency shell character, their equilibrium geometries can be rationalised in terms of the appropriate Walsh diagram[195]. For example, it would be expected that the A_2 component of the $\Delta_{(u)}$ state would bend far more than the B_2 component[6]. When there is a high degree of Rydberg character, the equilibrium geometry will approach that of the molecular ionic core.

Table 3.5.1 Correlation of the excited $(\pi^3\pi)$ electronic state representations between various point groups

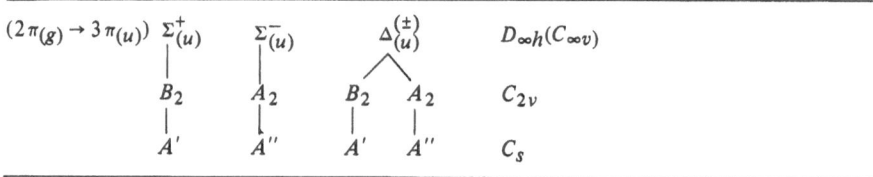

$(2\pi_{(g)} \to 3\pi_{(u)})$ $\Sigma_{(u)}^+$	$\Sigma_{(u)}^-$	$\Delta_{(u)}^{(\pm)}$		$D_{\infty h}(C_{\infty v})$
B_2	A_2	B_2	A_2	C_{2v}
A'	A''	A'	A''	C_s

A major attempt at providing a coherent analysis of the electronic transitions in 16-valence electron molecules was presented in 1971, by Rabalais, McDonald, Scherr and McGlynn[6], who confronted the experimental u.-v. and v.u.-v. spectral data with the predictions of approximate m.o. calculations based principally on the Mulliken-Wolfsberg-Helmholz technique. By looking for regular patterns of behaviour and correlating the spectral features of individual molecules within the iso-electronic group, they attempted a complete assignment of all the observed electronic transitions. In a more recent series of papers, King and his co-workers[196-200] have reviewed and revised some of the proposals made by Rabalais et al., since subsequent discussions by Mulliken[201], Winter et al.[202, 203], and M. Jungen[204] have shown that the earlier work paid insufficient attention to a proper assessment of the degree of Rydberg character in the composition of the excited molecular orbitals.

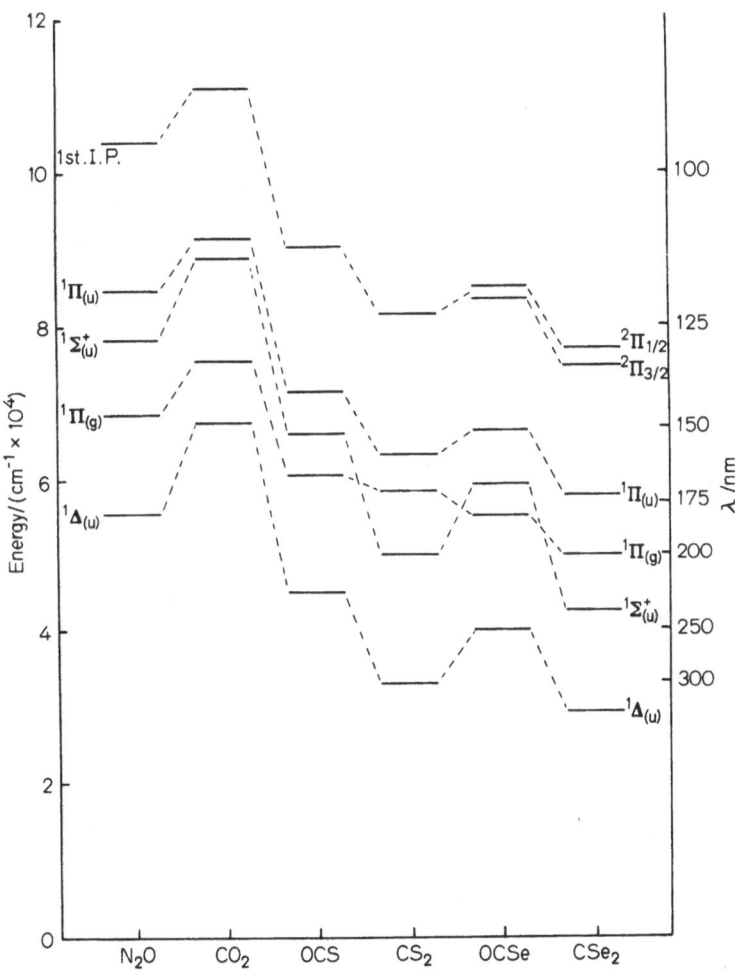

Fig. 3.5.1. Summary of the probable assignments and ordering of the low lying electronic states of CO_2, OCS, CS_2, OCSe, CSe_2 and N_2O

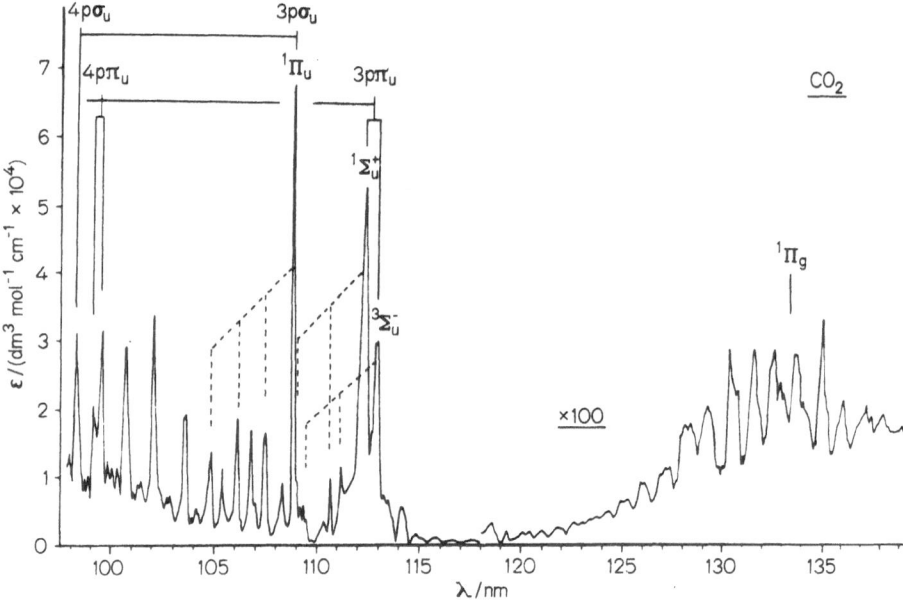

Fig. 3.5.2. Vacuum ultraviolet absorption spectrum of CO_2 (after Greening and King, Ref.[196])

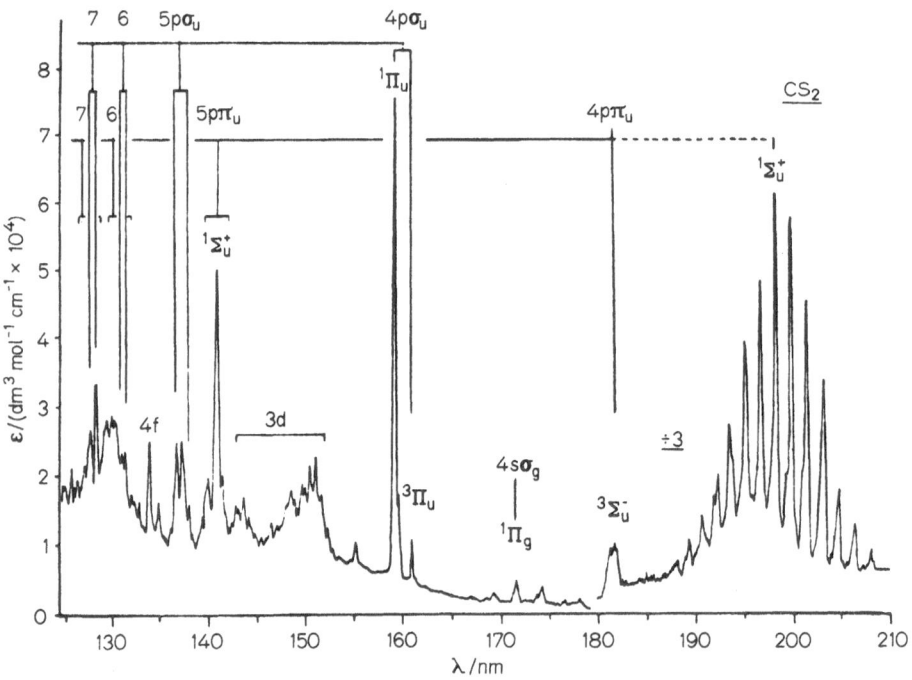

Fig. 3.5.3. Vacuum ultraviolet absorption spectrum of CS_2 (after Greening and King, Ref.[196])

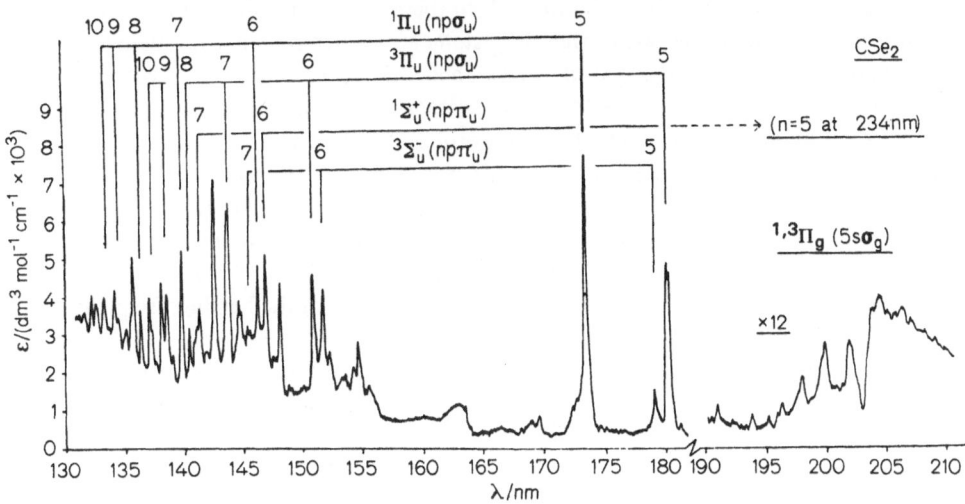

Fig. 3.5.4. Vacuum ultraviolet absorption spectrum of CSe_2 (after Greening and King, Ref.[198])

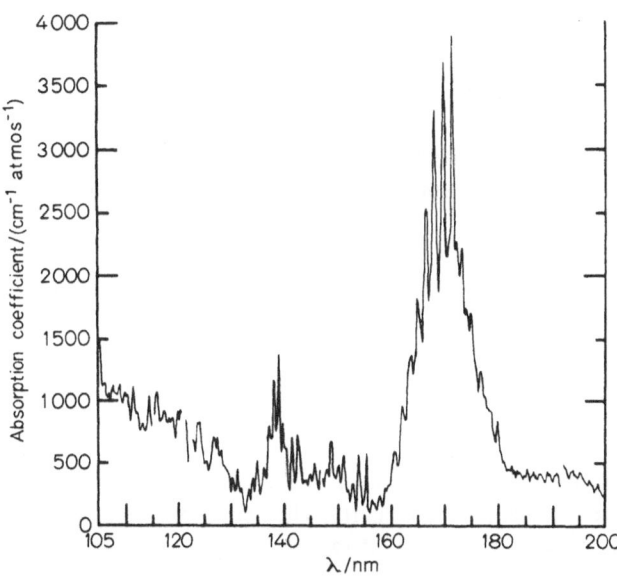

Fig. 3.5.5. Vacuum ultraviolet absorption spectrum of OCSe (after Black et al., Ref.[250])

An energy level diagram which summarises the most likely assignments and ordering of the lowest-lying electronic states in the vacuum u.-v. is presented in Fig. 3.5.1. It results from an assessment of the current published work but leans heavily on the conclusions reached by King and his co-workers[196–200]. Representative v.u.-v. absorption spectra are displayed in Figs. 3.5.2–3.5.5 together with the principal electronic assignments.

$^1\Delta_{(u)}^{(\pm)}$ and $^1\Sigma_{(u)}^-$ states. Excitation into each of these states is forbidden in the linear configuration, and, in the case of the $^1\Sigma_{(u)}^-$ state, even when non-linear, unless the point group symmetry is reduced to C_s (see Table 3.5.1); these states are expected to be bent and to possess intravalency shell character. In each molecule the $^1\Delta_{(u)} \leftarrow {}^1\Sigma_{(g)}^+$ transition excites a broad absorption band, which in N_2O (~ 180 nm)[6, 205, 206] is completely structureless. In CO_2 (~ 148 nm)[6, 196, 207] CS_2 (~ 310 nm)[6, 208] and CSe_2 (~ 385 nm)[200], the absorption displays a complex vibrational band structure which is complicated by the effects of Renner-Teller inter-actions in the bent upper B_2 and/or A_2 states. In OCS (~ 225 nm)[6] and OCSe (~ 250 nm)[197, 199], diffuse vibrational progressions in the bending frequency ν_2' may be identified.

Evidence for transitions into the $^1\Sigma_{(u)}^-$ state in absorption is less well established. In CO_2 (~ 190 nm) and N_2O (~ 270 nm), very weak, continuous regions of absorption have been attributed[6] to the $^1\Sigma_{(u)}^- \leftarrow {}^1\Sigma_{(g)}^+$ transition, though more recent cal-culations for N_2O[203, 209] and CO_2[202, 210] predict that this transition will lie at con-siderably shorter wavelength. In CS_2, both the very weak bands centred around 350 nm[6] and the weak absorption underlying transitions into the B_2 component of the $^1\Delta_u$ state[211] have been variously assigned to the $^1\Sigma_u^- \leftarrow {}^1\Sigma_g^+$ transition. How-ever, the recent work of Ch. Jungen et al.[208] found no evidence for the $^1\Sigma_u^-$ state in absorption. Similarly, in CSe_2, the absorption features at longer wavelength have been satisfactorily interpreted without recourse to a $^1\Sigma_u^-$ state[200]. The structured absorption systems around 200 nm in OCSe[197, 199] and around 230 nm in OCS[212] have been discussed in terms of excitation either to the bent $^1A''(^1\Sigma^-)$ electronic state, or to the $^1A'(^1\Delta)$ state. The latter assignment seems to us to be the more likely, in view of the earlier discussion and the expected weakness of transitions into the $^1\Sigma^-$ state.

$^1\Sigma_{(u)}^+$ states: Transitions into the $^1\Sigma_{(u)}^+$ state are fully allowed and absorption bands associated with them are very intense. In CO_2 (~ 112 nm)[196, 207], OCS (~ 153 nm)[6, 213] and OCSe (~ 170 nm)[197] the electronic transitions are very sharp, and lie at energies near enough to the lowest molecular ionisation potentials for the upper state to be regarded as predominantly Rydberg in character, with a major contribution from the $np\pi_{(u)}$ orbital function. Short progressions in the stretching frequencies of the upper state can be identified[6, 196, 197], consistent with a slightly stretched but linear or near-linear geometry. The corresponding band in N_2O (~ 128 nm) is, like the transition into the $^1\Delta$ state, entirely without struc-ture[6, 206, 214], reflecting very fast dissociation; the energy of the transition should favour a large Rydberg contribution, but one may speculate that as the N_2-O bond stretches, the orbital character evolves into one which is primarily intravalency anti-bonding without there being any significant potential barrier associated with the nuclear motion. In contrast, the electronic transitions into the $^1\Sigma^+$ states in CS_2 (~ 200 nm)[6, 196] and CSe_2 (~ 230 nm)[198] are far less energetic and exhibit a complex vibrational structure which includes the symmetric stretching frequency. They are best described as mixed intravalency/Rydberg transitions; a rotational ana-lysis of bands in the spectrum of CS_2 has identified the upper state as bent, $^1B_2$, correlating with $^1\Sigma_{(u)}^+$[215].

$^1\Pi_{(g)}$ and $^1\Pi_{(u)}$ states. In the centro-symmetric molecules the transitions into a linear $^1\Pi_g$ state are strongly forbidden; they are likely to remain relatively weak even when the symmetry is lowered. The $^1\Pi_u$ states, in contrast, are associated with intense absorption features, which correspond to well-defined members of the $2\pi_g \to np\,\sigma_u$ Rydberg series. Transitions of both types have been identified in CO_2 (~ 133 nm and ~ 108.8 nm)[196], CS_2 (~ 171 nm and ~ 159 nm)[196] and CSe_2 (~ 200 nm and ~ 173 nm)[198]; in each molecule the upper $^1\Pi_u$ states are believed to possess a high degree of Rydberg character[196, 198, 202]. In the non-centro-symmetric molecules, the two types are mixed through configuration interaction. In N_2O, the relatively weak banded system centred around 146 nm has been assigned to the $2\pi \to 3s\,\sigma$ ($^1\Pi \leftarrow {}^1\Sigma^+$) transition[6] with the more energetic $2\pi \to 3p\,\sigma$ Rydberg excitation occurring as a sharp peak at 117.9 nm[6, 205, 206, 214]. There has been some controversy over the assignment of the lower $^1\Pi$ state in OCS, where the banded system at 167 nm has been ascribed either to population of the lower $^1\Pi$[6] or to the $^1\Sigma^-$ state[199]. If the latter assignment were correct it would place the $^1\Sigma^-$ state surprisingly far above the $^1\Delta$ level (see Fig. 3.5.1), and, by analogy with OCSe, raise a question mark in regard to the whereabouts of the $^1\Pi$ level which is expected to occur in the same spectral region. Furthermore, if a $^1\Pi$ level does not lie in the region below $^1\Sigma^+$, difficulties arise in understanding the photochemical behaviour of OCS in the vacuum u.-v. (see later discussion). On balance, therefore, the assignment of the banded system ~ 167 nm in OCS to a $^1\Pi(4s\,\sigma)$ state is preferred, with the higher $^1\Pi(4p\,\sigma)$ state associated with the "Rydberg" absorption at ~ 140 nm. The analogous $^1\Pi(5s\,\sigma)$ and $^1\Pi(5p\,\sigma)$ absorption features in OCSe occur at ~ 180 nm and ~ 151 nm[197].

Triplet states. The lowest-lying triplet states associated with the intravalence $^1\Sigma^-_{(u)}$ and $^1\Delta_{(u)}$ states need not concern us since they lie at energies corresponding to absorption in the near u.-v. However, triplet transitions have been identified among the Rydberg systems in CS_2 and CSe_2[196, 198]. For example, the $^1\Pi_u \leftarrow {}^1\Sigma^+_g$ transition at 159.5 nm in CS_2 has a weak satellite band at a slightly longer wavelength[196], corresponding to an energy interval of the order of the spin-orbit interaction energy in the CS_2^+ ion. The transition into the $^3\Pi_u$ state is promoted by the (Ω_c, ω) coupling which becomes increasingly important as the quantum number n of the Rydberg state increases, so that in higher members of the series the $^1\Pi_u$ and $^3\Pi_u$ components reach comparable intensity[196]. Spin-orbit interaction is even more pronounced in the heavier atomic CSe_2, where the bands have similar intensities even in the lowest $^{1,3}\Pi_u \leftarrow {}^1\Sigma^+_g$ transition[198] (see Fig. 3.5.4), and a separation of 2125 cm^{-1}, which corresponds to the splitting in the ground state of CSe_2^+[198].

3.5.2 Photochemistry

CO_2. The photodissociation of CO_2 has been the subject of an enormous number of studies, partly for the sake of pure research and partly because of its importance as a planetary atmospheric constituent, especially of Mars. The literature up to the end of 1974 has been exhaustively reviewed by Filseth[216], but since that time a number of significant publications have helped to clarify some of the mysteries that clouded understanding of the molecular photochemistry.

Provided the absorbed photons lie at wavelengths longer than 108.2 nm the energy is insufficient to allow primary formation of electronically excited CO, and the only accessible dissociation product channels are

$$CO_2 \longrightarrow CO(X^1\Sigma^+) + O(^3P) \qquad\qquad \lambda < 227.5 \text{ nm} \qquad\qquad \text{(i)}$$
$$CO(X^1\Sigma^+) + O(^1D) \qquad\qquad \lambda < 167.1 \text{ nm} \qquad\qquad \text{(ii)}$$
$$CO(X^1\Sigma^+) + O(^1S) \qquad\qquad \lambda < 128.6 \text{ nm} \qquad\qquad \text{(iii)}$$

(Table 3.5.2 summarises the energetics of alternative primary processes at shorter wavelengths). In many of the earlier studies[217] a variable deficiency of O_2 was observed in the dissociation products compared with the expected stoichiometric ratio $O_2:CO = 0,5$ but more recent work has established that the discrepancy is apparent rather than real, and is caused by secondary heterogeneous reactions on the vessel wall[218]. The primary quantum yield of CO (detected via resonance fluorescence excitation) in the "spin-forbidden" channel has been measured as $\phi(CO) = 1.0 \pm 0.25$ over the range λ (174–210) nm[219].

At wavelengths $\lambda < 167$ nm, $O(^1D)$ becomes energetically accessible [reaction (ii)]. Its very long radiative lifetime (~ 100 s)[220] rules out direct fluorescence as a means of detection, but the high efficiency[220] of the fast[221] collisional deactivation step,

$$O(^1D) + CO_2 \rightarrow O(^3P) + CO_2, \quad k = 1.8 \times 10^{-10} \text{ cm}^3 \text{ molecule}^{-1} \text{ s}^{-1} \text{ [221]} \qquad \text{(iv)}$$

allows detection of the $O(^1D)$, produced by the spin allowed reaction (ii), using resonance fluorescence excitation of the $O(^3P)$ formed in 100% yield through reaction step (iv)[220]; (the technique cannot, of course, distinguish between $O(^3P)$ formed

Table 3.5.2 Threshold energies for primary product channels in the photodissociation of CO_2 (upper values, in brackets, are in electron volts, lower values are in nanometres)

CO ↓		$O \rightarrow {}^3P_2$ (0.00)	$^1D_2$ (1.97)[b]	$^1S_0$ (4.19)[b]
$X^1\Sigma^+$	(0.00)	(5.453)[c] 227.5	(7.42) 167.1	(9.64) 128.6
$a^3\Pi$	(6.01)[a]	(11.46) 108.2	(13.43) 92.3	(15.65) 79.2
$a'^3\Sigma^+$	(6.86)[a]	(12.31) 100.7	(14.28) 86.8	(16.50) 75.1
$d^3\Delta$	(7.52)[a]	(12.97) 95.6	(14.94) 80.9	(17.16) 70.7
$e^3\Sigma^-$	(7.90)[a]	(13.48) 92.0	(15.45) 80.3	(17.67) 70.2
$A^1\Pi$	(8.03)[a]	(13.59) 91.2	(15.56) 79.7	(17.78) 69.8

a Ref.[260].   c Ref.[3].
b Ref.[261].

via (iv) and the primary process (i)). Earlier investigations suggested a total quantum yield $\phi[O(^3P)] < 1$ in this wavelength range[222–224], a result not in conflict with a final O_2:CO ratio of 0.5, but when special attention was paid to minimising the contribution of heterogeneous reactions during photolysis, Slanger and Black[225] found unit quantum yields of total O atom production at (130.2–130.4) nm and at 147 nm. The rates of production of CO were also found to be proportional to the respective rates of generation of O atoms, implying a unit quantum yield for CO as well[225]. These results[224, 225], taken together with those of Ausloos[226], strongly suggest a unit quantum efficiency for the photodissociation of CO_2 at all wavelengths shorter than 167 nm, the threshold for production of $O(^1D)$.

Figure 3.5.6 shows measurements of the quantum yield of $O(^1S)$, determined by Slanger, Sharpless and Black[227] through monitoring the green auroral emission at 557.7 nm, associated with the transition $O(^1S) \rightarrow O(^1D)$. The onset of $O(^1S)$[227–229] formation closely follows the developing absorption associated with the transition into the $^1\Sigma_u^+$ state and reaches a maximum quantum efficiency of unity. However, the excitation spectrum exhibits a very sharp dip centred at 108.9 nm, which coincides exactly with the sharp Rydberg feature in the absorption spectrum associated with excitation into the $^1\Pi_u$ ($3p\sigma_u$) state (see Figs. 3.5.2 and 3.5.6). Within this narrow wavelength band, the preferred primary products were $CO\ (X^1\Sigma^+) + O\ (^1D)$ with $\phi_{max}[O\ (^1D)] \geqslant 0.65$[227]. The threshold wavelength for the primary production of electronically excited CO ($a^3\Pi$) (which probably accounts for the decline in $\phi\ [O\ (^1S)]$ at short wavelengths) lies below 108.9 nm (see Table 3.5.2), but since there is no observable molecular fluorescence, the two reaction channels (ii) and (iii) probably account for the entire primary process at $\lambda > 108.2$ nm.

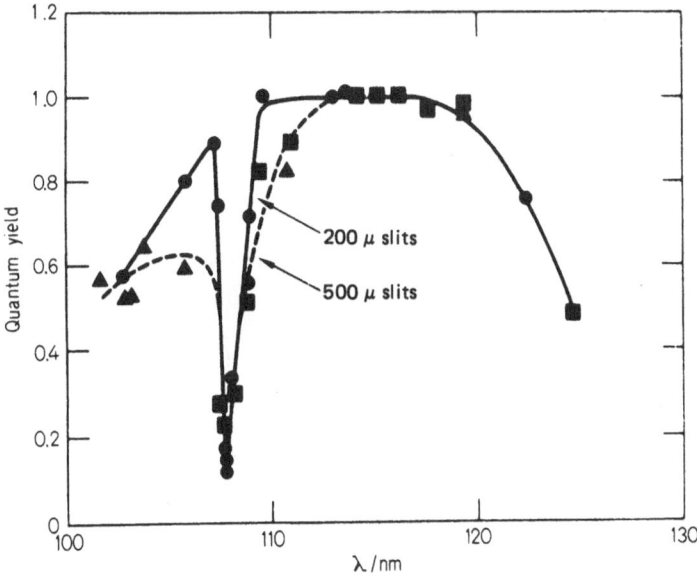

Fig. 3.5.6. Variation in $\phi[O(^1S)]$ from CO_2 photolysis over wavelength range 100–130 nm (after Slanger et al., Ref.[227])

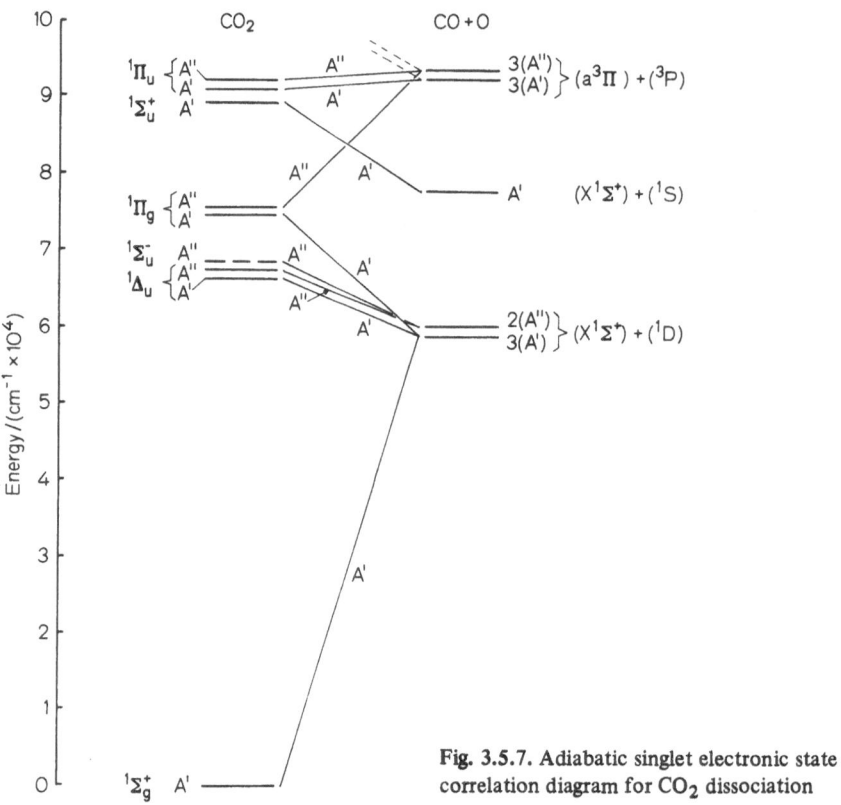

Fig. 3.5.7. Adiabatic singlet electronic state correlation diagram for CO_2 dissociation

The specificity of the dissociation path leading to $O(^1S)$ is easily understood on the basis of a minimum symmetry correlation diagram; similarly for $O(^1S)$ from N_2O, $S(^1S)$ from OCS and $Se(^1S)$ from OCSe (see later discussion). Figure 3.5.7 shows the correlations in the singlet manifold for CO_2 and it is clear that it is exclusively the lowest $^1\Sigma_u^+(^1A')$ state which correlates adiabatically with $CO(^1\Sigma^+) + O(^1S)$. The adjacent $^1\Pi_u$ state correlates with $CO(a^3\Pi) + O(^3P)$, but since formation of these products would be endothermic at 108.9 nm (see Table 3.5.2), $CO(^1\Sigma^+) + O(^1D)$ are formed instead, presumably through radiationless transfer on to a lower potential surface.

In a rare study aimed at clarifying the molecular dynamics of the primary process, Gilpin and Welge[230] measured the time-of-flight of metastable $O(^1S)$ atoms produced in reaction (iii) in the wavelength range λ (105–116) nm. The resulting time-of-flight distribution was both broad and delayed, reflecting considerable internal excitation (vibrational and/or rotational) of the $CO(^1\Sigma^+)$ molecular fragment. The most recent work[227] establishes that $O(^1S)$ is produced through excitation of the $^1\Sigma_u^+$ state in CO_2 which is largely Rydberg in character and is associated with a narrow, structured absorption system which includes a short progression in (probably) the symmetric stretching frequency, ν_1'. On the other hand, the direct correlation $^1\Sigma^+(A') \rightarrow {}^1\Sigma^+(A') + {}^1S(A')$ and the specificity of the primary process imply that although the upper state appears bound near the equilibrium conformation – one which

48

is close to that of the ground state – it readily evolves into a repulsive configuration. Pack[231] has developed a simple theory to account for this type of behaviour which allows the molecule "to have its cake and eat it". It assumes (a) a vertical electronic transition which prepares the excited molecule near the saddle point, and (b) an upper potential function that is repulsive along the "reaction coordinate" but attractive perpendicular to it, (i.e. respectively parallel to the antisymmetric (Q_3') and symmetric (Q_1') normal coordinates in the region of the saddle point). Under these conditions, the nett two-dimensional Franck-Condon factors permit the appearance of a diffuse, but none the less well-resolved, progression in the symmetric stretching frequency ν_1', as observed in $CO_2\,({}^1\Sigma_u^+)^{206)}$. The introduction of intravalency shell antibonding character with excitation of motion along Q_3' could account for the repulsion along the reaction coordinate.

The energy threshold calculated for the production of electronically excited CO ($\lambda < 108.2$ nm) has been verified experimentally by Lawrence[232], who measured the quantum yield of "Cameron" band emission from $CO(a^3\Pi)$ following photolysis in the range 85 nm $< \lambda <$ 109 nm. The yield increased monotonically from threshold to reach a maximum ~ 0.5 at 90.1 nm, the first ionisation potential. Inspection of the correlation diagram (Fig. 3.5.7) shows that a number of the states excited in this region can correlate with $CO(a^3\Pi) + O(^3P)$. An estimate of the branching ratios at the Ar resonance line, 106.7 nm[233], indicates $\phi[O(^1S)] = 0.60 \pm 0.06$, $\phi[CO(a^3\Pi)]$ $\sim 0.05^{232)}$ and by difference, $\phi[O(^1D)] = 0.35$, since the quantum yield of CO was found to be unity.

As the wavelength is reduced still further, several other electronically excited states of CO become energetically accessible (see Table 3.5.2) and visible emission from the triplet states $a'\,^3\Sigma^+$, $d^3\Delta$ and $e^3\Sigma^-$ has been detected following photolysis at 76.4 nm $< \lambda <$ 92.3 nm[234]. Judge and Lee[234] estimated a total yield of triplet CO at $\lambda > 85$ nm which was considerably greater than the earlier estimate reported by Lawrence[232]. At shorter wavelengths, fluorescence from $CO(A^1\Pi)$ can also be detected in the vacuum u.-v.[235–237] becoming increasingly dominant as the wavelength falls. Cross-sections for production of this fluorescence have been measured at wavelengths down to 70 nm[236, 237] and the results indicate that photodissociation from Rydberg states is competing against photoionisation and that $O(^3S)$ is not an important primary product.

Spectral analysis of the various emission bands of CO* allows an estimate of the vibrational energy disposal in the excited CO fragments, and Lawrence[232] and Lee and Judge[238, 239] have reported vibrational energy distributions in $CO(d^3\Delta, a'^3\Sigma^+$ and $a^3\Pi)^{238)}$ and $CO(A^1\Pi)^{239)}$. Their results for the d and a' states have been reproduced by Band and Freed[240, 241] using a model which combines Franck-Condon and half-collision considerations, but assumes direct photodissociation from an upper state continuum; the experimental results for the $a^3\Pi$ state could not be reproduced by their model. Berry has also applied a simplistic Golden Rule model to the analysis of the vibrational energy disposal[242].

OCS and OCSe. In marked contrast to the wealth of information in the literature regarding the vacuum u.-v. photodissociation of CO_2, the related molecules OCS and OCSe have received relatively little attention. The near u.-v. photochemistry of OCS,

Table 3.5.3. Threshold energies for primary product channels in the photodissociation of OCS (upper values, in brackets, are in electron volts, lower values are in nanometres)

CO ↓		$S \to {}^3P_2$ (0.00)	${}^1D_2$ $(1.15)^b$	${}^1S_0$ $(2.75)^b$
$X^1\Sigma^+$	(0.00)	$(3.16 \pm 0.03)^c$ 392.3	(4.31) 287.7	(5.91) 209.8
$a^3\Pi$	$(6.01)^a$	(9.17) 135.2	(10.32) 120.1	(11.92) 104.0
$a'^3\Sigma^+$	$(6.86)^a$	(10.02) 123.7	(11.17) 111.0	(12.77) 97.1
$d^3\Delta$	$(7.52)^a$	(10.68) 116.1	(11.83) 104.8	(13.43) 92.3
$e^3\Sigma^-$	$(7.90)^a$	(11.06) 112.1	(12.21) 101.5	(13.81) 89.8
$A^1\Pi$	$(8.03)^a$	(11.19) 110.8	(12.34) 100.5	(13.94) 88.9

CS ↓		$O \to {}^3P_2$ (0.00)	${}^1D_2$ $(1.97)^d$	${}^1S_0$ $(4.19)^d$
$X^1\Sigma^+$	(0.00)	$(7.30 \pm 0.16)^k$ 169.8	(9.27) 133.7	(11.49) 107.9
$a^3\Pi$	$(3.42)^e$	(10.72) 115.6	(12.69) 97.7	(14.91) 83.2
$a'^3\Sigma^+$	$(3.86)^f$	(11.16) 111.1	(13.13) 94.4	(15.35) 80.8
$d^3\Delta$	$(4.42)^g$	(11.72) 105.8	(13.69) 90.6	(15.91) 77.9
$e^3\Sigma^-$	$(4.79)^h$	(12.09) 102.6	(14.06) 88.2	(16.28) 76.2
$A^1\Pi$	$(4.81)^j$	$(12.11 \pm 0.16)^k$ 102.4	(14.08) 88.1	(16.30) 76.1

a Ref.[260].
b Ref.[262].
c Ref.[263].
d Ref.[261].
e Ref.[264–267].
f Ref.[267, 268].
g Ref.[268].
h Ref.[265, 268].
j Ref.[264, 265, 268, 269].
k Ref.[248].

at wavelengths $\lambda > 200$ nm where parallel production of both $S({}^3P)$ and $S({}^1D)$ has been inferred by product analysis, has been reviewed by Filseth[216] and by Gunning and Strausz[243].

At wavelengths $\lambda > 165$ nm, the flash photolysis of OCS excites laser emission on vibrationally excited $CO(X^1\Sigma^+)$ and up to thirty vibration-rotation transitions have been identified, ranging from $\Delta v(13 \to 12)$ to $\Delta v(7 \to 6)$[244]. Laser emission could not be detected when the photolysis light was restricted to $\lambda > 200$ nm. Although the dissociations

$$\nu^4_{max}$$

$$OCS + h\nu(\lambda > 165 \text{ nm}) \rightarrow CO(X^1\Sigma^+)_v + S(^3P) \qquad 18 \qquad \text{(v)}$$

$$\rightarrow CO(X^1\Sigma^+)_v + S(^1D) \qquad 13 \qquad \text{(vi)}$$

$$\rightarrow CO(X^1\Sigma^+)_v + S(^1S) \qquad 6 \qquad \text{(vii)}$$

are all energetically accessible (see Table 3.5.3), the correlation diagram favours channel (vi) (see Fig. 3.5.8). The triplet channel (v) was discounted on the grounds that
no laser emission was detected from $v > 13$; but this would only be valid if the
$CO(X)_v$ products from dissociation channel (v) also had an inverted vibrational population at $v \geqslant 14$. Furthermore, the absence of laser emission from $v \leqslant 6$, consistent with a non-inverted population of the lower vibrational levels in CO, probably
reflects a contribution from channel (vii), the quantum efficiency of which has been
measured as ~ 0.2 at 165–170 nm[245].

Black et al.[245] have measured the quantum yield of $S(^1S)$ as a function of wavelength as part of the continuing search for suitable laser amplifier systems based on
the ^1S metastable states of elements in Group 6. In the range 110 nm $< \lambda <$ 170 nm,
they found $\phi[S(^1S)] \approx 1.0$ between 142 nm and 160 nm, which precisely mirrors the

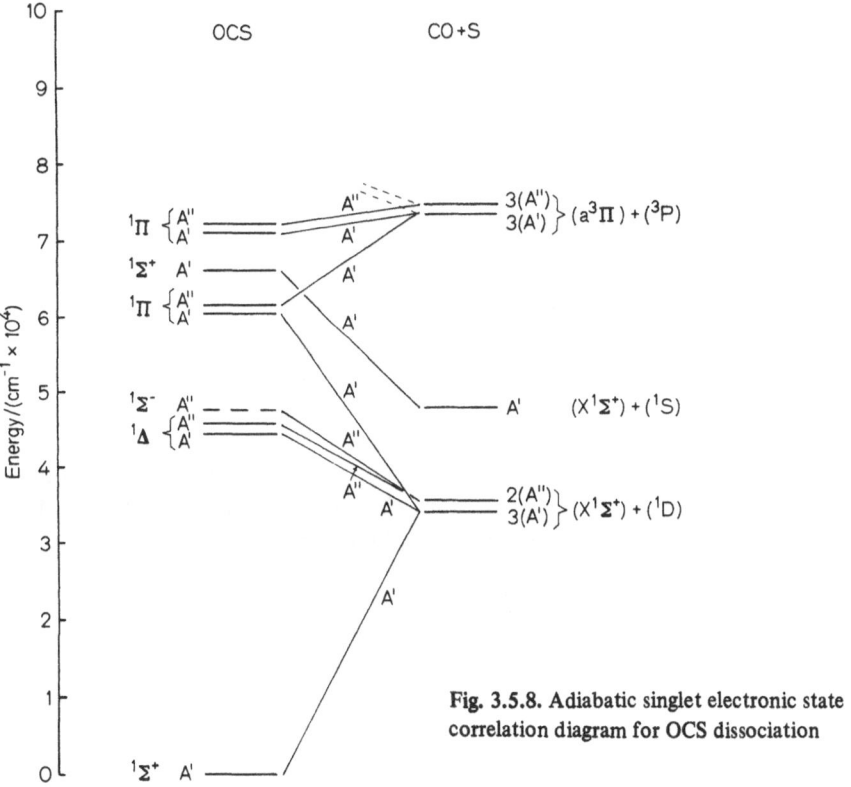

Fig. 3.5.8. Adiabatic singlet electronic state
correlation diagram for OCS dissociation

---

4 Assuming that the total available energy is taken up in vibrational excitation of $CO(X^1\Sigma^+)$,
$\omega_e = 2170.2 \text{ cm}^{-1}$ [130]

region of intense absorption assigned to population of the $^1\Sigma^+$ state in OCS (see Fig. 3.5.1). As in CO_2, the correlation diagram (Fig. 3.5.8) confirms that population of the $^1\Sigma^+$ state should indeed lead preferentially to $CO(X^1\Sigma^+) + S(^1S)$.

The first of the alternative dissociation channels involving cleavage of the O–CS bond becomes energetically accessible at $\lambda < 170$ nm (see Table 3.5.3) but experiments to date suggest this route to be of minor importance only. Donovan[246] observed transient absorption from CS following flash photolysis of OCS at $\lambda > 110$ nm, but $< 10\%$ of the photodissociation yield could be attributed to primary production of CS. The relative contributions of SC–O and S–CO cleavage have also been estimated by using resonance fluorescence techniques to measure the initial concentrations of $O(^3P)$ and $S(^3P)$ following flash photolysis of OCS in an atmosphere of Ar at $\lambda > 105$ nm[247]. An upper limit of $\phi[O(^3P)]: \phi[S(^3P] \sim 2\%$ was estimated; however, the experimental observations were complicated by a long-lived molecular fluorescence emission in the near and vacuum u.-v. and by the inability to compensate the result for the $O(^1S)$ and $S(^1S)$ which would also have been produced.

Vibrational energy disposal in $CS(A^1\Pi)$ produced via the vacuum u.-v. photodissociation of OCS at $\lambda < 99.2$ nm has been investigated through spectral analysis of the $CS(A \rightarrow X)$ emission[248]. The experimental results approximated a Poisson distribution over vibrational states, though this may need revision in the light of more recent studies on the photodissociation of CS_2 (see later discussion). The results[248] also provided the first accurate experimental estimate of the dissociation energy D_0^0 (O-CS) by identifying the threshold wavelength for production of $CS(A \rightarrow X)$ fluorescence following the spin forbidden process

$$OCS + h\nu \rightarrow CS(A^1\Pi) + O(^3P) \qquad\qquad \lambda < 102.4 \text{ nm} \qquad\qquad \text{(viii)}$$

Although there have been very few photochemical studies on OCSe the potential utility of $Se(^1S)$ atoms as an energy store in laser amplification has encouraged investigation of its v.u.-v. photochemistry. Excited $Se(^1D_2)$ atoms have been detected in absorption following the flash photolysis of OCSe at $\lambda > 200$ nm[249] and the quantum yield of $Se(^1S)$ has been measured over the range $110\,\text{nm} < \lambda < 200\,\text{nm}$ by monitoring the $Se(^1S_0 \rightarrow {}^1D_2)$ emission intensity at 777 nm[250]. As in the related molecules CO_2, N_2O and OCS, the excitation spectrum for $Se(^1S)$ closely follows the broad outline of the $(^1\Sigma^+ \leftarrow {}^1\Sigma^+)$ absorption system in OCSe and the quantum yield reaches unity at $\lambda \sim 170$ nm. This behaviour once again satisfies the predictions of the minimum symmetry correlation diagram which is based on the assignment of the band systems centred around 167 nm and 180 nm to transitions into the lowest excited $^1\Sigma^+$ and $^1\Pi$ states, respectively (see Fig. 3.5.9).

An unexpected by-product of the study of OCSe was the detection of strong emission from $Se_2 (B^3\Sigma_u^- \rightarrow X^2\Sigma_g^-)$ as the pressure of the parent molecule increased[250], which was ascribed to the fast reaction of $Se(^1S)$ with the parent OCSe molecule $(k = (1.6 \pm 0.2) \times 10^{-10} \text{ cm}^3 \text{ molecule}^{-1} \text{ s}^{-1}$ [250])*. The short wavelength threshold of the molecular emission (~ 380 nm) allowed a determination of the dissociation energy D_0^0 (OC-Se) $\leqslant 259 \pm 5$ kJ mol^{-1} [250], from which a value for the heat of for-

---

* *Note added in proof:* subsequent work has demonstrated that this assignment is incorrect; an unknown higher excited state of Se_2 is formed initially[368].

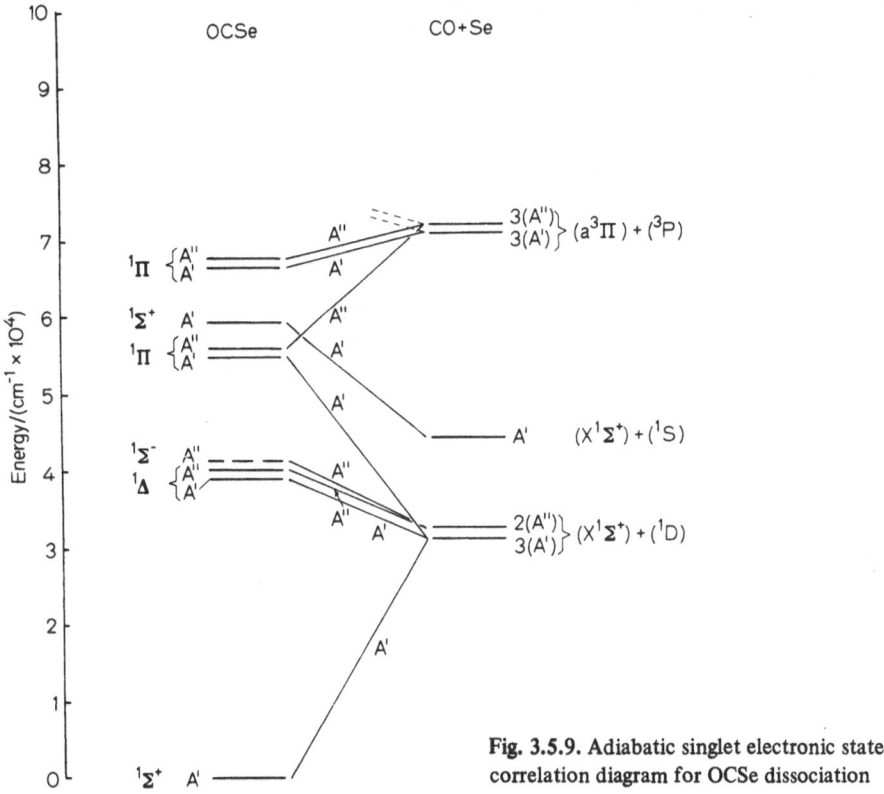

Fig. 3.5.9. Adiabatic singlet electronic state correlation diagram for OCSe dissociation

mation $\Delta H^0_{f_0}$ (OCSe) = -132 ± 5 kJ mol^{-1} and the dissociation energy at the alternative bond D^0_0 (O–CSe) = 740 ± 8 kJ mol^{-1} can be derived (see Table 3.5.4). Dissociation at the C–O bond has not been reported so far.

CS_2. Photodissociation of CS_2 at wavelengths $\lambda > 185$ nm could lead to formation of both S(^3P) and S(^1D). The principle absorption system ~ 200 nm populates the $^1\Sigma^+_u(^1A')$ electronic state, which lies *below* the lowest $^1\Pi_u$ state (in contrast to all the other molecules in this group apart from CSe_2), and in consequence it correlates with CS($X^1\Sigma^+$) + S(^1D) (see Fig. 3.5.10). Flash photolysis of CS_2 at $\lambda > 185$ nm produced vibrationally excited CS($X^1\Sigma^+$) and S(^3P) as the only detectable transients[251] in absorption, but the detection of S(^1D) would have been prevented through rapid quenching by the diluent gas, N_2. At the pressures employed, S(^1D) atoms would decay within a few nanoseconds, since the rate coefficient for quenching by N_2 is $\sim 1.5 \times 10^{-11}$ cm^3 molecule^{-1} s^{-1} [252].

In contrast to the molecules discussed so far, the quantum yield of metastable ^1S atoms is negligible (< 0.05), throughout the entire wavelength range 110 nm $< \lambda < 210$ nm[253], though CS($a^3\Pi$) is produced in high yield, particularly between 125 nm and 140 nm. It is probable that the absolute yield in this region is of the order unity, following a steady rise from the threshold at 158 nm[253]. This behaviour is readily explained by the fact that the dissociation channel producing

Table 3.5.4. Threshold energies for primary product channels in the photodissociation of OCSe (upper values, in brackets, are in electron volts, lower values are in nanometres)

CO ↓		Se → $^3P_2$ (0.00)	$^1D_2$ $(1.19)^b$	$^1S_0$ $(2.78)^b$
$X^1\Sigma^+$	(0.00)	$(2.69 \pm 0.05)^c$ 460.8	(3.88) 319.4	(5.47) 226.6
$a^3\Pi$	$(6.01)^a$	(8.70) 142.5	(9.89) 125.3	(11.48) 108.0
$a'^3\Sigma^+$	$(6.86)^a$	(9.55) 129.8	(10.74) 115.1	(12.33) 100.5
$d^3\Delta$	$(7.52)^a$	(10.21) 121.4	(11.40) 108.7	(12.99) 95.4
$e^3\Sigma^-$	$(7.90)^a$	(10.59) 117.0	(11.78) 105.2	(13.37) 92.7
$A^1\Pi$	$(8.03)^a$	(10.72) 115.6	(11.91) 104.1	(13.50) 91.8
CSe ↓		O → $^3P_2$ (0.00)	$^1D_2$ $(1.97)^f$	$^1S_0$ $(4.19)^f$
$X^1\Sigma^+$	(0.00)	$(7.68 \pm 0.09)^g$ 161.4	(9.65) 128.5	(11.87) 104.4
$a^3\Pi$	$(2.99)^d$	(10.67) 116.2	(12.64) 98.1	(14.86) 83.4
$D^1\Pi$	$(4.36)^e$	(12.04) 103.0	(14.01) 88.5	(16.23) 76.4

a	Ref.260).	c	Ref.250).	e	Ref.272).
b	Ref.272).	d	Ref.271).	f	Ref.261).

S(^1S) is energetically accessible only when $\lambda < 172$ nm, and by the correlation diagram shown in Fig. 3.5.10. As mentioned earlier, CS$_2$ (and CSe$_2$) are distinguished by having their lowest $^1\Pi_u$ state lying above the all important $^1\Sigma_u^+$ level and because of this, the dissociation products CS($X^1\Sigma^+$) + S(^1S) correlate with the higher lying $^1\Pi_g$ state. Transitions into this state are weak, since the transition is parity forbidden in the linear conformation and in any case the absorption bands are centred slightly below the energy threshold at 172 nm; excitation into the $^1\Sigma_u^+$ state occurs ~ 200 nm. On the other hand, the intense Rydberg features lying to shorter wavelengths correlate directly with CS($a^3\Pi$) + S(^3P) and once the threshold at 158 nm has been crossed, it is in this region that the quantum yield of CS($a^3\Pi$) rises steeply.

Measurements of the photofluorescence thresholds have been very useful in the determination of bond dissociation energies[254] and the value for D_0^0 (SC–S) estimated from the threshold for CS($a^3\Pi$) luminescence[253] was in very good agreement with an earlier estimate reported by Okabe[255] (see Table 3.5.5) which was derived from the threshold for CS($A \to X$) fluorescence, excited via the (spin-forbidden) dissociation

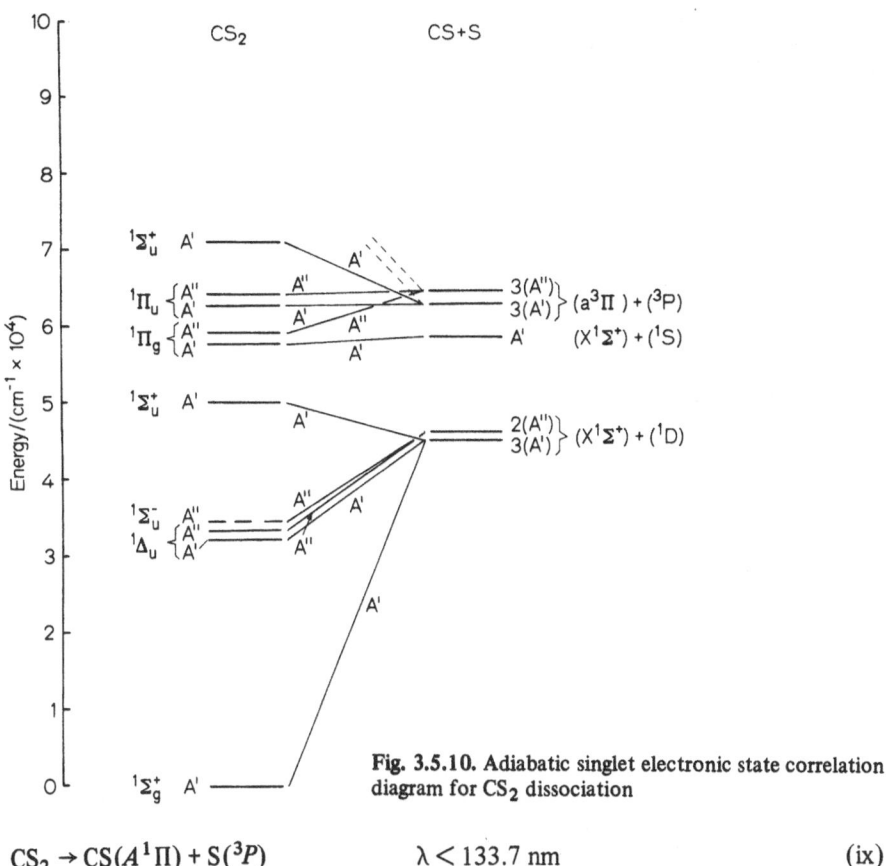

Fig. 3.5.10. Adiabatic singlet electronic state correlation diagram for CS_2 dissociation

$$CS_2 \rightarrow CS(A^1\Pi) + S(^3P) \qquad\qquad \lambda < 133.7 \text{ nm} \qquad\qquad \text{(ix)}$$

The peaks in the excitation spectrum of the $CS(A \rightarrow X)$ fluorescence correspond to the higher members of the Rydberg progressions populating the Σ_u^+ $(np\pi_u)$ and Π_u $(np\sigma_u)$ states, though it is not clear whether the quantum yields on the triplet bands are larger than those on the singlet components. It is probable that the dissociation proceeds through intersystem crossing in the excited state, since the quantum yield of $CS(A^1\Pi)$ remains small ($\lesssim 0.07$) and declines with the onset of photoionisation at wavelengths $\lambda \lesssim 124$ nm[248, 255].

The vibrational energy disposal in $CS(A^1\Pi)$ was first studied by Lee and Judge[248] who reported relative populations approximating Poisson distributions following photodissociation at 123.9 nm and 92.3 nm, with strong population inversion between $v = 0$ and 1. However, more recent studies[256] of the $CS(A \rightarrow X)$ fluorescence spectrum, conducted at considerably higher spectral resolution (0.05 nm rather than ~ 0.3 nm) at $\lambda = 130.4$, 129.5, 123.6 and 121.6 nm, have established that the vibrational populations actually decay monotonically from $v = 0$, with $\sim 30\%$ of the excess energy appearing in vibration at the two shorter wavelengths. The Franck-Condon factors for the $CS(A \rightarrow X)$ transition determined from the relative emission intensities[256] were also in excellent agreement with values recently computed by Coxon et al.[257]. An underlying sequence of emission bands attributed to the decay of $CS(d^3\Delta_i \rightarrow X)$, produced in a competing primary process,

Table 3.5.5. Threshold energies for primary product channels in the photodissociation of CS_2 (upper values, in brackets, are in electron volts, lower values are in nanometres)

CS ↓		$S \to {}^3P_2$ (0.00)	${}^1D_2$ (1.15)[f]	${}^1S_0$ (2.75)[f]
$X^1\Sigma^+$	(0.00)	(4.463)[g] 277.7	(5.61) 220.9	(7.21) 171.9
$a^3\Pi$	(3.42)[a]	(7.88) 157.3	(9.03) 137.3	(10.63) 116.6
$a'^3\Sigma^+$	(3.86)[b]	(8.32) 149.0	(9.47) 130.9	(11.07) 112.0
$d^3\Delta$	(4.42)[c]	(8.88) 139.6	(10.03) 123.6	(11.63) 106.6
$e^3\Sigma^-$	(4.79)[d]	(9.25) 134.0	(10.40) 119.2	(12.00) 103.3
$A^1\Pi$	(4.81)[e]	(9.27)[g] 133.7	(10.42) 119.0	(12.02) 103.1

a   Ref.264–267.   c   Ref.268).   e   Ref.264, 265, 268, 269).   g   Ref.255).
b   Ref.267, 268).   d   Ref.265, 268).   f   Ref.262).

$$CS_2 \to CS(d^3\Delta_i) + S(^3P) \tag{x}$$

has been observed[256] following photodissociation at $\lambda \leqslant 130$ nm. The emission almost certainly corresponds to the unidentified luminescence ($\tau_r \sim 8$ μs) observed in the same region by Black, Sharpless and Slanger[253].

CSe_2. The vacuum u.-v. photochemistry of CSe_2 is largely unexplored, apart from a preliminary study in the authors' laboratory[256]. From the point of view of laser amplification the molecule is unlikely to be a prolific source of $Se(^1S)$, since the ordering of the states in the correlation diagram closely parallels that in CS_2 (see Fig. 3.5.1). The products $CSe(X^1\Sigma^+) + Se(^1S)$ correlate with the $^1\Pi_g$ state (see Fig. 3.5.11) but, as in CS_2, the threshold for dissociation to $Se(^1S)$ at 191.3 nm (see Table 3.5.6) lies above the excitation energies both of the $^1\Pi_g$ state (~ 200 nm) and the $^1\Sigma_u^+$ state (~ 230 nm). Photodissociation at 149.4 nm or 147 nm excites the fluorescence of $CSe(D^1\Pi \to X^1\Sigma^+)^{[256]}$ (analogous with the $A \to X$ emission of CS from CS_2) and strong emission bands which overlap this system can be assigned to the $^3\Delta_i$ and/or $^3\Sigma^-$ states of CSe. As in CS_2, the $CSe(D^1\Pi)$ fragments are vibrationally excited, though in this case with a small population inversion between $v = 0$ and 1, which is consistent with the relative changes in the value of the equilibrium bond lengths in CSe compared with CS. The vibrational band intensities have been analysed to estimate Franck-Condon factors for the $D \to X$ transitions[256]. Assuming that all vibrational levels energetically accessible in $CSe(D^1\Pi)$ are populated (as in $CS(A)$ from CS_2) an estimate of the dissociation energy D_0^0 (SeC—Se) can be derived[256] in good agreement with the current literature value[258, 259] (see Table 3.5.6).

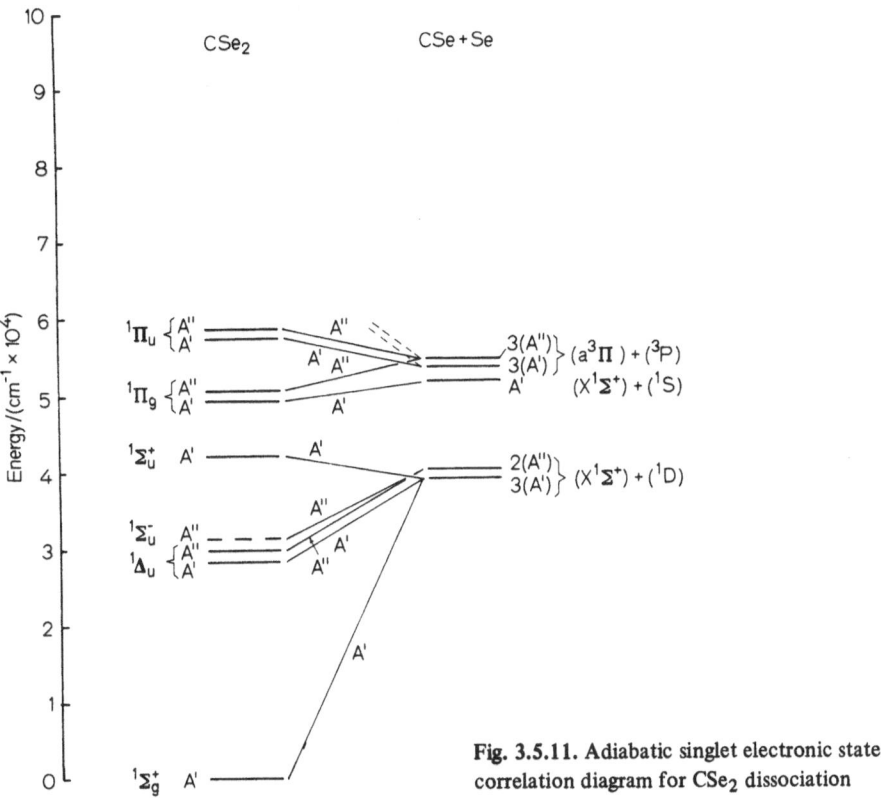

Fig. 3.5.11. Adiabatic singlet electronic state correlation diagram for CSe_2 dissociation

Table 3.5.6 Threshold energies for primary product channels in the photodissociation of CSe_2 (upper values, in brackets, are in electron volts, lower values are in nanometres)

CSe ↓		$Se \rightarrow {}^3P_2$ (0.00)	${}^1D_2$ (1.19)[c]	${}^1S_0$ (2.78)[c]
$X^1\Sigma^+$	(0.00)	$(3.70 \pm 0.05)^d$ 335.1	(4.89) 253.5	(6.48) 191.3
$a^3\Pi$	(2.99)[a]	(6.69) 185.3	(7.88) 157.3	(9.47) 130.9
$D^1\Pi$	(4.36)[b]	(8.06) 153.8	(9.25) 134.0	(10.84) 114.4

a Ref.271).   c Ref.270).
b Ref.272);   d Ref.256, 258, 259).

N₂O. Early studies of the vacuum u.-v. photolysis of N_2O concentrated on the identification of final molecular products and the wavelength dependence for their production[10] but more recent work has been directed towards detection of the metastable atomic and molecular species of oxygen and nitrogen formed as primary products. These are of considerable importance in stratospheric photochemistry, where

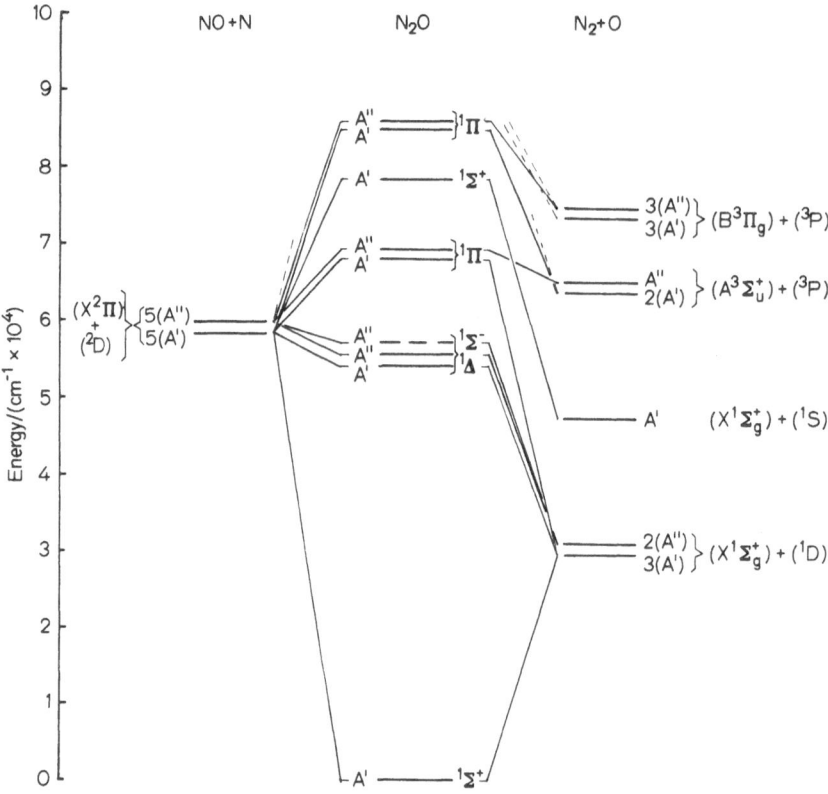

Fig. 3.5.12. Adiabatic singlet electronic state correlation diagram for N_2O dissociation

N_2O provides the main source of NO (through its reaction with $O(^1D)$ atoms). In the laboratory, the vacuum u.-v. photodissociation of N_2O has been used as a source of $O(^1D)^{273-276}$, $O(^1S)^{230, 277-279}$, $N(^2D)^{280-283}$, $N(^2P)^{282, 283}$, $N_2(A^3\Sigma_u^+)^{284-285}$ and $N_2(B^3\Pi_g)^{284, 286}$, and a wealth of kinetic information on their rates of decay has been derived, following collisions with a variety of quenching gases.

Preston and Barr[287] reported that production of atomic nitrogen was insignificant at $\lambda > 185$ nm and proposed that the dissociation

$$N_2O \rightarrow N_2(X^1\Sigma_g^+) + O(^1D) \qquad \lambda > 185 \text{ nm} \qquad \text{(xi)}$$

was the only important primary process in the range 185 nm $< \lambda <$ 230 nm. Excitation in this region populates the $^1\Delta$ state in N_2O, which correlates with the observed products (see Fig. 3.5.12). At shorter wavelengths, the relative quantum yields for production of $O(^1S)^{286, 288}$, $N(^2D)^{288}$ and $N_2(A^3\Sigma_u^+)^{288}$ have been measured between 110 nm and 150 nm, which covers the regions in which the $^1\Pi(3p\sigma)$ and $^1\Sigma^+(3p\pi)$ states are populated (together with higher Rydberg states below 120 nm). The results, which are shown in Fig. 3.5.13, confirm earlier measurements of $\phi[O(^1S)]$ at 121.6 nm$^{278, 289}$ and 123.6 nm^{278}, and of all three relative quantum yields at 147 nm[273, 278, 290]. The wavelength dependence of $\phi[O(^1S)]$ closely mirrors the absorption profile of the $^1\Sigma^+ \leftarrow {}^1\Sigma^+$ transition, with $\phi \approx 1$ at 129 nm[286], as expected from the symmetry correlation diagram (see Fig. 3.5.12). At $\lambda \geqslant 136$ nm, the nett quantum yield shown in Fig. 3.5.13 falls below unity and the major fragment that has been identified following absorption at 147 nm is $O(^1D)$, with a quantum yield of 0.55 ± 0.03[273]; hence the reasonable suggestion that the deficit in the nett quantum yield at longer wavelengths is associated with formation of $O(^1D)$[288]. Excitation at 147 nm populates the $^1\Pi(3p\sigma)$ state in the parent molecule and should lead to $O(^1D)$ (see Fig. 3.5.12).

At shorter wavelengths, $\lambda < 125.6$ nm, the photodissociation of N_2O excites the "first positive" band system of N_2[286, 291], which corresponds to the transition $N_2(B^3\Pi_g \rightarrow A^3\Sigma_u^+)$. Two excitation paths have been identified[286], via photodissociation directly into the B state and by the decay of more highly excited triplet states of N_2 formed in alternative photodissociation channels; an unassigned luminescence with a radiative lifetime of (27 ± 5) μs was also observed[286], and might be associated with the second channel. Analysis of the $N_2(B \rightarrow A)$ fluorescence excited by photodissociation at 116.5 nm and 123.6 nm[291], and time-of-flight spectroscopy of the metastable $O(^1S)$ and $N_2(A)$[230, 292] produced through polychromatic flash photolysis between 105 nm and 131 nm, both indicate high levels of internal excitation in the molecular fragments $N_2(X^1\Sigma_g^+)$, and $N_2(A$ and $B)$. A more sophisticated experiment which uses a monochromatic flash photolysis lamp[293] has confirmed the

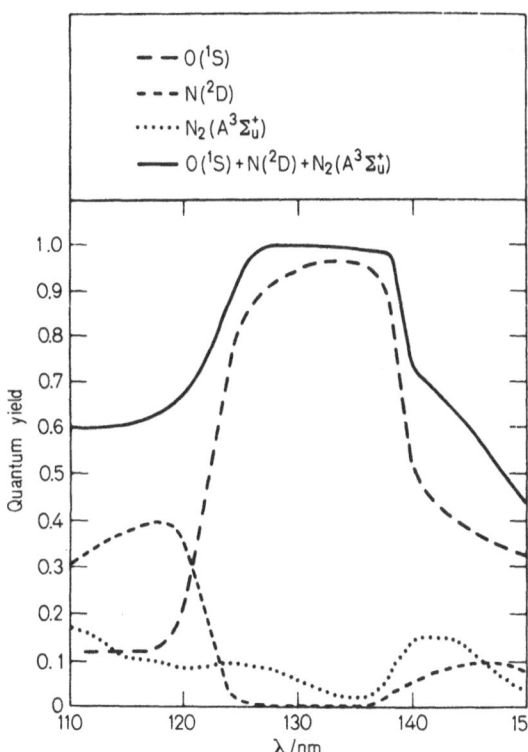

Fig. 3.5.13. Variation in product quantum yields from photodissociation of N_2O in the wavelength range 110–150 nm (after Black et al., Ref.[288])

Legend for figure:
— — $O(^1S)$
- - - $N(^2D)$
...... $N_2(A^3\Sigma_u^+)$
—— $O(^1S) + N(^2D) + N_2(A^3\Sigma_u^+)$

Quantum yield (vertical axis: 0, 0.1, 0.2, 0.3, 0.4, 0.5, 0.6, 0.7, 0.8, 0.9, 1.0)
λ/nm (horizontal axis: 110, 120, 130, 140, 150)

high internal excitation in $N_2(X)$ consistent with population of vibrational levels $v \lesssim 6$ following photodissociation at 130 nm[5]. The experiments[292, 293] also confirm that $N_2(X) + O(^1S)$ are products of a "parallel" transition in the parent molecule, as expected for the $^1\Sigma^+ \leftarrow {}^1\Sigma^+$ system, while the earlier experiments[292] showed that triplet $N_2(A \text{ and/or } B)$ was produced through a mixture of parallel and perpendicular parent molecular transitions, also consistent with the expected behaviour (see Fig. 3.5.12).

3.6. ICN, BrCN and ClCN

3.6.1 Spectroscopy

The vacuum ultraviolet absorption spectra of the cyanogen halides[176, 295] are shown in Figs. 3.6.1–3.6.3. The molecules are isovalent with those discussed in Section 3.5, and in the ground state their electronic configuration is

$$\ldots (1\pi)^4 (6\sigma)^2 (2\pi)^4; {}^1\Sigma^+$$

Similarities may be expected between corresponding pairs of isovalent molecules, e.g. SCO and ClCN, SeCO and BrCN.

The absorption spectra begin with weak, broad continua, which in ICN and BrCN can be resolved into two components, designated the A and α-systems. These were first discussed in detail by King and Richardson[296] primarily by drawing comparisons with other halide molecules rather than with the isovalent series of triatomic molecules: the lowest lying continua were therefore attributed to the electron promotion $2\pi \rightarrow 7\sigma(^1\Pi \leftarrow \tilde{X}^1\Sigma^+)$. Rabalais et al.[6] adopted the alternative strategy and assigned the A systems to the transition $2\pi \rightarrow 3\pi (^1\Sigma^-(^1A'') \leftarrow \tilde{X})$, forbidden under $C_{\infty v}$ point group symmetry, populating a bent upper state, with the α continua (lying to shorter wavelengths) being assigned to the bent $^1A'$ and/or $^1A''$ components of the $2\pi \rightarrow 3\pi(^1\Delta \leftarrow \tilde{X}^1\Sigma^+)$ transition (cf. Section 3.5.1). Rabalais et al.[6] were unconvinced by King and Richardson's conclusion that the lowest lying banded systems (the B and C systems) should be assigned to singlet and triplet components of the leading member of the Rydberg series, $2\pi \rightarrow ns\sigma(^{1, 3}\Pi \leftarrow X)$, preferring instead to assign one or other of the systems to the intravalency $^1\Sigma^+$ state generated by the $2\pi \rightarrow 3\pi$ electron promotion. Subsequent analysis indicates that King and Richardson were not mistaken in their original assignment of the B and C systems, at least in BrCN and ICN, where the $2\pi \rightarrow ns\sigma$ Rydberg series can be identified, terminating on the two spin-orbit components $^2\Pi_{3/2}$ and $^2\Pi_{1/2}$ of the molecular ion[297, 298] (see Figs. 3.6.1 and 3.6.2). The analysis of ClCN is a little more involved since there are additional absorption bands which precede and overlap the B and

---

5   This result conflicts with a theoretical calculation of the vibrational energy disposal[294], which while reproducing the $^1\Sigma^+ \leftarrow {}^1\Sigma^+$ absorption profile, predicts very little excitation of $N_2(X)_{v>0}$ unless the parent N_2O is vibrationally excited.

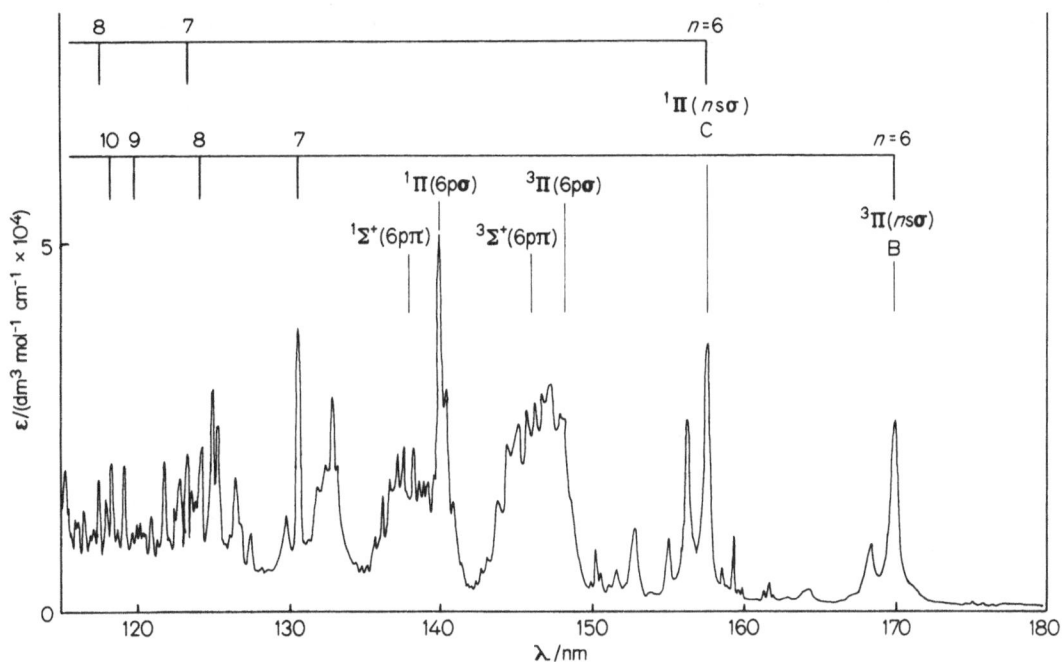

Fig. 3.6.1. Vacuum ultraviolet absorption spectrum of ICN (after West and Berry, Refs.[176] and [299])

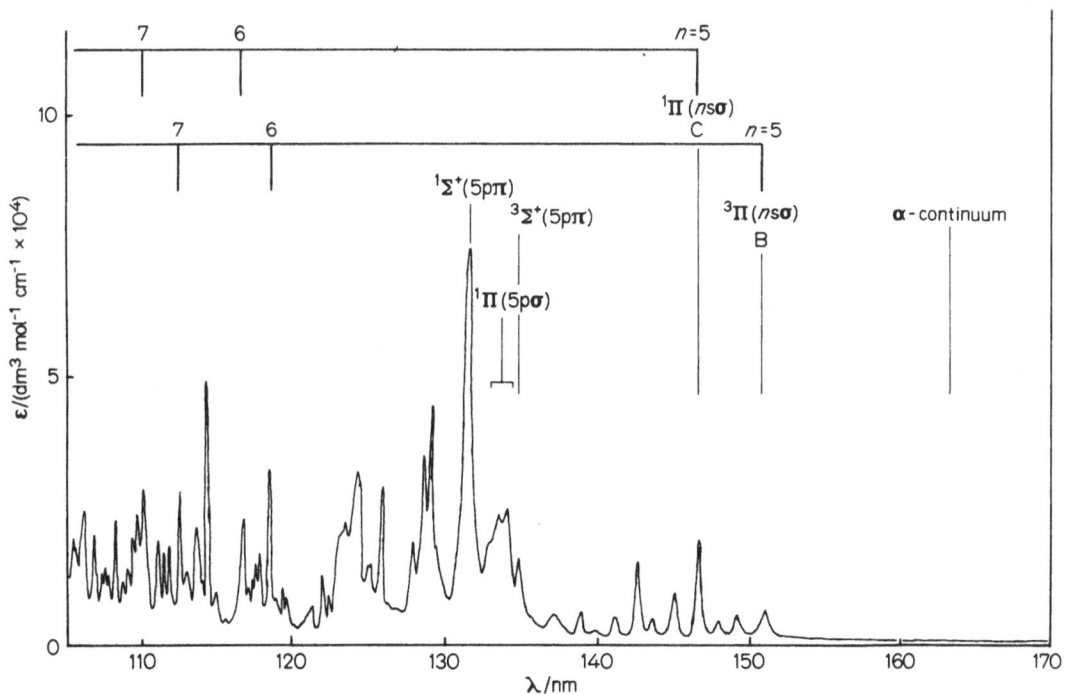

Fig. 3.6.2. Vacuum ultraviolet absorption spectrum of BrCN (after West, Ref.[176])

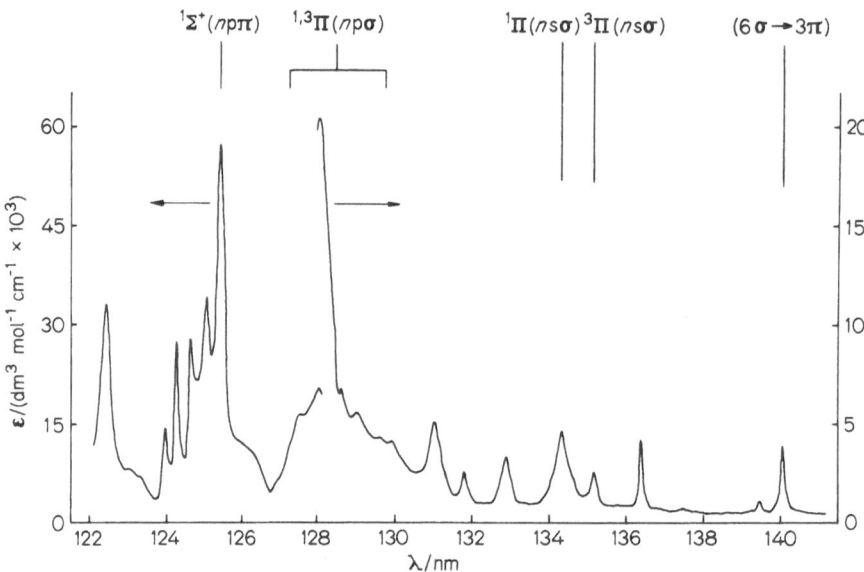

Fig. 3.6.3. Vacuum ultraviolet absorption spectrum of ClCN (after Macpherson and Simons, Ref. [317])

Table 3.6.1 Probable assignments for some of the shorter wavelength absorption features of the cyanogen halides*

Molecule	ν/cm^{-1}	T_n/cm^{-1}	δ	Upper state assignment	Ionization potentials[a]
ICN	58890	28785	1.05	$B, ^3\Pi(ns\sigma)$	
	63470	29130	1.06	$C, ^1\Pi(ns\sigma)$	
	67500	20175	0.67	$^3\Pi(np\sigma)$	$ICN^+(^2\Pi_{3/2}) = 87675\ cm^{-1}$
	71400	21200	0.72	$^1\Pi(np\sigma)$	
	67960 (adiab.)	19500	0.63	$^3\Sigma^+(np\pi)$	$ICN^+(^2\Pi_{1/2}) = 92600\ cm^{-1}$
	72400 (adiab.)	20200	0.67	$^1\Sigma^+(np\pi)$	
BrCN	66250	29250	1.06	$B, ^3\Pi(ns\sigma)$	
	68240	28760	1.05	$C, ^1\Pi(ns\sigma)$	
	72920 (adiab.)	22580	0.80	$^3\Pi(np\sigma)$	$BrCN^+(^2\Pi_{3/2}) = 95500\ cm^{-1}$
	74600 (adiab.)	22400	0.79	$^1\Pi(np\sigma)$	
	74130	21370	0.73	$^3\Sigma^+(np\pi)$	$BrCN^+(^2\Pi_{1/2}) = 97000\ cm^{-1}$
	75980	21020	0.72	$^1\Sigma^+(np\pi)$	
ClCN	73965	25485	0.93	$^3\Pi(ns\sigma)$	
	74460	25270	0.93	$^1\Pi(ns\sigma)$	$ClCN^+(^2\Pi_{3/2}) = 99450\ cm^{-1}$
	78065	21540	0.74	$^{1,\,3}\Pi(np\sigma)$	$ClCN^+(^2\Pi_{1/2}) = 99730\ cm^{-1}$
	79730	19870	0.65	$^{1,\,3}\Sigma^+(np\pi)$	

a Ref. [301].

* *Note added in proof:* for a fuller discussion and more recent work, see Ref. [317].

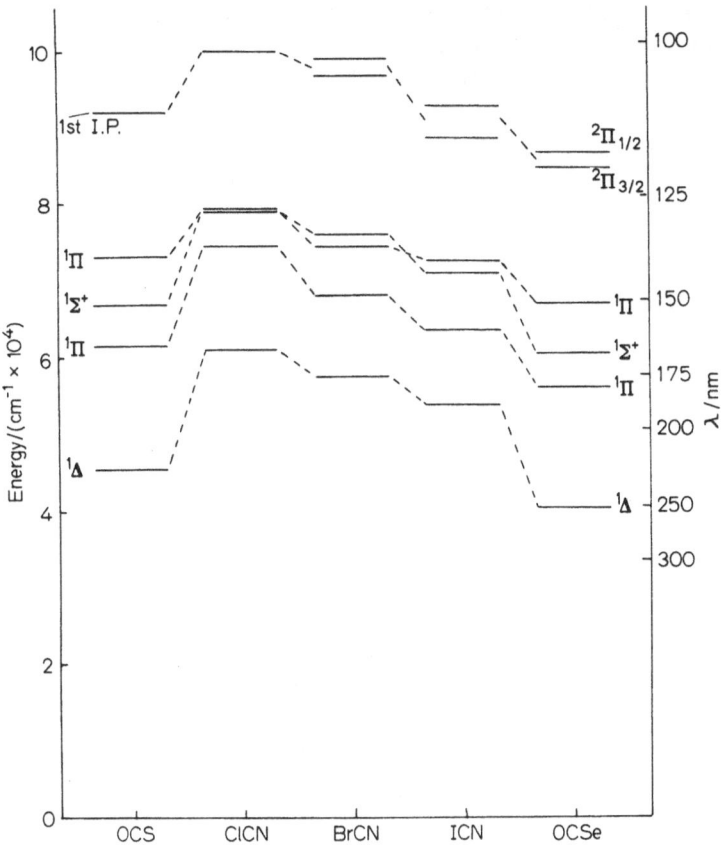

Fig. 3.6.4. Summary of the probable assignments and energetic ordering of the low lying electronic states in ICN, BrCN and ClCN; the corresponding electronic states in the isovalent molecules OCSe and OCS are shown for comparison

C systems[298]), some of which did not appear in the original spectrum of King and Richardson.

By reviewing the spectroscopy of the cyanogen halides in the light of the more recent work on isovalent molecules discussed in Section 3.5.1 and anticipating some of of the results of photodissociation spectroscopy studies discussed in Section 3.6.2, the most probable assignments may be summarised, as shown in Table 3.6.1 and Figs. 3.6.1 to 3.6.3. The orderings of the lower lying states are correlated with those in OCS and OCSe in Fig. 3.6.4.

$^1\Sigma^-$ and $^1\Delta$ *states.* The absorption continua at long wavelengths are assigned to transitions into the $\Sigma^-(A'')$ and $\Delta(A', A'')$ states generated by the intravalence configuration . . . $(2\pi)^3 (3\pi)$. Diffuse vibrational structure can be resolved on the long wavelength side of the continuum in ClCN. The spacings are irregular, though a principal progression with an average spacing ~ 700 cm^{-1} can be distinguished consistent with excitation of the C—Cl stretching frequency ν_1' [298]). By analogy with the

63

corresponding transitions in the isovalent molecules discussed earlier, it is reasonable to associate the perturbed structure with Renner-Teller interactions in a $^1\Delta$ upper state. Measurements of rotational energy disposal in excited $CN(B^2\Sigma^+)$ fragments produced via photodissociation in the α-continua, imply that the degree of nonlinearity in the photoexcited parent molecule increases in the order $ICN < BrCN < ClCN$[194].

The weak continua, termed the A systems[296], in ICN and BrCN, are probably associated with transitions into components of the $^3\Delta$ and $^3\Sigma^-$ configurations (see Sect. 3.6.2).

$^{1,3}\Pi(ns\sigma)$ *states.* These states are by far the best characterised, and in BrCN and ICN the term values and the relative spacings between the B and C band systems are consistent with their assignment to the first members of the $2\pi \rightarrow ns\,(^{1,3}\Pi\,(ns\sigma) \leftarrow \tilde{X}^1\Sigma^+)$ Rydberg series[296, 298, 299]. Spin-orbit coupling splits the molecular core into $^2\Pi_{3/2}$ and $^2\Pi_{1/2}$ components, and under (Ω_c, ω) coupling the "singlet" and "triplet" states correlate with components of the $(1/2, 1/2)$ and $(3/2, 1/2)$ states, respectively[130]. Assignment of the B and C bands to the leading members of the $\Pi(ns\sigma)$ series places them lower than the first $^1\Sigma^+$ level, in harmony with the ordering found in OCS and OCSe (see Sect. 3.5.1).

Similar arguments also apply to ClCN, but the two $^{1,3}\Pi$ states lie at such short wavelengths, because of the higher molecular ionisation potential, that an additional band system is exposed, with an origin at ~ 140 nm (see Fig. 3.6.3). The second band in this system, which lies at 136.4 nm and can be associated with the excitation of one quantum of the C-N stretching frequency $v_3' \sim 1900$ cm^{-1}, was incorrectly assigned to the origin of the B system by King and Richardson[296]. The extra transition does not seem to have any analogue in the spectra of BrCN or ICN but, by comparing the absorption spectra of ClCN and HCN and noting the relative energies of the 1π, 6σ and 2π valence m.o's in the three cyanogen halides (determined from the respective photoelectron spectra)[300, 301], we assign this additional band system to the analogue of the $(\tilde{C} \leftarrow \tilde{X})$ transition in HCN (see Sect. 3.4.1). The latter transition lies in the same spectral region and is associated with a $\sigma \rightarrow \pi^*$ inner electron promotion[3, 172] analogous with the transition $6\sigma \rightarrow 3\pi\,(^1\Pi \leftarrow \tilde{X}^1\Sigma^+)$ in the cyanogen halides. The ordering of the valence orbital energies revealed in the photoelectron spectra of the cyanogen halides[300, 301] suggests that in BrCN and ICN the corresponding transition lies buried beneath the intense absorption at shorter wavelengths.

$^{1,3}\Sigma^+$ and $^{1,3}\Pi(np\sigma)$ *states.* In OCS and OCSe, the transitions into the $^1\Sigma^+$ state are very intense and sharp and are believed to be predominantly Rydberg $(np\,\pi)$ in character rather than intravalency . . . $(2\pi)^3\,(3\pi)$[197]. The term values for the intense bands indicated in Figs. 3.6.2 and 3.6.3 are consistent with the same assignment, and the doublet splittings associated with $^2\Pi_{3/2}$ and $^2\Pi_{1/2}$ states in the molecular ion[300, 301] are reflected in the excited state multiplet separations (see Table 3.6.1); assignment of this spectral region in ICN is complicated by the effects of increased spin-orbit coupling[298]. It is not clear if and where a conjugate intravalence $^1\Sigma^+$ state should appear.

The weaker band systems attributed to the $^{1,3}\Pi\,(np\sigma)$ states are interlaced with the neighbouring $^1\Sigma^+$ systems, and lie at comparable energies. The transitions display a diffuse vibrational structure and although no analysis has been reported, the spacings may indicate some excitation of the bending frequency. In BrCN, for example, where the overlapping is less severe, the diffuse bands are separated by $\sim 340 \text{ cm}^{-1}$. Like the $^1\Sigma^+$ systems, the transitions have term values consistent with their assignment to the first members of the $2\pi \rightarrow np$ Rydberg series[317].

3.6.2 Photochemistry

Virtually all the photochemical studies of the cyanogen halides have been concerned with the determination of electronic branching ratios and energy disposal among the

Table 3.6.2 Threshold energies for primary product channels in the photodissociation of ICN (upper values, in brackets, are in electron volts, lower values are in nanometres)

CN ↓		$I \rightarrow {}^2P_{3/2}$ (0.00)	${}^2P_{1/2}$ $(0.94)^c$
$X^2\Sigma^+$	(0.00)	(3.11) 398.6	(4.05) 306.1
$A^2\Pi_i$	$(1.13)^a$	(4.24) 292.4	(5.18) 239.3
$B^2\Sigma^+$	$(3.20)^a$	(6.31) 196.5^d	(7.25) 171.0
$a^4\Sigma^+$	$(4.09)^b$	(7.20) 172.0	(8.14) 152.0
$D^2\Pi_i$	$(6.69)^a$	(9.80) 126.5	(10.74) 115.4

a Ref.130). c Ref.296).
b Ref.324). d Ref.184).

Table 3.6.3 Threshold energies for primary product channels in the photodissociation of BrCN (upper values, in brackets, are in electron volts, lower values are in nanometres)

CN ↓		$Br \rightarrow {}^2P_{3/2}$ (0.00)	${}^2P_{1/2}$ $(0.46)^c$
$X^2\Sigma^+$	(0.00)	(3.77) 328.9	(4.23) 293.1
$A^2\Pi_i$	$(1.13)^a$	(4.90) 253.0	(5.36) 231.3
$B^2\Sigma^+$	$(3.20)^a$	(6.97) 178.0^d	(7.43) 166.9
$a^4\Sigma^+$	$(4.09)^b$	(7.86) 158	(8.32) 149
$D^2\Pi_i$	$(6.69)^a$	(10.46) 118.5	(10.92) 113.5

a Ref.130). c Ref.296).
b Ref.324). d Ref.184).

Table 3.6.4 Threshold energies for primary product channels in the photodissociation of ClCN (upper values, in brackets, are in electron volts, lower values are in nanometres)

CN ↓		$Cl \rightarrow {}^2P_{3/2}$ (0.00)	${}^2P_{1/2}$ (0.07)[c]
$X^2\Sigma^+$	(0.00)	(4.34) 285,7	(4.41) 281.1
$A^2\Pi_i$	(1.13)[a]	(5.47) 226.6	(5.54) 223.8
$B^2\Sigma^+$	(3.20)[a]	(7.54) 164.5[d]	(7.61) 162.9
$a^4\Sigma^+$	(4.09)[b]	(8.43) 147	(8.50) 146
$D^2\Pi_i$	(6.69)[a]	(11.03) 112.4	(11.10) 111.7

a   Ref.[130].   c   Ref.[296].
b   Ref.[324].   d   Ref.[184].

primary products of photodissociation. The threshold wavelengths at which the alternative channels become energetically accessible are summarised in Tables 3.6.2–3.6.4. The experimental results have sometimes been in conflict, particularly those for ICN, but they have stimulated a vigorous effort in the theoretical analysis of the dynamics of photodissociation, both semi-classical and quantum mechanical, which is still very much in evidence.

The photochemistry at the longer wavelengths proceeds through absorption in the broad continua whose spectral composition remains uncertain. There can be little doubt that components of the intravalence $(2\pi)^3$ (3π), ${}^1\Delta$ state contribute to the α-continuum, but the number and identity of the states involved over the whole range of continuous absorption, especially in ICN where spin-orbit coupling should be strong, is not known. It might be expected that identification of the electronic states of the primary photofragments, their angular distributions with respect to the polarization vector of the incident photon, and judicious application of the appropriate correlation diagrams, could help in assigning the initially photoexcited state, and this information has been extensively sought in ICN. At the time of writing, the picture remains confused, but it is well worth summarising (despite our being led temporarily out of the vacuum u.-v. into the gentler realms of the near u.-v.). A brief historical review of the photochemistry of ICN and a theoretical analysis have been presented recently by Beswick and Jortner[302].

The latest experiments have used tunable laser induced fluorescence techniques to monitor the $CN(X^2\Sigma^+)$ fragments produced through photodissociation of ICN by a frequency quadrupled Nd:YAG laser ($\lambda = 266.2$ nm)[303] or a flash lamp ($\lambda \geqslant 220$ nm)[304]. They each concluded that virtually all the $CN(X)$ fragments are formed without vibrational excitation ($N_{v=0}:N_{v=1} \geqslant 1.00:0.02$) but that they are rotationally excited with a distribution characterised approximately by a rotational temperature of ~ 3000 K. While earlier time-of-flight photofragment spectroscopy experiments[305] had been rationalised by assuming that $CN(X^2\Sigma^+)$ and $CN(A^2\Pi)$

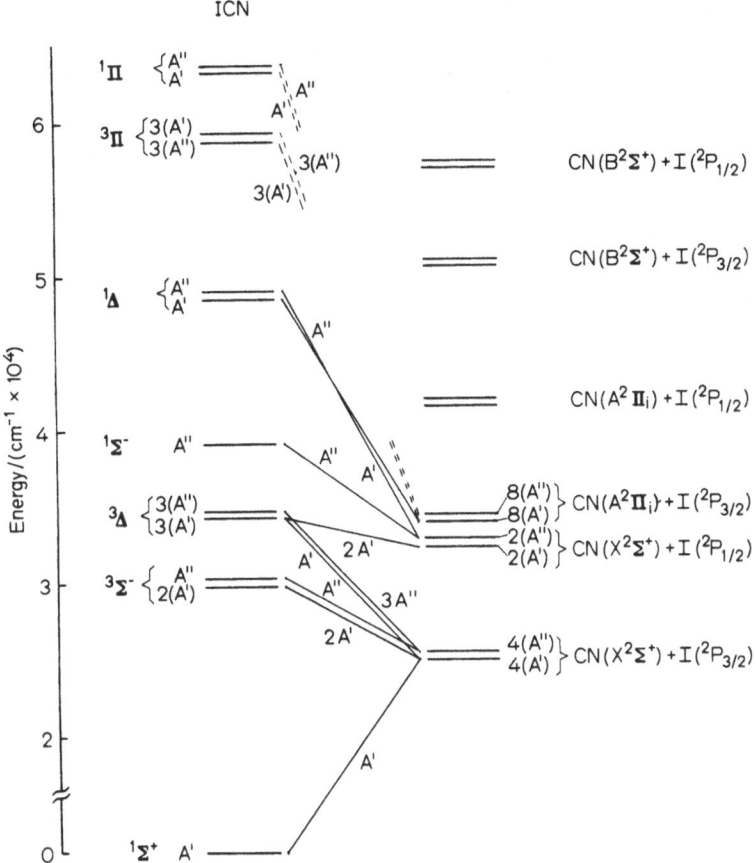

Fig. 3.6.5. Hund's case (c) correlation diagram for the dissociation of ICN

were both primary products at 266.2 nm[306], the more recent fluorescence studies[303] positively exclude production of CN(*A*) at this wavelength, although the excited fragment is energetically accessible (see Table 3.6.2). This absence of any measurable CN(*A* → *X*) emission has been confirmed by West and Berry's report that the CN(*A* → *X*) fluorescence excitation spectrum which can be monitored following excitation in the vacuum u.-v., falls to zero when λ > 200 nm[299]. Its absence at the longer wavelengths could indicate the operation of a dynamical constraint but it can also be understood in terms of the minimum symmetry correlation diagram for case (c) coupling (see Fig. 3.6.5). This assumes that the weak *A*-continuum in ICN at λ > 200 nm is composed principally of transitions into spin-orbit components of the bent $^3\Sigma^-$ and $^3\Delta$ states, and that the $^1\Delta$ state, which is the first to correlate with CN(*A*), only becomes optically accessible when λ < 200 nm. The correlation diagram also requires that branching into $I(^2P_{3/2})$ and $I(^2P_{1/2})$ occurs in the long wavelength continuum; this has been confirmed by Baronavski and McDonald[303] who observed $I(^2P_{1/2} \rightarrow ^2P_{3/2})$ chemiluminescence at 1.315 μm following photodissociation at 266.2 nm. These results suggest that the "fast" and "slow" compo-

67

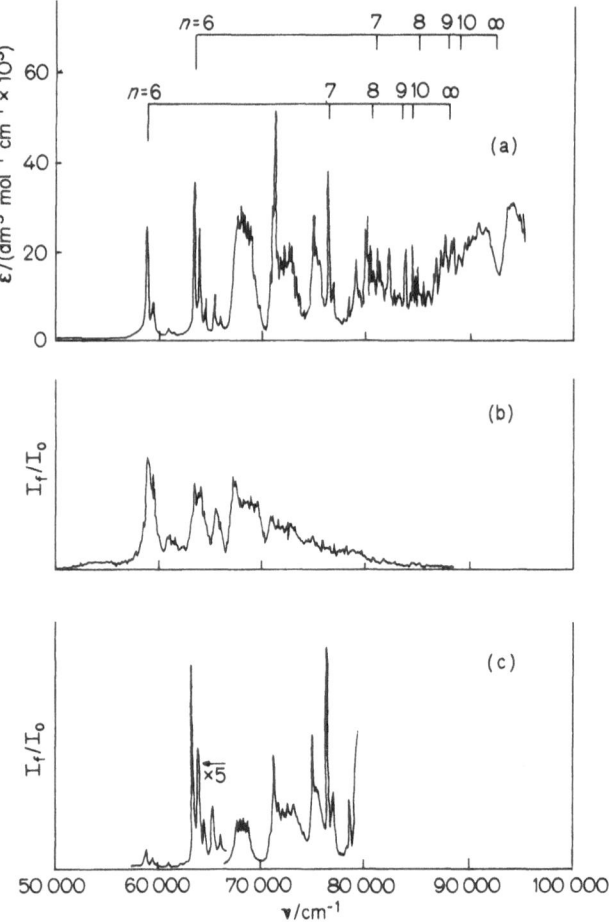

Fig. 3.6.6. Photofragment excitation spectra of (b) CN($A\,^2\Pi_i$) and (c) CN($B\,^2\Sigma^+$) from ICN, with (a) the absorption spectrum of ICN included for comparison (after West and Berry, Ref.[299]) and Macpherson and Simons, Ref.[317])

nents in the time-of-flight photofragment spectrum of ICN at the same wavelength[306] are associated with branching in the I atom rather than the CN radical. The preferential polarization of the photofragment spectrum parallel to the C-I axis[306] can only be reconciled with an initial electronic transition into the Σ^+ spin-orbit component of the $^3\Sigma^-$ state. Since the absorbed photon excites the long wavelength edge of the continuum, this assignment is quite plausible, though it would be very helpful if polarization data were available at other wavelengths in the continuum. With the advent of intense laser sources based on Raman-shifted rare gas halide lasers [e.g. KrF at 249 nm (unshifted) and ArF at 193 nm (unshifted)], these measurements can be expected in the very near future.

A number of theoretical discussions of the photodissociation of ICN within the A-continuum, i.e. at $\lambda \geqslant 200$ nm, have been presented, most of which have reproduced the experimental data current at the time[302, 307-312]. It is easy to be critical

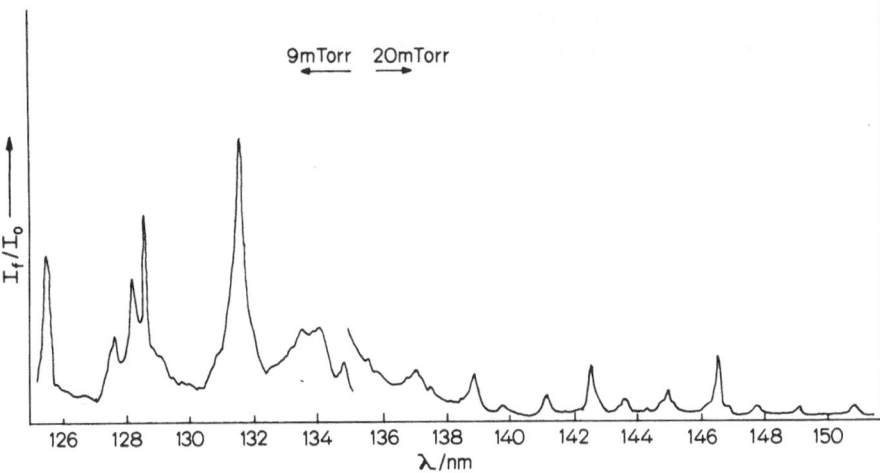

Fig. 3.6.7. Photofragment excitation spectrum of $CN(B^2\Sigma^+)$ from BrCN (after Macpherson and Simons, Ref.[317])

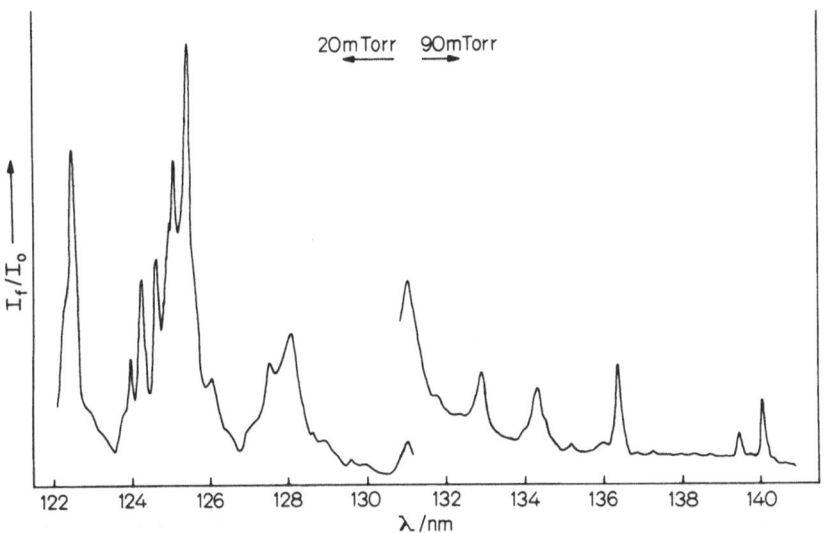

Fig. 3.6.8. Photofragment excitation spectrum of $CN(B^2\Sigma^+)$ from ClCN (after Macpherson and Simons, Ref.[317])

with the benefit of hindsight, and since one of the authors was caught in this way, the point will not be pursued! The most recent quantum mechanical treatments include those of Beswick and Jortner[302] and Morse et al.[312], who calculated excitation spectra for the dissociation of ICN into $CN(X)$ and $I(^2P_{3/2})$ or $I(^2P_{1/2})$, which approximately reproduced the near u.-v. absorption profile (where the $CN(X)$ fragments are produced predominantly in their ground vibrational state). Morse et al.[312] extended earlier treatments of the dynamics of photodissociation to include the bending and rotational degrees of freedom and were able to reproduce the broad features of the rotational and vibrational energy disposal in $CN(X)$ obtained by Baronavski and McDonald[303] (assuming production of $I(^2P_{3/2})$). Their discussion emphasised the constraint imposed by the simultaneous requirements of energy and angular momentum conservation, which had been noted earlier[310, 313, 314]; when the departing atom is as heavy as iodine, the constraint prevents anything more than a few degrees departure from linearity during dissociation[310, 312, 315]*.

Studies of the photodissociation of cyanogen halides in the vacuum ultraviolet can be separated into two broad types: (i) those which employed polychromatic flash photolysis and which were primarily concerned with the detection of electronic and/or vibrational laser action in the CN fragments, and (ii) those which employed continuous monochromatic photolysis sources and were primarily concerned with the branching ratios and energy disposal following dissociation from well-defined initially populated levels.

Excitation spectra of $CN(A \rightarrow X)$ emission have been recorded from ICN[299] and of $CN(B \rightarrow X)$ emission from all three cyanogen halides (see Figs. 3.6.6–3.6.8)[316, 317]. In each case the spectral features closely resemble the absorption spectrum, though changes in their relative intensities reveal sensitivity to the nature of the electronic and/or vibronic levels being populated. In ClCN, for example, the two features assigned to the B ($^3\Pi(ns\sigma)$) system which are well defined in the absorption spectrum are very weak in the $CN(B \rightarrow X)$ fluorescence excitation spectrum, though the corresponding features in the C ($^1\Pi(ns\sigma)$) system are undiminished[317] (see Figs. 3.6.3 and 3.6.8). This distinctive behaviour provided a helpful clue in analysing the spectrum of ClCN in this region.

The $CN(A \rightarrow X)$ excitation spectrum from ICN rises from a threshold near 200 nm, follows the contour of the α-continuum and the $^3\Pi(ns\sigma)$ (B) and $^1\Pi(ns\sigma)$ (C) band systems (where vibronic bands involving excitation of ν'_3 ($C\equiv N$ stretch) are relatively more intense), but begins to fade out at shorter wavelengths[299] (see Fig. 3.6.6). The intense peak attributed to the $^1\Sigma^+$ $(np\pi)$ state is almost entirely suppressed and the yield of $CN(A)$ from the C state is markedly less than from the B state; the latter phenomenon probably reflects competition from the alternative channel leading to $CN(B)$ since the relative intensities of the two systems are inverted in the $CN(B \rightarrow X)$ excitation spectrum, cf. ClCN (see Fig. 3.6.6). The $CN(B \rightarrow X)$ fluorescence can be excited at all wavelengths from threshold at ~ 196 nm[184] to wavelengths at least as short as 120 nm.

Electronic laser action on the $CN(A)_{v'=0} \rightarrow CN(X)_{v''=2}$ levels has been excited following flash photolysis of BrCN at $\lambda > 155$ nm[183], i.e. within the α-continuum. The fluorescence spectrum of $CN(B)$ has also been recorded at a number of wavelengths in the same region, and in each dissociation channel the CN fragments are produced with a low level of vibrational excitation[194]. Population of $CN(X)_{v=0,1}$ either directly or through relaxation of $CN(B)$, prevents laser action on the lower vibrational levels, but following flash photolysis at $\lambda \geqslant 155$ nm, vibrational laser emission to higher vibrational levels in $CN(X)$ can be observed in all three cyanogen halides[183]. Near resonant, collisionally induced intersystem crossing from $CN(A)_{v'=0} \rightarrow CN(X)_{v''=4}$ provides an efficient route for introducing vibrational population inversion in the ground electronic state[183, 304].

---

* *Note added in proof:* for a more recent treatment see later work by Freed and his coworkers[369].

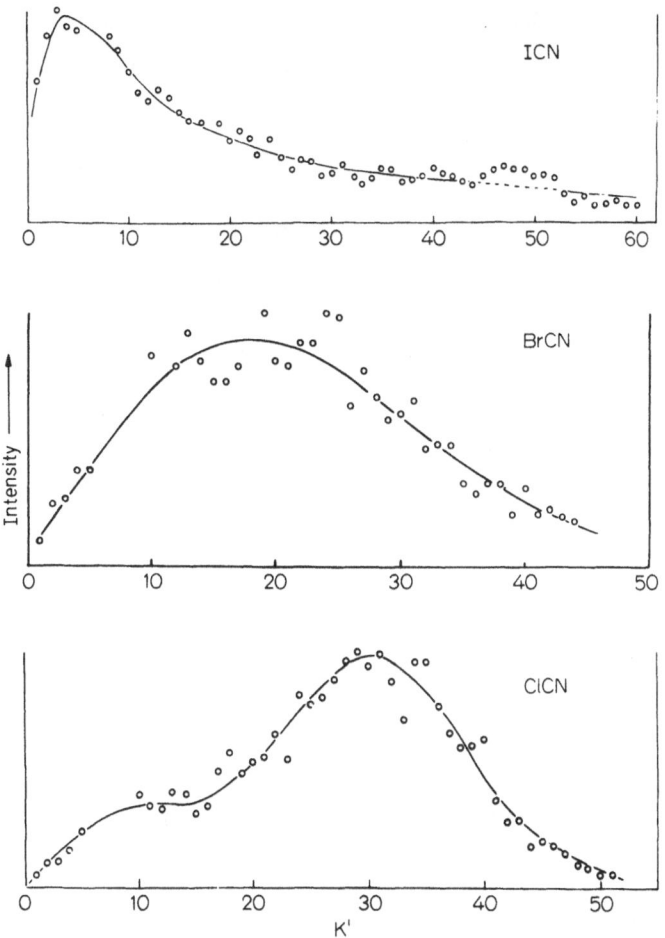

Fig. 3.6.9. Rotational distributions in CN $(B)_{v=0}$ from cyanogen halide photodissociation within the α-continuum absorption region. (a) ICN photolysis at 174.4 nm, $\epsilon_{avl} \sim 6{,}500$ cm^{-1}. (b) BrCN photolysis at 158.0 nm, $\epsilon_{avl} \sim 7{,}100$ cm^{-1}. (c) ClCN photolysis at 147.0 nm, $\epsilon_{avl} \sim 7{,}250$ cm^{-1}. (after Ashfold and Simons, Ref.[194])

A systematic study of energy disposal in the photodissociation of the cyanogen halides has utilised the intense fluorescence from CN(B) as a probe of the molecular dynamics and hence the topography of the photoexcited molecular potential energy surface[194, 317]. Photodissociation at a number of wavelengths in the α-continuum generates CN(B) fragments which carry very little energy in vibration, but which have very different levels of rotational excitation, decreasing in the order ClCN > BrCN > ICN[194, 315] (see Fig. 3.6.9). In the case of ClCN, the rotational distributions are bimodal. The CN(X) fragments produced in the photodissociation of ClCN within the α-continuum, at $\lambda \geqslant 166$ nm, carry even higher rotational excitation, with the distribution peaking at $K' \sim 66\text{--}70$[318], but again very little energy appears in vibration. The distributions can be rationalised in terms of recent theo-

71

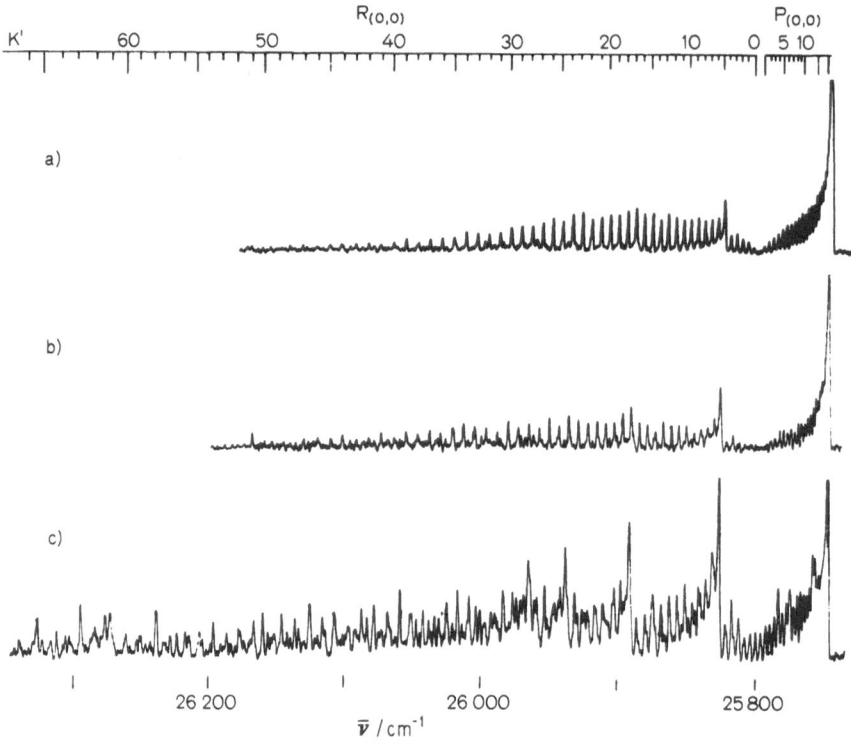

Fig. 3.6.10. Fluorescence spectra of CN$(B^2\Sigma^+)$ following photolysis of BrCN at: (a) 156.0 and 165.4 nm (α-continuum). (b) 149.4 nm (B-state, 1_0^1). (c) 147.0 nm (C-state, 0_0^0 and B-state 3_0^1). (after Ashfold and Simons, Ref.[194])

retical discussions of rotational energy disposal[312, 313, 315, 369] which identify the crucial influence of (i) Franck-Condon factors for the radiative or radiationless transfer onto the final repulsive surface, (ii) the constraints imposed by energy and angular momentum conservation, and (iii) final state interactions between the separating fragments. The results imply the retention of near collinearity in the photoexcited state of ICN, but increasing nonlinearity as the mass of the halogen atom is reduced[194, 315]. This is consistent with the assignment of the α-continuum to bent components of the ... $(2\pi)^3$ (3π); $^1\Delta$ state[194].

The disposal of energy into vibration may be very sensitive to the nature of the excited state initially populated, and photodissociation at wavelengths in the banded region of the absorption spectra of the cyanogen halides promotes increasingly higher levels of vibrational excitation in CN(B)[194, 298, 319, 320]. At the same time, rotational perturbations appear in the $(B \rightarrow X)_{0 \rightarrow 0}$ band, which arise from the parallel formation of CN fragments in the $(A^2\Pi)$ state, excited into vibrational levels $v' \geq 10$[319]. A number of rotational levels in CN$(B)_{v=0}$ and CN$(A)_{v=10}$ are nearly isoenergetic[321–323], and if the pressure is sufficiently high, collisionally induced intersystem crossing can modify the vibrational and rotational populations observed

72

in the CN($B \rightarrow X$) fluorescence spectrum. For example, Luk and Bersohn[190] observed non-exponential decay of CN radical fluorescence following flash photodissociation of ICN at ~ 158 nm. This could be understood in terms of contributions from CN($B \rightarrow X$) and CN($A \rightarrow X$) emission, in relative proportions dependent on the total pressure. The early results of Mele and Okabe[185] have been revised, following the discovery that the estimated vibrational distributions in CN(B) from BrCN were pressure dependent and did not reflect the distributions produced under collision free conditions[319]. The general conclusion is that the appearance of A-state perturbations, which reveals the population of high vibrational levels in the CN(A) state, correlates well with the onset of increased disposal of vibrational energy into the CN(B) fragments. In each case, the observed levels of excitation reflect changes in the parent molecular dimensions and force constants on excitation to a common initial photoexcited electronic state. A dramatic illustration of the influence of the upper state is shown in Fig. 3.6.10, where a slight reduction in the wavelength from 149.4 nm, which excites BrCN in the B ($^3\Pi(n s\sigma)$) absorption system, to 147.0 nm, which excites the (0,0,1) band of the B-system and (weakly) the (0,0,0) band in the $C(^1\Pi(n s\sigma))$ system[370], leads to a large increase in the proportion of the available energy released into both vibration and rotation. At the same time, the rotational perturbations in the CN($B \rightarrow X$) fluorescence associated with the population of CN(A)$_{v'=10}$ which are absent at 149.4 nm, are clearly displayed at the shorter wavelength. The photofragment spectral analysis provides an alternative method for studying the nature of the parent molecule in regions where predissociation broadens the structure in the molecular absorption spectrum.

Further information can be obtained from measurements of the polarization of the photofragment fluorescence excitation spectrum. The directions and magnitudes of the fluorescence polarization, which can be measured following photodissociation at any wavelength in the excitation spectrum, reflect the orientation of the electric dipole transition moment in the parent molecule and its lifetime with respect to predissociation into the observed products[316]*.

3.7 (CN)$_2$ and the Cyanoacetylenes

3.7.1 Spectroscopy

The electronic configuration of the linear ground electronic state of cyanogen may be written

$$\ldots (2\,\sigma_u)^2\,(1\pi_u)^4\,(3\sigma_g)^2\,(1\pi_g)^4;\ ^1\Sigma_g^+.$$

The electronic spectrum thus displays many of the absorption features that are to be expected as a result of $\pi \rightarrow \pi^*$ ($1\pi_g \rightarrow 2\pi_{g,\,u}$) and $n_N \rightarrow \pi^*$ ($3\sigma_g \rightarrow 2\pi_{g,u}$) excitations.

---

* *Note added in proof:* a theoretical analysis of the polarization of fluorescence from the diatomic fragments formed in the photodissociation of triatomic molecules has been developed by Macpherson, Simons and Zare[371].

In the near ultraviolet the very weak absorption features centred at ~ 300 nm and ~ 250 nm have been attributed to the forbidden $^3\Sigma_u^+ \leftarrow {}^1\Sigma_g^+$ [325] and $^3\Delta_u \leftarrow {}^1\Sigma_g^+$ [326] transitions respectively. These are followed to shorter wavelengths by two slightly stronger regions of absorption centred around 220 nm ($^1\Sigma_u^- \leftarrow {}^1\Sigma_g^+$)[327] and 207 nm ($^1\Delta_u \leftarrow {}^1\Sigma_g^+$)[328].

There have been relatively few studies of the rather complex absorption spectrum of cyanogen in the vacuum ultraviolet[328-330] (see Fig. 3.7.1). The comprehensive experimental and theoretical studies of Connors et al.[330] have confirmed that the strong absorption bands in the region 145–170 nm involve excitation to a $^1\Pi_u$ upper state in accord with the earlier conclusions of Bell et al.[328]. A well-defined progression in ν_1' (σ_g^+) (the C≡N symmetric stretching mode, spacing $\sim 2,100$ cm^{-1}) has been discerned in the sharp but complex vibrational structure associated with this system[330]. The strongest region of absorption in the spectrum of cyanogen occurs at $\lambda \leqslant 132$ nm, and consists of a diffuse absorption feature which exhibits a short progression also in ν_1' [330]. CNDO/CI calculations predict transitions to both $^1\Sigma_u^+(1\pi_g \to 2\pi_u)$ and $^1\Pi_u$ ($3\sigma_g \to 2\pi_u$) states in this region[330] and, consequently, it has been proposed that excitation to both of these upper states contributes to this rather diffuse region of absorption[330]. The intravalence nature of all of these transitions in cyanogen has been demonstrated by their insensitivity to the addition of high pressures of helium[330], but this is not surprising in view of the high value for the ionisation potential of this molecule of 13.36 eV.[331]. A few Rydberg bands have however been reported at wavelengths below 110 nm[328, 329]. The vacuum ultraviolet absorption spectra of the cyanoacetylenes HC≡C.CN, NC.C≡C.CN and NC.C≡C–C≡C.CN have also been recorded by Connors et al.[330] (q.v.). As with

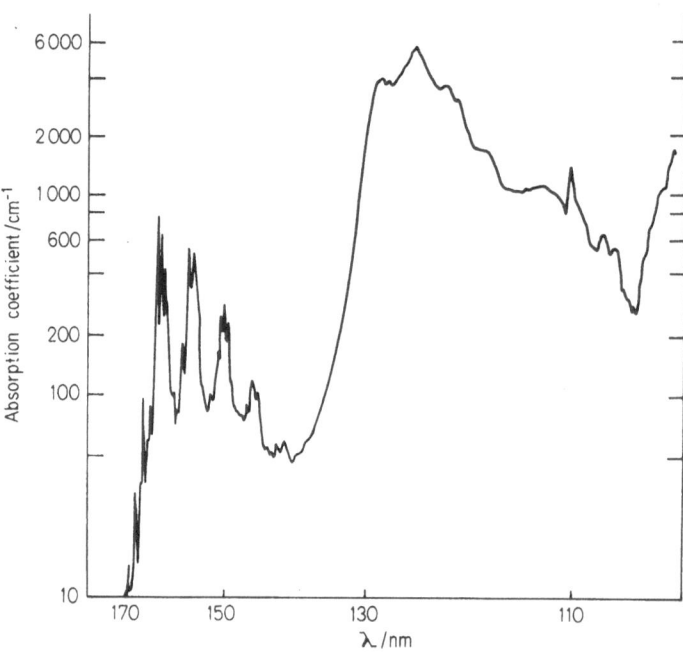

Fig. 3.7.1. Vacuum ultraviolet absorption spectrum of $(CN)_2$ (after Connors et al., Ref.[330])

cyanogen, the intense $^1\Sigma^+_{(u)} \leftarrow {}^1\Sigma^+_{(g)}$ intravalence transition is the outstanding feature in the absorption spectrum of each molecule though it shifts to longer wavelength as the degree of conjugation increases. The less intense $n \to \pi^* ({}^1\Pi_{(u)} \leftarrow {}^1\Sigma^+_{(g)})$ intravalence transitions occur at ~ 167 nm in each molecule[330] and all three cyanoacetylenes display well defined Rydberg series which terminate at the first ionisation potential[330].

3.7.2 Photochemistry

The first few energetically accessible dissociation channels for $(CN)_2$ are listed below, with their thermochemical thresholds:

$$(CN)_2 \longrightarrow 2CN(X^2\Sigma^+) \qquad\qquad \lambda < 220 \text{ nm}^{[184]} \qquad\qquad (i)$$

$$\longrightarrow CN(X^2\Sigma^+) + CN(A^2\Pi_i) \qquad \lambda < 183 \text{ nm} \qquad\qquad (ii)$$

$$\longrightarrow 2CN(A^2\Pi_i) \qquad\qquad \lambda < 157 \text{ nm} \qquad\qquad (iii)$$

$$\longrightarrow CN(X^2\Sigma^+) + CN(B^2\Sigma^+) \qquad \lambda < 141 \text{ nm}^{[184]} \qquad\qquad (iv)$$

Using kinetic absorption spectroscopy, Basco et al.[332] observed a high level of vibrational excitation in the CN(X) fragments produced during the flash photolysis of $(CN)_2$; laser emission on the $CN(X^2\Sigma^+)_{v''=4\to3}$ transition at $5.2\ \mu$ has subsequently been reported by Pollack[333] and by West and Berry[183] who also observed $CN(A^2\Pi)_{v'=0} \to CN(X^2\Sigma^+)_{v''=2}$ electronic laser emission following the flash photodissociation of $(CN)_2$ at $\lambda > 155$ nm. Cody, Sabety-Dzvonik and Jackson[334, 335] have used laser induced fluorescence of the $CN(B \to X)$ system to monitor the internal energy partitioning in the $CN(X^2\Sigma^+)$ fragments produced by polychromatic flash dissociation of $(CN)_2$ at ~ 160 nm. The ground state fragments were observed to carry a high level of rotational excitation which approximated a Boltzmann distribution over rotational states (T_{rot} ($v'' = 0$) ~ 1400 K) but only a low level of vibrational excitation, with 95% of all the CN(X) radicals produced in $v'' = 0$ and 1 [335]. The parallel primary production of $CN(A)_{v'=0}$ fragments was demonstrated by the enhancement of the population of $CN(X^2\Sigma^+)_{v''=4}$ as the pressure of added buffer was increased[335]. On the basis of these collisionally induced energy transfer studies and correlation arguments it has been proposed[183, 335] that the dominant primary process operating at $\lambda \sim 160$ nm is

$$(CN)_2 \to CN(X^2\Sigma^+)_{v''=0,1} + CN(A^2\Pi)_{v'=0} \qquad\qquad (v)$$

However, West[176] has reported a value of only 0.1 for $\phi[CN(A^2\Pi)]$ at this wavelength.

A low level of vibrational excitation has also been observed in the $CN(B^2\Sigma^+)$ fragments produced through the photolysis of $(CN)_2$ at 130.4 nm, 129.5 nm and 123.6 nm, with f_v (the fraction of excess energy going into vibration of the CN(B) fragment) in the range 7–12%[336]. The (0, 0) band of the $CN(B \to X)$ fluorescence displays none of the enhanced rotational intensities associated with the parallel formation of

$CN(A)_{v' \geq 10}$, from which it can be inferred that the low level of fragment vibrational excitation extends to dissociation channel (ii) as well.

There have been very few photochemical studies involving the cyanoacetylenes; Okabe and Dibeler[337] reported the $CN(B \to X)$ fluorescence excitation spectrum for $HC \equiv C.CN$, which faithfully reproduces the Rydberg absorption features at wavelengths below the observed production threshold for $CN(B^2 \Sigma^+)$, presumably via the reaction:

$$HC \equiv CCN \longrightarrow CN(B^2 \Sigma^+) + HC_2(X^2 \Sigma^+) \quad \lambda < 131.8 \text{ nm} \qquad \text{(vi)}$$

Cody, Sabety-Dzvonik and Jackson[338, 339] used laser induced fluorescence to measure the product state distributions in $CN(X)$ following the flash photolysis of $HC \equiv C.N$ at $\lambda > 135$ nm[338] and of $NC.C \equiv C.CN$ at $\lambda \sim 160$ nm[339] (these excitation wavelengths effectively overlap the intense, intravalence $^1\Sigma_u^+ \leftarrow {}^1\Sigma_g^+$ absorption, for which a linear upper state has been characterised, in both molecules[330]). Again, the $CN(X)$ fragments were rotationally excited (e.g. for C_4N_2, T_{rot} $(v = 0) \sim 1400$ K, T_{rot} $(v = 1) \sim 1100$ K) but carried little energy in vibration; dissociation channels leading to $CN(A^2 \Pi)$ products were found to be of very minor importance at these wavelengths.

3.8 CH_3CN, CH_3NC and CF_3CN

3.8.1 Spectroscopy

The electronic configuration for these molecules in their ground state may be represented:

$$\ldots \ldots (6a)^2 \, (1e)^4 \, (7a)^2 \, (2e)^4; \quad \tilde{X}^1 A_1$$

Ab initio calculations[340, 341] and photoelectron spectroscopy[342] agree that in CH_3CN the highest occupied molecular orbital is the $2e(\pi_{C-N})$ bonding orbital. In contrast, the relative ordering of the $7a$ and $2e$ molecular orbitals is reversed in CH_3NC. The first photoelectron band of CH_3NC at 11.24 eV displays short progressions attributable to the $N \equiv C$ stretching and CH_3 deformation frequencies in the molecular ion[343], indicating that the $7a$ m.o., although derived from the carbon lone pair orbital, contains considerable σ_{N-C} bonding character. In CF_3CN it has been proposed[344] that the $7a$ and $2e$ molecular orbitals are nearly isoenergetic since the photoelectron spectrum reveals just one broad peak in the appropriate energy region.

CH₃CN. The vacuum ultraviolet absorption spectrum of CH_3CN (see Fig. 3.8.1) displays a weak, diffuse feature centred at ~ 165 nm[345] which has been variously assigned to an intravalence $\pi \to \pi^*$ or $n_N \to \pi^*$ transition[1]. To shorter wavelengths, a more intense region of continuous absorption at ~ 135 nm[346] (which is thought to include the $\pi(2e) \to 3s$ Rydberg transition[1]) precedes a sharp band at 129.2 nm[346, 347]. This peak (term value 21,230 cm^{-1} [1]) has been attributed to the

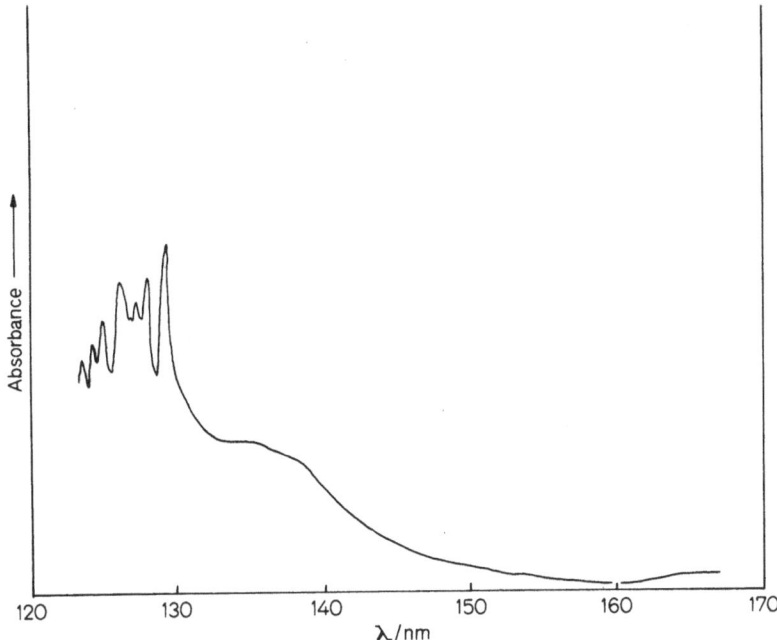

Fig. 3.8.1. Vacuum ultraviolet absorption spectrum of CH_3CN (after West and Berry, Ref.[295])

first member of the $\pi(2e) \to np\sigma$ Rydberg series, $n = 3$[1, 346, 347] and progressions based on quanta of ν_1' (C–H stretch, 2724 cm^{-1}), ν_2' (C≡N stretch, 2000 cm^{-1}) and ν_3' (CH$_3$ deformation, 1240 cm^{-1}) have been identified[337, 346].

CH$_3$NC. The leading members of two Rydberg series have been identified in the vacuum ultraviolet absorption spectrum of CH_3NC[344] (see Fig. 3.8.2). The progression starting at 133 nm has been assigned as the $\pi(2e) \to 3s$ member[344] of the Rydberg series with $\delta \sim 0.93$, terminating at the second ionisation potential, 12.46 eV[343]. The $\sigma(7a) \to 3s$ Rydberg band, which would be expected at longer wavelengths, has been assigned to the intense continuous absorption centred around 160 nm[344, 348]. A second, broadly structured feature, displaying a short progression (spacing ~ 950 cm^{-1} attributable to the C–N–C bending mode[344]), which is superimposed on the short wavelength side of the $\sigma(7a) \to 3s$ band, has been assigned to the $\pi(2e) \to \pi^*(3e)$ intravalence transition[344], and the rather intense absorption feature at ~ 185 nm may be due to the intravalence $\sigma(7a) \to \pi^*(3e)$ excitation. Figure 3.8.3 attempts to correlate these various electronic transitions with those observed in CH$_3$CN, HCN and CF$_3$CN.

CF$_3$CN. Although the vacuum ultraviolet absorption spectrum of CF$_3$CN (see Fig. 3.8.4) exhibits very little resolvable structure[176], its major electronic features have been interpreted[344] by comparison with those displayed by HCN in the same spectral region (see Fig. 3.8.3). On this basis, the weak diffuse banded absorption at ~ 140 nm has been assigned to the intravalency $\sigma(7a) \to \pi^*(3e)$ transition leading

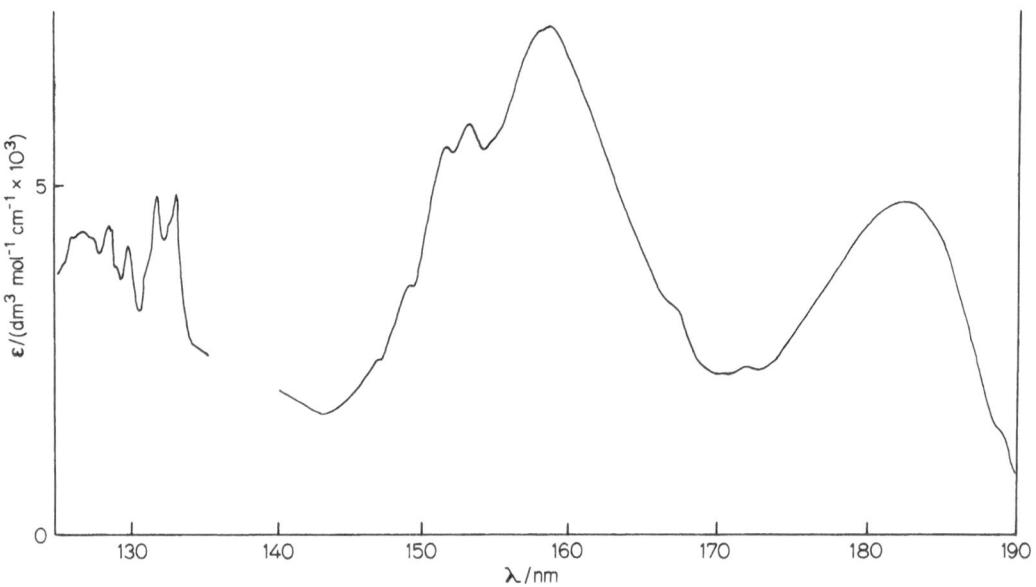

Fig. 3.8.2. Vacuum ultraviolet absorption spectrum of CH_3NC (after West and Berry, Ref.[295])

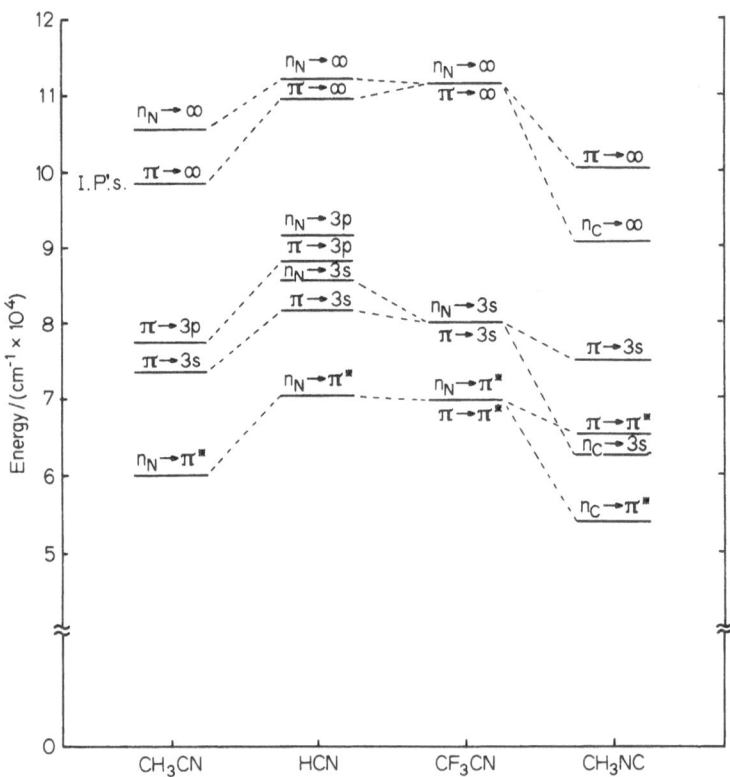

Fig. 3.8.3. Electronic transition energy correlation for CH_3CN, HCN, CF_3CN and CH_3NC (after Ashfold and Simons, Ref.[344])

to population of a bent upper state, and the intense region of continuous absorption at $\lambda \lesssim 125$ nm has been attributed to overlapping $\sigma(7a) \rightarrow 3s$ and $\pi(2e) \rightarrow 3s$ Rydberg transitions[344]. The resultant term value of $\sim 31{,}500$ cm^{-1} is typical of an $(n, 3s)$ Rydberg transition in a simple perfluorinated molecule[1].

3.8.2 Photochemistry

Two primary processes have been identified in the photolysis of CH_3CN at 184.9 nm:

$$CH_3CN \rightarrow H + CH_2CN \qquad \text{(i)}$$

$$\rightarrow CH_3 + CN \qquad \text{(ii)}$$

Final product analysis suggested that route (i) is probably the more important dissociation channel at this wavelength[349].

A host of intense molecular electronic $CN(A^2\Pi_{3/2})_{v=0} \rightarrow CN(X^2\Sigma^+)_{v=0, 1, 2}$ laser transitions have been identified following the flash photolysis of CH_3NC at $\lambda > 155$ nm[183, 348], and Knudtson and Berry[348] have identified a highly non-statistical product state distribution arising from this photodissociation, in which the formation of $CN(A^2\Pi_i)_{v'=0}$ accounts for $\sim 90\%$ of the CN radicals from channel (ii), and correlation arguments have been proposed to account for the high specifici-

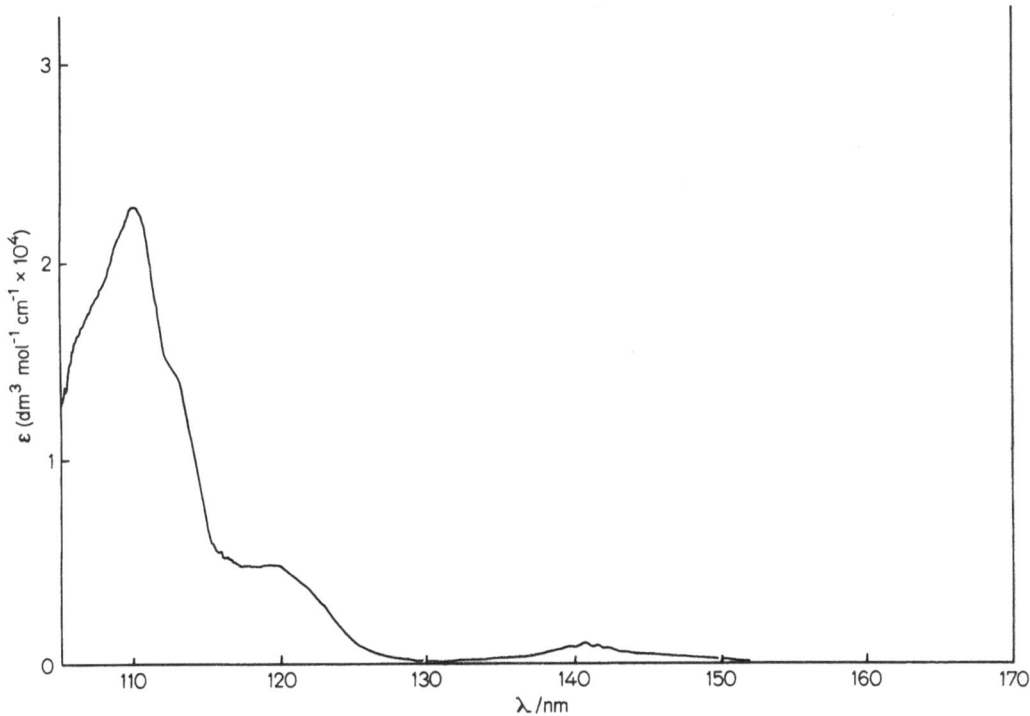

Fig. 3.8.4. Vacuum ultraviolet absorption spectrum of CF_3CN (after West, Ref.[176])

Table 3.8.1 Threshold energies for product channels in the photodissociation of CH_3CN, CH_3NC and CF_3CN (after Ref.[344])

	$CX_3(\tilde{X}) + CN(B)$ Threshold/nm	$CX_3(\tilde{X}) + CN(A)$ Threshold/nm	$D(CX_3-CN)$	
			nm	kJ mol^{-1}
CH_3CN	146 ± 1 [a, b, c]	< 192	< 234	517 ± 6
CH_3NC	162 ± 2 [a]	< 221	< 278	430 ± 10
CF_3CN	154 ± 1 [a, c]	< 206	< 255	479 ± 6

a Ref.[344].   c Ref.[183].
b Ref.[337].

ty[348]. Vibration-rotation laser transitions on $CN(X^2\Sigma^+)$ were not detected and their absence was attributed to rapid "leakage" *via* the atom abstraction:

$$CN^\dagger + CH_3NC \rightarrow HCN + CH_2NC \qquad \text{(iii)}$$

which is expected to be efficient[183]. In CF_3CN, both electronic and vibrational CN laser transitions can be observed following photodissociation at $\lambda > 155$ nm, (though surprisingly neither electronic nor vibrational laser emission was observed from CH_3CN dissociation under the same experimental conditions)[183].

The $CN(B \rightarrow X)$ photofragment fluorescence excitation spectrum from the photodissociation of CH_3CN qualitatively reproduces the absorption spectrum at wavelengths below the threshold, estimated at ~ 146 nm[183, 337, 350]. A similar value has been derived through an analysis of the rotational energy disposal in the $CN(B^2\Sigma^+)$ photofragments from CH_3CN[344]. Table 3.8.1 lists the threshold wavelengths determined for production of $CN(B)$ from each of the three parent molecules, together with the extrapolated threshold values for $CN(A^2\Pi)$ and $CN(X^2\Sigma^+)$ formation. The vibrational and rotational energy partitioning within the $CN(B^2\Sigma^+)$ fragments has been estimated from analysis of the $CN(B \rightarrow X)$ fluorescence[344] and interpreted in terms of statistical energy partitioning, limited by the constraints imposed by the requirements of energy and angular momentum conservation[344].

4 Towards the Future

Our review has concentrated on relating the photochemistry of simple polyatomic molecules in the vacuum ultraviolet to current knowledge of their absorption spectroscopy, in order to penetrate the elusive secrets of the photodissociative state. We have not discussed current experimental techniques since this aspect has recently been reviewed elsewhere[21, 25], but as writing progressed the technological constraints which have limited comparison of experimental observations with theoretical analyses became increasingly evident. Unless the experimentalist is blessed with sources of monochromatic vacuum ultraviolet light of adequate intensity to permit photofragment excitation spectra to be recorded with resolution comparable to that of the absorption spectrum, their correlation is inevitably blurred.

Many of the photochemical advances that have occurred during the past two decades have followed from the development of microwave discharge vacuum ultraviolet light sources, emitting either intense monochromatic atomic resonance radiation at fixed wavelengths (determined by nature rather than for utility) or over broad continua produced by the fluorescent decay of rare gas excimers which provide tunable sources after passage through a vacuum monochromator (but at the cost of reduced intensity).

Laser technology has fortunately now invaded the vacuum ultraviolet region, and promises that, in future, research will be less impeded by experimental limitations. These are being swept away in the flood of photons generated by intense laser sources which can be tuned over wide ranges of the vacuum ultraviolet, and which are becoming available in increasing numbers. For example, pulsed tunable vacuum u.-v. laser radiation can be generated directly from rare gas excimers (Xe_2^*, 173 nm; Kr_2^*, 146 nm; Ar_2^*, 126 nm[351−354]) from rare gas halide exciplexes (e.g. ArF*, 193 nm[355]), which has already been used for multiphoton dissociation[356]; ArCl*, 175 nm[357]) and by three and four wave mixing in potentially non-linear media such as metal vapours (e.g. by using two dye lasers to provide second harmonic photons in the near u.-v. and mixing with fundamental photons in the visible). A review of the latter technique has been presented by Stoicheff and Wallace[22], and the probability of producing c.w. vacuum ultraviolet laser radiation by this method has also been proposed[358]. Pulsed tunable lasers are also finding increasing application in the detection of primary photofragments by using laser induced fluorescence (l.i.f.)[359] or multiphoton ionization (m.p.i.)[360−362] techniques to monitor their internal energy distributions. The latter technique promises to provide a more universal detection method for monitoring the transient products of vacuum ultraviolet photochemistry, since the method is not restricted to fluorescent species though there are difficulties in relating signal strengths to concentrations. Vacuum ultraviolet time-of-flight experiments are also a practical possibility and measurements of the temporal evolution of the photodissociating molecule can be expected in the near future.

An ingenious tunable vacuum u.-v. "photofragment spectrometer" has been constructed by Jackson et al.[363]. The photolysis source is provided by a flash lamp, the electrodes of which constitute the entrance slit of a vacuum monochromator; product state distributions from selectively excited parent molecules have been monitored using laser induced fluorescence.

Vacuum ultraviolet photochemistry can also be studied by using near ultraviolet lasers of high intensity to excite 2-photon transitions in the parent molecule. An elegant early experiment was that of Wang and Davis[364], which utilised a frequency-doubled dye laser at ~ 300 nm to excite simultaneously the biphotonic dissociation of H_2O (effectively at ~ 150 nm, in the $A \leftarrow \tilde{X}$ continuum) and the single photon ($A \rightarrow X$) fluorescence of the resultant OH fragments found along the axis of the focussed laser beam.

Another fruit of modern technology is the advent of synchrotron radiation from electron storage rings[22, 365]. This produces a pulsed, polarized tunable continuum source extending into the X-ray region, having a typical pulse duration ~ 0.4 ns and a repetition ~ (1−10) MHz, and is an ideal source for time-resolved spectroscopy in the vacuum ultraviolet. The expected intensity of the system at Daresbury, for example, will be $\sim 10^{14}$ photons nm^{-1} s^{-1} $mrad^{-1}$, sufficient for it to be a very useful photolysis source. With all these systems at his disposal the experimentalist should have a thoroughly enjoyable career between now and the end of the century, indeed he may well believe the millenium has already arrived.

5 References

1. Robin, M. B.: Higher excited states of polyatomic molecules, Vols. I and II. New York: Academic Press 1974
2. Duncan, A. B. F.: Rydberg series in atoms and molecules. New York: Academic Press 1971
3. Herzberg, G.: Molecular spectra and molecular structure, Vol. III. Electronic spectra and electronic structure of polyatomic molecules. Princeton, New Jersey: Van Nostrand Reinhold 1966
4. Zaidel, A. N., Schreider, E. Ya.: Vacuum ultraviolet spectroscopy. Ann Arbor: Ann Arbor-Humphrey 1970
5. Milazzo, G., Cecchetti, G.: Vacuum ultraviolet spectroscopy – A review. Appl. Spectrosc. *23*, 197 (1969)
6. Rabalais, J. W., McDonald, J. M., Scherr, V., McGlynn, S. P.: Electronic spectroscopy of isoelectronic molecules, II. Linear triatomic groupings containing sixteen valence electrons. Chem. Rev. *71*, 73 (1971)
7. Innes, K. K.: Geometries of molecules in excited electronic states, in: Excited states, Vol. 2, p. 1. Lim, E. C. (ed.). New York: Academic Press 1975
8. Duxbury, G.: The electronic spectra of triatomic molecules and the Renner-Teller effect, in: Molecular spectroscopy Vol. III, p. 497. (Chem. Soc. Spec. Period. Rep., 1975)
9. Wittel, K., McGlynn, S. P.: The orbital concept in molecular spectroscopy. Chem. Rev. *77*, 745 (1977)
10. McNesby, J. R., Okabe, H.: Vacuum ultraviolet photochemistry, in: Advances in photochemistry, Vol. 3, p. 157. New York-London: Interscience 1964
11. Zare, R. N., Herschbach, D. R.: Atomic and molecular fluorescence excited by photodissociation. Appl. Optics. (Suppl. 2), 193 (1965)
12. Calvert, J. G., Pitts, Jr. J. N.: Photochemistry. New York: Wiley 1966
13. Simons, J. P.: Photochemistry and spectroscopy. New York: Wiley 1971
14. McNesby, J. R., Braun, W., Ball, J.: Vacuum ultraviolet techniques in photochemistry, in: Creation and detection of the excited state, p. 503. Lamola, A. A. (ed.). New York: Dekker 1971
15. Welge, K. H.: Photolysis of O_x, HO_x, CO_x and SO_x compounds. Can. J. Chem. *52*, 1424 (1974)
16. Phillips, D.: Gas-phase photoprocesses, in: Photochemistry Vol. 8, p. 105. (Chem. Soc. Spec. Period. Rep., 1977)
17. Simons, J. P.: The dynamics of photodissociation, in: Gas kinetics and energy transfer Vol. 2, p. 58. (Chem. Soc. Spec. Period. Rep., 1977)
18. Gelbart, W. M.: Photodissociation dynamics of polyatomic molecules. Ann. Rev. Phys. Chem. *28*, 323 (1977)
19. Samson, J. A. R.: Techniques of vacuum ultraviolet spectroscopy. New York: Wiley 1967
20. Damany, N., Vodar, B., Romand, J. (eds.): Some aspects of vacuum ultraviolet radiation physics. Oxford: Pergamon 1974
21. Samson, J. A. R.: Nuclear and atomic spectroscopy. Far ultraviolet region, in: Methods of experimental physics, 1976, Vol. 13 (A-Spectroscopy), p. 204.
22. Stoicheff, B. P., Wallace, S. C.: Tunable coherent VUV radiation, in: Springer Ser. Opt. Sci. 1976, 3 (Tunable Lasers Appl., Proc. Loen Conf.), p. 1
23. Ewing, J. J., Brau, C. A.: High efficiency UV lasers, in: Springer Ser. Opt. Sci. 1976, 3 (Tunable Lasers Appl., Proc. Loen Conf.), p. 21

24. Bradley, D. J., Hutchinson, M. H. R., Ling, C. C.: Tunable VUV excimer laser systems, in: Springer Ser. Opt. Sci. 1976, 3 (Tunable Lasers Appl., Proc. Loen Conf.), p. 40
25. West, M. A.: Developments in instrumentation and techniques, in: Photochemistry, Vol. 8, p. 3. (Chem. Soc. Spec. Period. Rep. 1977)
26. Buenker, R. J., Peyerimhoff, S. D.: Theoret. Chim. Acta 35, 33 (1975)
27. Baird, N. C.: Pure and Applied Chemistry 49, 223 (1977)
28. Runau, R., Peyerimhoff, S. D., Buenker, R. J.: J. Mol. Spect. 68, 253 (1977)
29. Walsh, A. D.: J. Chem. Soc. 2260 (1953)
30. Johns, J. W. C.: Rotational structure in the Rydberg series of diatomic molecules, in: Molecular spectroscopy, Vol. 2, p. 513. (Chem. Soc. Spec. Period. Rep., 1974)
31. Mulliken, R. S.: Phys. Rev. 61, 277 (1942)
32. Wang, H.-t., Felps, W. S., McGlynn, S. P.: J. Chem. Phys. 67, 2614 (1977)
33. Mulliken, R. S.: J. Chem. Phys. 3, 506 (1935)
34. Gürtler, P., Saile. V., Koch, E. E.: Chem. Phys. Lett. 51, 386 (1977)
35. Brundle, C. R., Turner, D. W.: Proc. Roy. Soc. A, 307, 27 (1968)
36. Potts, A. W., Price, W. C.: Proc. Roy. Soc. A, 326, 181 (1972)
37. Karlsson, L., Mattsson, K., Jadrny, R., Albridge, R. G., Pinchas, S., Bergmark, T., Siegbahn, K.: J. Chem. Phys. 62, 4745 (1975)
38. Dixon, R. N., Duxbury, G., Rabalais, J. W., Åsbrink, L.: Mol. Phys. 31, 423 (1976)
39. Winter, N. W., Goddard III, W. A., Bobrowiez, F. W.: J. Chem. Phys. 62, 4325 (1975)
40. Macpherson, M. T., Simons, J. P.: Chem. Phys. Lett. 51, 261 (1977)
41. Buenker, R. J., Peyerimhoff, S. D.: Chem. Phys. Lett. 29, 253 (1974)
42. Yeager, D., McKoy, V., Segal, G. A.: J. Chem. Phys. 61, 755 (1974)
43. Wight, G. R., Brion, C. E.: J. Elec. Spec. Rel. Phen. 4, 25 (1974)
44. Price, W. C.: J. Chem. Phys. 4, 147 (1936)
45. Johns, J. W. C.: Can. J. Phys. 41, 209 (1963)
46. Bell, S.: J. Mol. Spectrosc. 16, 205 (1965)
47. Katayama, D. H., Huffmann, R. E., O'Bryan, C. L.: J. Chem. Phys. 59, 4309 (1973)
48. Ishiguro, E., Sasanuma, M., Morioka, H., Nakamura, M.: J. Phys. B. 11, 993 (1978)
49. Knoop, F. W. E., Brongersma, H. H., Oosterhoff, L. J.: Chem. Phys. Lett. 13, 20 (1972)
50. Trajmar, S., Williams, W., Kuppermann, A.: J. Chem. Phys. 54, 2274 (1971); 58, 2521 (1973)
51. Tsurubuchi, S.: Chem. Phys. 10, 335 (1975)
52. Dixon, R. S.: Radiation Res. Rev. 2, 237 (1970)
53. Oldershaw, G. A., in: Gas kinetics and energy transfer, Vol. 2, p. 96. (Chem. Soc. Spec. Per. Rep., 1977)
54. Black, G., Porter, G.: Proc. Roy. Soc. A. 266, 185 (1962)
55. Stuhl, F., Welge, K. H.: J. Chem. Phys. 47, 332 (1967)
56. Stief, L. J., Payne, W. A., Klemm, R. B.: J. Chem. Phys. 62, 4000 (1975)
57. Cottin, M., Masanet, J., Vermeil, C.: J. Chim. Phys. 63, 959 (1966)
58. Masanet, J., Vermeil, C.: J. Chim. Phys. 66, 1248 (1969)
59. Ung, A. Y.-M.: Chem. Phys. Lett. 28, 603 (1974)
60. Rebbert, R. E., Lilly, R. L., Ausloos, P.: Abstracts of Papers, 164th National ACS Meeting, New York, 1972
61. Chou, C. C., Lo, J. G., Rowland, F. S.: J. Chem. Phys. 60, 1208 (1974)
62. Ung, A. Y.-M., Back, R. A.: Can. J. Chem. 42, 753 (1964)
63. Gangi, R. A., Bader, R. F. W.: J. Chem. Phys. 55, 5369 (1971)
64. Miller, K. J., Mielczarek, S. R., Krauss, M.: J. Chem. Phys. 51, 26 (1969)
65. Welge, K. H., Stuhl, F.: J. Chem. Phys. 46, 2440 (1967)
66. McNesby, J. R., Tanaka, I., Okabe, H.: J. Chem. Phys. 36, 605 (1962)
67. Chamberlain, G. A., Simons, J. P.: Chem. Phys. Lett. 32, 355 (1975)
68. Vinogradov, I. P., Vilesov, F. I.: Opt. Spectrosc. 40, 32 (1976)
69. Flouquet, F., Horsley, J. A.: J. Chem. Phys. 60, 3767 (1974)
70. Flouquet, F.: Disc. Faraday Soc. No. 62, 143 (1977)
71. Caplan, C. F., Child, M. S.: Mol. Phys. 23, 249 (1972)
72. Carrington, T.: J. Chem. Phys. 41, 2012 (1964)

M. N. R. Ashfold, M. T. Macpherson, and J. P. Simons

73. Kley, D.: Ph. D. Thesis, Bonn, 1967
74. Becker, K. H., Haaks, D.: Z. Naturforsch. *28a*, 249 (1973)
75. Yamashita, I.: J. Phys. Soc. Jap. *39*, 205 (1975)
76. Tanaka, I., Carrington, T., Broida, H. P.: J. Chem. Phys. *35*, 750 (1961)
77. Sutherland, R. A., Anderson, R. A.: J. Chem. Phys. *58*, 1226 (1973)
78. Akamatsu, R., O-ohata, K.: J. Phys. Soc. Jap. *43*, 264 (1977)
79. Akamatsu, R., O-ohata, K.: J. Phys. Soc. Jap. *44*, 589 (1978)
80. Heller, E. J.: J. Chem. Phys. *62*, 1544 (1975)
81. O-ohata, K., Liu, B.: IBM Technical Reports (to be published)
82. Möhlmann, G. R., Beenakker, C. I. M., de Heer, F. J.: Chem. Phys. *13*, 375 (1976)
83. Simons, J. P., Tasker, P. W.: Mol. Phys. *27*, 1691 (1974)
84. Lee, L. C., Judge, D. L.: Bull. Amer. Phys. Soc. *23*, 391 (1977)
85. Masanet, J., Vermeil, C.: J. Chim. Phys. *74*, 795 (1977)
86. Watanabe, K., Zelikoff, M.: J. Opt. Soc. Amer. *43*, 753 (1953)
87. Goodeve, C. F., Stein, N. O.: Trans. Faraday Soc. *27*, 393 (1931)
88. Watanabe, K., Jursa, A. S.: J. Chem. Phys. *41*, 1650 (1964)
89. Thompson, S. D., Carroll, D. G., Watson, F., O'Donnell, M., McGlynn, S. P.: J. Chem. Phys. *45*, 1367 (1966)
90. Nakamoto, K.: Infrared spectra of inorganic and coordination compounds, 3rd. ed., New York: Wiley 1977
91. Karlsson, L., Mattsson, L., Jadrny, R., Bergmark, T., Siegbahn, K.: Physica Scripta *13*, 229 (1976)
92. Murrell, J. N., Conway, A., Harada, Y.: Mol. Phys. *20*, 161 (1971)
93. Shih, S.-k., Peyerimhoff, S. D., Buenker, R. J.: Chem. Phys. *17*, 391 (1976)
94. Polezzo, S., Stabitini, M. P., Simonetta, M.: Mol. Phys. *17*, 609 (1969)
95. Hillier, I. H., Saunders, V. R.: Chem. Phys. Lett. *5*, 384 (1970)
96. Clark, L. B., Simpson, W. T.,: J. Chem. Phys. *43*, 3666 (1965)
97. Rosenfeld, J. S., Moscowitz, A.: J. Amer. Chem. Soc. *94*, 4797 (1972)
98. Price, W. C., Teegan, J. P., Walsh, A. D.: Proc. Roy. Soc. A, *201*, 600 (1950)
99. Gallo, A. R., Innes, K. K.: J. Mol. Spec. *54*, 472 (1975)
100. Bell, S.: reported in Ref. 92
101. Carroll, D. G., Armstrong, A. T., McGlynn, S. P.: J. Chem. Phys. *44*, 1865 (1966)
102. McDiarmid, R.: J. Chem. Phys. *68*, 945 (1978)
103. Donovan, R. J., Little, D. J., Konstantatos, J.: J. Chem. Soc. Faraday II, *68*, 1812 (1972)
104. Porter, G.: Disc. Faraday Soc. *9*, 60 (1950)
105. Ramsay, D. A.: J. Chem. Phys. *20*, 1920 (1952)
106. Darwent, B. de B., Roberts, R.: Proc. Roy. Soc. A, *216*, 344 (1953)
107. Gann, R. G., Dubrin, J.: J. Chem. Phys. *47*, 1867 (1967)
108. Compton, L. E., Gole, J. L., Martin, R. M.: J. Phys. Chem. *73*, 1158 (1969)
109. Sturm, G. P., White, J. M.: J. Chem. Phys. *50*, 5035 (1969)
110. Compton, L. E., Martin, R. M.: J. Chem. Phys. *52*, 1613 (1970)
111. Davis, D. D., Braun, W., unpublished results, reported in: Kurylo, M. J., Peterson, N. C., Braun, W.: J. Chem. Phys. *54*, 943 (1971)
112. Haaks, D.: Ph. D. Thesis, Bonn (1972)
113. Flouquet, F.: Chem. Phys. *13*, 257 (1976)
114. Dobson, D. C., James, F. C., Safarik, I., Gunning, H. E., Strausz, O. P.: J. Phys. Chem. *79*, 771 (1975)
115. Watanabe, K.: J. Chem. Phys. *22*, 1564 (1954)
116. Walsh, A. D., Warsop, P. A.: Trans. Faraday Soc. *57*, 345 (1961)
117. Douglas, A. E.: Discuss. Faraday Soc. *35*, 158 (1963)
118. Herzberg, G., Longuet-Higgins, H. C.: Discuss. Faraday Soc. *35*, 77 (1963)
119. Rianda, R., Frueholz, R. P., Goddard III, W. A.: Chem. Phys. *19*, 131 (1976)
120. Douglas, A. E., Hollas, J. M.: Can. J. Phys. *39*, 479 (1961)
121. Harshbarger, W. R.: J. Chem. Phys. *54*, 2504 (1971)

122. Branton, G. R., Frost, D. C., Herring, F. G., McDowell, C. A., Stenhouse, I. A.: Chem. Phys. Lett. *3*, 581 (1969)
123. Humphries, C. M., Walsh, A. D., Warsop, P. A.: Discuss. Faraday Soc. *35*, 148 (1963)
124. Walsh, A. D., Warsop, P. A.: Ultraviolet spectra of hydrides of group V Elements, in: Advances in molecular spectroscopy Vol. 2, p. 582. Mangini, A. (ed.). Oxford: Pergamon 1962
125. McConaghie, V. M., Nielson, H. H.: J. Chem. Phys. *21*, 1836 (1953)
126. Claxton, T. A., Smith, N. A.: Trans. Farad. Soc. *66*, 1825 (1970)
127. Maier, J. P., Turner, D. W.: J. Chem. Soc. Farad. II, *68*, 711 (1972)
128. Aarons, L. J., Guest, M. F., Hall, M. P., Hillier, I. H.: J. Chem. Soc. Farad. II, *69*, 643 (1973)
129. Masanet, J., Gilles, A., Vermeil, C.: J. Photochem. *3*, 417 (1974/5)
130. Herzberg, G.: Molecular spectra and molecular structure, Vol. I. Spectra of diatomic molecules, Princeton, New Jersey: Van Nostrand 1950
131. Okabe, H., Lenzi, M.: J. Chem. Phys. *47*, 5241 (1967)
132. Bayes, K. D., Becker, K. H., Welge, K. H.: Z. Naturforsch, *17a*, 676 (1962)
133. Groth, W. E., Schurath, U., Schindler, R. N.: J. Phys. Chem. *72*, 3914 (1968)
134. Schurath, U., Tiedemann, P., Schindler, R. N.: J. Phys. Chem. *73*, 456 (1969)
135. Di Stefano, G., Lenzi, M., Margani, A., Nguyen Xuan, C.: J. Chem. Phys. *67*, 3832 (1977)
136. Branton, G. R., Frost, D. C., Herring, F. G., McDowell, C. A., Stenhouse, I. A.: Chem. Phys. Lett. *5*, 1 (1970)
137. Watanabe, K., Sood, S. P.: Sci. Light (Tokyo) *14*, 36 (1965)
138. (a) Koda, S., Back, R. A.: Can. J. Chem. *55*, 1380 (1977)
    (b) Back, R. A., Koda, S.: Can. J. Chem. *55*, 1387 (1977)
139. Koda, S.: Bull. Chem. Soc. Japan *50*, 1683 (1977)
140. Lenzi, M., McNesby, J. R., Mele, A., Nguyen Xuan, C.: J. Chem. Phys. *57*, 319 (1972)
141. Halpern, J. B., Hancock, G., Lenzi, M., Welge, K. H.: J. Chem. Phys. *63*, 4808 (1975)
142. Gelernt, B., Filseth, S. V., Carrington, T.: Chem. Phys. Lett. *36*, 238 (1975)
143. Masanet, J., Fournier, J., Vermeil, C.: Can. J. Chem. *51*, 2946 (1973)
144. Lilly, R. L., Rebbert, R. E., Ausloos, P.: J. Photochem. *2*, 49 (1973/4)
145. Becker, K. H., Welge, K. H.: Z. Naturforsch *18a*, 600 (1963)
146. Hansen, I., Höinghaus, K. K., Zetzsch, C., Stuhl, F.: Chem. Phys. Lett. *42*, 370 (1976)
147. Beyer, K. D., Welge, K. H.: Z. Naturforsch, *12a*, 1161 (1957)
148. Becker, K. H., Welge, K. H.: Z. Naturforsch, *19a*, 1006 (1964)
149. Kawasaki, M., Hirata, Y., Tanaka, I.: J. Chem. Phys. *59*, 648 (1973)
150. Kawasaki, M., Tanaka, I.: J. Phys. Chem. *78*, 1784 (1974)
151. Gilles, A., Masanet, J., Vermeil, C.: Chem. Phys. Lett. *25*, 346 (1974)
152. Zetzsch, C., Stuhl, F.: Chem. Phys. Lett. *33*, 375 (1975)
153. Gelernt, B., Filseth, S. V., Carrington, T.: J. Chem. Phys. *65*, 4940 (1976)
154. Ramsay, D. A.: Nature (London) *178*, 374 (1956)
155. Norrish, R. G. W., Oldershaw, G. A.: Proc. Roy. Soc. *A262*, 1 (1961)
156. Kley, D., Welge, K. H.: Z. Naturforsch. *20a*, 124 (1965)
157. Berthou, J. M., Pascat, B., Guenebaut, H., Ramsay, D. A.: Can. J. Phys. *50*, 2265 (1972), and references therein
158. Lee, J. H., Michael, J. V., Payne, W. A., Whytock, D. A., Stief, L. J.: J. Chem. Phys. *65*, 3280 (1976)
159. Di Stefano, G., Lenzi, M., Margani, A., Mele, A., Nguyen Xuan, C.: J. Photochem. *7*, 335 (1975)
160. Di Stefano, G., Lenzi, M., Margani, A., Nguyen Xuan, C.: J. Chem. Phys. *68*, 959 (1978)
161. Price, W. C., Passmore, T. R.: Discuss. Faraday Soc. *35*, 232 (1963)
162. Jordan, P. C.: J. Chem. Phys. *41*, 1442 (1964)
163. Cade, P. E.: Can. J. Phys. *46*, 1989 (1968)
164. Rostas, J., Cossart, D., Bastien, J. B.: Can. J. Phys. *52*, 1274 (1974)
165. Suchard, S. N. (ed.): Spectroscopic Data 1, part B. IFI-Plenum, New York 1975, p. 815
166. Balfour, W. J., Douglas, A. E.: Can. J. Phys. *46*, 2277 (1968)
167. Legay, F.: Can. J. Phys. *38*, 797 (1960)

168. Walsh, A. D.: J. Chem. Soc. 2288, (1953)
169. Absar, I., McEwen, K. L.: Can. J. Chem. *50*, 653 (1972)
170. Hilgendorff, H. J.: Z. Physik *95*, 781 (1935)
171. Herzberg, G., Innes, K. K.: Can. J. Phys. *35*, 842 (1957)
172. Schwenzer, G. M., O'Neill, S. V., Schaefer, H. F. III, Baskin, C. P., Bender, C. F.: J. Chem. Phys. *60*, 2787 (1974)
173. Perić, M., Peyerimhoff, S. D., Buenker, R. J.: Can. J. Chem. *55*, 3664 (1977)
174. Schwenzer, G. M., Bender, C. F., Schaefer, H. F. III: Chem. Phys. Lett. *36*, 179 (1975)
175. Herzberg, G., Innes, K. K.: unpublished results, reported in Ref. 3
176. West, G. A.: Ph. D. Thesis, Wisconsin (1975)
177. Macpherson, M. T., Simons, J. P.: J. Chem. Soc. Faraday II, *74*, 1965, (1978)
178. Tam, W.-C., Brion, C. E.: J. Elec. Spect. Rel. Phen. *3*, 281 (1974)
179. Fridh, C., Åsbrink, L.: J. Elec. Spect. Rel. Phen. *7*, 119 (1975)
180. Chutjian, A., Tanaka, H., Wicke, B. G., Srivastava, S. K.: J. Chem. Phys. *67*, 4835 (1977)
181. Åsbrink, L., Fridh, C., Lindholm, E.: Chem. Phys. *27*, 159 (1978)
182. Price, W. C., Teegan, J. P., Walsh, A. D.: Proc. Roy. Soc. *201A*, 600 (1950)
183. West, G. A., Berry, M. J.: J. Chem. Phys. *61*, 4700 (1974)
184. Davis, D. D., Okabe, H.: J. Chem. Phys. *49*, 5526 (1968)
185. Mele, A., Okabe, H.: J. Chem. Phys. *51*, 4798 (1969)
186. Tereschenko, E. N., Dodonova, N. Ya.: Opt. Spectrosc. *41*, 286 (1976)
187. Cook, T. J., Levy, D. H.: J. Chem. Phys. *57*, 5059 (1972)
188. Tatematsu, S., Kuchitsu, K.: Bull. Chem. Soc. Japan *50*, 2896 (1977)
189. Ashfold, M. N. R., Macpherson, M. T., Simons, J. P.: Chem. Phys. Lett. *55*, 84 (1978)
190. Luk, C. K., Bersohn, R.: J. Chem. Phys. *58*, 2153 (1973)
191. Jackson, W. M.: J. Chem. Phys. *61*, 4177 (1974)
192. Mohamed, K. A., King, G. C., Read, F. H.: J. Elec. Spec. Rel. Phen. *12*, 229 (1977)
193. See Refs. 17 and 189
194. Ashfold, M. N. R., Simons, J. P.: J. Chem. Soc. Faraday II *74*, 280 (1978)
195. Walsh, A. D.: J. Chem. Soc. 2266 (1953)
196. Greening, F. R., King, G. W.: J. Mol. Spect. *59*, 312 (1976)
197. Finn, E. J., King, G. W.: J. Mol. Spect. *56*, 52 (1975)
198. Greening, F. R., King, G. W.: J. Mol. Spect. *61*, 459 (1976)
199. Finn, E. J., King, G. W.: J. Mol. Spect. *56*, 39 (1975)
200. King, G. W., Srikameswaran, K.: J. Mol. Spect. *31*, 269 (1969)
201. Mulliken, R. S.: Chem. Phys. Lett. *25*, 305 (1974)
202. Winter, N. M., Bender, C. F., Goddard, W. A.: Chem. Phys. Lett. *20*, 489 (1973)
203. Winter, N. M.: Chem. Phys. Lett. *33*, 300 (1975)
204. Jungen, M.: Chem. Phys. Lett. *27*, 256 (1974)
205. Zelikoff, M., Watanabe, K., Inn, E. C. Y.: J. Chem. Phys. *21*, 1643 (1953)
206. Tanaka, Y., Jursa, A. S., Le Blanc, F. J.: J. Chem. Phys. *32*, 1205 (1960)
207. Nakata, R. S., Watanabe, K., Matsunaga, F. M.: Sci. Light (Tokyo) *14*, 54 (1965)
208. Jungen, Ch., Malm, D. N., Merer, A. J.: Can. J. Phys. *51*, 1471 (1973)
209. Peyerimhoff, S. D., Buenker, R. J.: J. Chem. Phys. *49*, 2473 (1968)
210. England, W. B., Ermler, W. C., Wahl, A. C.: J. Chem. Phys. *66*, 2336 (1977)
211. Bajema, L., Gouterman, M., Meyer, B.: J. Phys. Chem. *75*, 2004 (1971)
212. Breckenridge, W. H., Taube, H.: J. Chem. Phys. *52*, 1713 (1970)
213. Matsunaga, F. M., Watanabe, K.: J. Chem. Phys. *46*, 4457 (1967)
214. Huebner, R. H., Celotta, R. J., Mielczarek, S. R., Kuyatt, C. E.: J. Chem. Phys. *63*, 4490 (1975)
215. Douglas, A., Zanon, I.: J. Chem. Phys. *42*, 627 (1964)
216. Filseth, S. V.: Vapour phase photochemistry of the neutral oxides and sulphides of carbon, in Advances in photochemistry Vol. 10, p. 1. New York: Interscience 1977
217. See for example the reviews of: Preston, K. F., Cvetanović, R. J.: Comprehensive chemical kinetics *4*, 47 (1972); Welge, K. H.: Ref. 15
218. Slanger, T. G.: J. Chem. Phys. *45*, 4127 (1966)

219. Inn, E. C. Y., Heimerl, J.: J. Atmos. Sci. 28, 838 (1971)
220. Slanger, T. G., Black, G.: J. Chem. Phys. 54, 1889 (1971)
221. Chemical kinetic and photochemical data for modelling atmospheric chemistry. Hampson, R. F., Jr., and Garvin, D. (ed.). NBS Technical Note 866, p. 21 (1975)
222. Sach, R. S.: Int. J. Radiat. Phys. Chem. 3, 45 (1971)
223. Inn, E. C. Y.: J. Geophys. Res. 77, 1991 (1972)
224. Slanger, T. G., Sharpless, R. L., Black, G., Filseth, S. V.: J. Chem. Phys. 61, 5022 (1974)
225. Slanger, T. G., Black, G.: J. Chem. Phys. 68, 1844 (1978)
226. Ausloos, P.: unpublished results (reported in Ref. 225)
227. Slanger, T. G., Sharpless, R. L., Black, G.: J. Chem. Phys. 67, 5317 (1977)
228. Lawrence, G. M.: J. Chem. Phys. 57, 5616 (1972)
229. Koyano, I., Wauchop, T. S., Welge, K. H.: J. Chem. Phys. 63, 110 (1975)
230. Welge, K. H., Gilpin, R.: J. Chem. Phys. 54, 4224 (1971)
231. Pack, R. T.: J. Chem. Phys. 65, 4765 (1976)
232. Lawrence, G. M.: J. Chem. Phys. 56, 3435 (1972)
233. Slanger, T. G., Black, G.: J. Chem. Phys. 68, 989 (1978)
234. Judge, D. L., Lee, L. C.: J. Chem. Phys. 58, 104 (1973)
235. Phillips, E., Lee, L. C., Judge, D. L.: J. Chem. Phys. 65, 3118 (1976)
236. Gentieu, E. P., Mentall, J. E.: J. Chem. Phys. 58, 4803 (1973)
237. Lee, L. C., Carlson, R. W., Judge, D. L., Ogawa, M.: J. Chem. Phys. 63, 3987 (1975)
238. Lee, L. C., Judge, D. L.: Can. J. Phys. 51, 378 (1973)
239. Phillips, E., Lee, L. C., Judge, D. L.: J. Chem. Phys. 66, 3688 (1977)
240. Band, Y. B., Freed, K. F.: J. Chem. Phys. 64, 4329 (1976)
241. Freed, K. F., Band, Y. B.: Product energy distributions in the dissociation of polyatomic molecules, in: Excited states, Vol. 3, p. 109. Lim, E.C. (ed.). New York: Academic Press 1978
242. Berry, M. J.: Chem. Phys. Lett. 29, 329 (1974)
243. Gunning, H. E., Strausz, O. P.: Reactions of sulphur atoms, in: Advances in photochemistry, Vol. 4, p. 143. New York: Interscience 1966
244. Lin, M. C.: Chem. Phys. 7, 433 (1975)
245. Black, G., Sharpless, R. L., Slanger, T. G., Lorents, D. C.: J. Chem. Phys. 62, 4274 (1975)
246. Donovan, R. J.: Trans. Farad Soc. 65, 1419 (1969)
247. Klemm, R. B., Glicker, S., Stief, L. J.: Chem. Phys. Lett. 33, 512 (1975)
248. Lee, L. C., Judge, D. L.: J. Chem. Phys. 63, 2782 (1975)
249. Donovan, R. J., Little, D. J.: Chem. Phys. Lett. 53, 394 (1978)
250. Black, G., Sharpless, R. L., Slanger, T. G.: J. Chem. Phys. 64, 3985 (1976)
251. Callear, A. B.: Proc. Roy. Soc. A276, 401 (1963)
252. Little, D. J., Dalgleish, A., Donovan, R. J.: Disc. Farad. Soc. 53, 211 (1972)
253. Black, G., Sharpless, R. L., Slanger, T. G.: J. Chem. Phys. 66, 2113 (1977)
254. Okabe, H. in: Chemical spectroscopy and photochemistry in the vacuum ultraviolet, p. 513. Sandorfy, C., Ausloos, P. J., and Robin, M. B. (eds.). Boston, Mass.: D. Reidel Publishing Co. 1974
255. Okabe, H.: J. Chem. Phys. 56, 4381 (1972)
256. Ashfold, M. N. R., Simons, J. P.: unpublished results
257. Coxon, J. A., Marcoux, P. J., Setser, D. W.: Chem. Phys. 17, 403 (1976)
258. Smoes, S., Drowart, J.: J. Chem. Soc. Farad. II 73, 1746 (1977)
259. Marquart, J. R., Bedford, R. L., Fraenkel, H. A.: Int. J. Chem. Kin. 9, 671 (1977)
260. Krupenie, P. H.: The band spectrum of carbon monoxide, NSRDS-NBS 5, U.S. Govt. Printing Office 1966
261. Weise, W. L., Smith, M. W., Glennon, B. M.: Atomic transition probabilities, Vol. 1, Hydrogen through Neon, NSRDS-NBS 4, U.S. Govt. Printing Office 1966
262. Weise, W. L., Smith, M. W., Miles, B. M.: Atomic transition probabilities, Vol. II, Sodium through Calcium, NSRDS-NBS 22, U. S. Govt. Printing Office 1969
263. Dibeler, V. H., Walker, J. A.: J. Opt. Soc. Amer. 57, 1007 (1967)
264. Tewarson, A., Palmer, H. B.: J. Mol. Spect. 27, 246 (1968)

265. Field, R. W., Bergeman, T. H.: J. Chem. Phys. *54*, 2936 (1971)
266. Taylor, G. W., Setser, D. W., Coxon, J. A.: J. Mol. Spect. *44*, 108, (1972)
267. Cossart, D., Horani, M., Rostas, J.: J. Mol. Spect. *67*, 283 (1977)
268. Cossart, D., Bergeman, T. H.: J. Chem. Phys. *65*, 5462 (1976)
269. Lagerqvist, A., Westerlund, H., Wright, C. V., Barrow, R. F.: Ark. Fys. *14*, 387 (1958)
270. Moore, C. E.: Atomic energy levels, Vol. I, NBS-Circ. 467 (1949)
271. Lebreton, J., Bosser, G., Marsigny, L.: J. Phys. B. *6*, L226 (1973)
272. Laird, R. K., Barrow, R. F.: Proc. Phys. Soc. A., *66*, 836 (1953)
273. Young, R. A., Black, G., Slanger, T. G.: J. Chem. Phys. *49*, 4769 (1968)
274. Boxall, C. R., Simons, J. P., Tasker, P. W.: Disc. Farad. Soc. *53*, 182 (1972)
275. Paraskevopoulos, G., Symonds, V. B., Cvetanovic, R. J.: Can. J. Chem. *50*, 1838 (1972)
276. Chamberlain, G. A., Simons, J. P.: J. Chem. Soc. Farad. II *71*, 402 (1975)
277. Young, R. A., Black, G., Slanger, T. G.: J. Chem. Phys. *50*, 309 (1969)
278. Hampson, R. F., Jr., Okabe, H.: J. Chem. Phys. *52*, 1930 (1970)
279. Slanger, T. G., Wood, B. J., Black, G.: Chem. Phys. Lett. *17*, 401 (1972)
280. Black, G., Slanger, T. G., St. John, G. A., Young, R. A.: J. Chem. Phys. *51*, 116 (1969)
281. Slanger, T. G., Wood, B. J., Black, G.: J. Geophys. Res. *76*, 8430 (1971)
282. Husain, D., Kirsch, L. J., Wiesenfeld, J. R.: Disc. Farad. Soc. *53*, 201 (1972)
283. Husain, D., Mitra, S. K., Young, A. N.: J. Chem. Soc. Farad. II *70*, 1721 (1974)
284. Young, R. A., Black, G., Slanger, T. G.: J. Chem. Phys. *50*, 303 (1969)
285. Slanger, T. G., Wood, B. J., Black, G.: J. Photochem. *2*, 63 (1973)
286. McEwan, M. J., Lawrence, G. M., Poland, H. M.: J. Chem. Phys. *61*, 2857 (1974)
287. Preston, K. F., Barr, R. F.: J. Chem. Phys. *54*, 3347 (1971)
288. Black, G., Sharpless, R. L., Slanger, T. G., Lorents, D. C.: J. Chem. Phys. *62*, 4266 (1975)
289. Ridley, B. A., Atkinson, R., Welge, K. H.: J. Chem. Phys. *58*, 3878 (1973)
290. Dodge, M. C., Heicklen, J.: Int. J. Chem. Kin. *3*, 269 (1971)
291. Young, R. A., Black, G., Slanger, T. G.: J. Chem. Phys. *48*, 2067 (1968)
292. Fisher, C. H., Welge, K. H.: 4th International Symposium on Molecular Beams, Peymeinada, France 1973
293. Stone, E. J., Lawrence, G. M., Fairchild, C. E.: J. Chem. Phys. *65*, 5083 (1976)
294. Shapiro, M.: Chem. Phys. Lett. *46*, 442 (1977)
295. West, G. A., Berry, M. J.: unpublished results
296. King, G. W., Richardson, A. W.: J. Mol. Spect. *21*, 339, 353 (1966)
297. Myer, J. A., Samson, J. A. R.: J. Chem. Phys. *52*, 266 (1970)
298. Macpherson, M. T., Simons, J. P.: to be published (see Ref.[317])
299. West, G. A., Berry, M. J.: Chem. Phys. Lett. *56*, 423, (1978)
300. Lake, R. F., Thompson, H.: Proc. Roy. Soc. A. *317*, 187 (1970)
301. Potts, A. W.: Ph. D. Thesis, University of London, 1969
302. Beswick, J. A., Jortner, J.: Chem. Phys. *24*, 1 (1977)
303. Baronavski, A. P., McDonald, J. R.: Chem. Phys. Lett. *45*, 172 (1977)
304. Sabety-Dzvonik, M. J., Cody, R. J.: J. Chem. Phys. *66*, 125 (1977)
305. Busch, G. E., Cornelius, J. F., Mahoney, R. T., Morse, R. I., Schlesser, D. W., Wilson, K. R.: Rev. Sci. Instr. *41*, 1066 (1970)
306. Ling, J. H., Wilson, K. R.: J. Chem. Phys. *63*, 101 (1975)
307. Mitchell, R. C., Simons, J. P.: Faraday Discuss. *44*, 208 (1967)
308. Holdy, K. E., Klotz, L. C., Wilson, K. R: J. Chem. Phys. *52*, 4588 (1970)
309. Shapiro, M., Levine, R. D.: Chem. Phys. Lett. *5*, 499 (1970)
310. Simons, J. P., Tasker, P. W.: Mol. Phys. *26*, 1267 (1973)
311. Halavee, U., Shapiro, M.: Chem. Phys. *21*, 105 (1977)
312. Morse, M. D., Band, Y. B., Freed, K. F.: Chem. Phys. Lett. *44*, 125 (1976)
313. Florida, D., Rice, S. A.: Chem. Phys. Lett. *33*, 207 (1975)
314. Rice, S. A.: Dynamics of primary photochemical processes, in: Excited states, Vol. 2, p. 111. Lim, E.C. (ed.). New York: Academic Press 1975
315. Ashfold, M. N. R., Simons, J. P.: J. Chem. Soc. Faraday II, *73*, 858 (1977)
316. Chamberlain, G. A., Simons, J. P.: J. Chem. Soc. Faraday II, *71*, 2043 (1975)

317. Macpherson, M. T., Simons, J. P.: J. Chem. Soc. Faraday II (1979), in press
318. Sabety-Dzvonik, M. S., Cody, R. J.: J. Chem. Phys. *64*, 4794 (1976)
319. Ashfold, M. N. R., Simons, J. P.: Chem. Phys. Lett. *47*, 65 (1977)
320. Tatematsu, S., Kondow, T., Nakagawa, T., Kuchitsu, K.: Bull. Chem. Soc. Japan *50*, 1056 (1977)
321. Wager, A. T.: Phys. Rev. *64*, 18 (1943)
322. Radford, H. E.: Phys. Rev. A. *136*, 1571 (1964)
323. Meakin, P., Harris, D. O.: J. Mol. Spect. *44*, 219 (1972)
324. Schaefer, H. F. III, Heil, T. G.: J. Chem. Phys. *54*, 2573 (1971)
325. Callomon, J. H., Davey, A. B.: Proc. Phys. Soc. Lond. *82*, 335 (1963)
326. Cartwright, G. J., O'Hare, D., Walsh, A. D., Warsop, P. A.: J. Mol. Spect. *39*, 393 (1971)
327. Fish, G. B., Cartwright, G. J., Walsh, A. D., Warsop, P. A.: J. Mol. Spect. *41*, 20 (1972)
328. Bell, S., Cartwright, G. J., Fish, G. B., O'Hare, D., Ritchie, R. K., Walsh, A. D., Warsop, P. A.: J. Mol. Spect. *30*, 162 (1969)
329. Price, W. C., Walsh, A. D.: Trans. Farad. Soc. *41*, 381 (1945)
330. Connors, R. E., Roebber, J. L., Weiss, K.: J. Chem. Phys. *60*, 5011 (1974)
331. Baker, C., Turner, D. W.: Proc. Roy. Soc. A. *308*, 19 (1968)
332. Basco, N., Nicholas, J. E., Norrish, R. G. W., Vickers, W. H. J.: Proc. Roy. Soc. A. *272*, 147 (1963)
333. Pollack, M.: Appl. Phys. Lett. *9*, 230 (1966)
334. Jackson, W. M., Cody, R. J.: J. Chem. Phys. *61*, 4183 (1974)
335. Cody, R. J., Sabety-Dzvonik, M. J., Jackson, W. M.: J. Chem. Phys. *66*, 2145 (1977)
336. Ashfold, M. N. R., Simons, J. P.: unpublished results.
337. Okabe, H., Dibeler, V. H.: J. Chem. Phys. *59*, 2430 (1973)
338. Cody, R. J., Sabety-Dzvonik, M. J: 12th Informal Conference on Photochemistry (Natl. Bureau of Standards, Gaithersburg, U.S.A., 1976)
339. Sabety-Dzvonik, M. J., Cody, R. J., Jackson, W. M.: Chem. Phys. Lett. *44*, 131 (1976)
340. Snyder, L. C., Basch, H.: Molecular wave functions and properties: Tabulated from SCF calculations in a Gaussian basis set. New York: Wiley 1972
341. Ha, T.-K.: J. Molecular Structure, *11*, 185 (1972)
342. Turner, D. W., Baker, A. D., Baker, C., Brundle, C. R.: Molecular photoelectron spectroscopy, p. 346. New York: Wiley 1970
343. Lake, R. F., Thompson, H.: Spectrochimica Acta *27*, 783 (1971)
344. Ashfold, M. N. R., Simons, J. P.: J. Chem. Soc. Faraday II *74*, 1263 (1978)
345. Herzberg, G., Scheibe, G.: Z. physik. Chem. *B7*, 390 (1930)
346. Cutler, J. A.: J. Chem. Phys. *16*, 136 (1948)
347. Stradling, R. S., Loudon, A. G.: J. Chem. Soc. Faraday II, *73*, 623 (1977)
348. Knudtson, J. T., Berry, M. J.: J. Chem. Phys. *68*, 4419, (1978)
349. McElcheran, D. C., Wijnen, M. H. J., Steacie, D. W. R.: Can. J. Chem. *36*, 321 (1958)
350. Vinogradov, I. P., Vilesov, F. I.: Opt. Spectrosc. *40*, 179 (1976)
351. Gerardo, J. B., Johnson, A. W.: IEEE J. Quant. Electron. QE-9, 746 (1973)
352. Hoff, P., Swingle, J. C., Rhodes, C K.: Appl. Phys. Lett. *23*, 246 (1973)
353. Wallace, S. C., Hodgson, R. T., Dreyfus, R. W.: Appl. Phys. Lett. *23*, 672 (1973)
354. Hughes, W. M., Shannon, J., Hunter, R.: Appl. Phys. Lett. *24*, 488 (1974)
355. Hoffman, J. M., Hays, A. K., Tisone, G. C.: Appl. Phys. Lett. *28*, 538 (1976)
356. Jackson, W. M., Halpern, J. B., Lin, C.-S.: Chem. Phys. Lett. *55*, 254 (1978)
357. Waynant, R. W.: Appl. Phys. Lett. *30*, 234 (1977)
358. Bjorklund, G. C., Bjorkholm, J. E., Freeman, R. R., Liao, P. F.: Appl. Phys. Lett. *31*, 330 (1977)
359. For a recent review of l.i.f. see: Kinsey, J. L., Ann. Rev. Phys. Chem. *28*, 349 (1977)
360. Johnson, P. M., Berman, M., Zakheim, D.: J. Chem. Phys. *62*, 2500 (1975)
361. Feldman, D. L., Lengel, R. N., Zare, R. N.: Chem. Phys. Lett. *52*, 413 (1977)
362. Vaida, V., Turner, R. E., Casey, J. L., Colson, S. D.: Chem. Phys. Lett. *54*, 25 (1978)
363. Miller, G. E., Halpern, J. B., Jackson, W. M.: Appl. Optics, *17*, 2821 (1978)
364. Wang, C. C., Davis, L. I.: J. Chem. Phys. *62*, 53 (1975)

M. N. R. Ashfold, M. T. Macpherson, and J. P. Simons

365. Vacuum ultraviolet physics. (eds. Koch, E. E., Haensel, R., and Kunz, C. (eds.). New York: Pergamon 1974
366. Masuki, H., Morioka, Y., Nakamura, M., Ishiguro, E., Sasanuma, Can. J. Phys. *57*, 745 (1979)
367. Nieman, G. C., Colson, S. D.: J. Chem. Phys. in press
368. Bischel, W. K., Black, G., Hawkins, R. T., Kligler, D. J., Rhodes, C. K.: J. Chem. Phys. *70*, 5589 (1979)
369a. Morse, M. D., Freed, K. F., Band, Y. B.: J. Chem. Phys. (1979), in press
    b. Freed, K. F., Morse, M. D., Band, Y. B.: J. Chem. Soc. Faraday Discuss. *69* (1979), in press.
370. Georgion, A. S., Quinton, A. M., Simons, J. P.: to be published
371. Macpherson, M. T., Simons, J. P., Zare, R. N.: Mol. Phys. (1979), in press.

Received October 26, 1978

Far-Ultraviolet Absorption Spectra
of Organic Molecules:
Valence-Shell and Rydberg Transitions

Camille Sandorfy

Département de Chimie, Université de Montréal, C. P. 6210, Succ. A.,
Montréal, Québec, Canada H3C 3VI

Table of Contents

Introduction

Far ultraviolet spectroscopy is nearly as old as ultraviolet spectroscopy; Herzberg's volume III is the authoritative reference. For the more chemically minded spectroscopist Robin's two volumes offer a framework based on chemical categories and nearly complete coverage.

The scope of the present chapter is limited in many ways: only optical absorption spectra of organic molecules are considered in the region of the spectrum extending roughly from 200 to 120 nm or 6 to 10 eV or 50000 to 80000 cm^{-1}. The spectra that can be obtained from molecules of this size are mostly of moderate resolution but with the help of photoelectron spectroscopy and quantum chemistry a general and fairly solid understanding can be achieved and this despite the complete lack of support from rotational structure and only a modest degree of support from vibrational structure. While an attempt is made to state the present position relating to the spectra of certain typical organic molecules no completeness is sought. The far ultraviolet spectra of organic molecules fall naturally into certain categories: spectra due to transitions of bonding σ electrons; spectra dominated by the transitions of lone pair electrons; spectra characterized by the transitions of π or both π and lone pair electrons. Examples are given for each of these categories. The first sections are meant for the less specialized Reader.

This chapter follows three leitmotives:
1. The fundamental fact and problem of far ultraviolet spectroscopy is the Rydberg versus valence-shell distinction;
2. This distinction must have photochemical consequences;
3. Therefore there is a need for creating the framework for Rydberg photochemistry and, in general, to work towards a more complete junction of molecular spectroscopy and photochemistry.

The arguments developed along these lines are of a personal and tentative nature. The writer is hopeful that while many of his answers might be wrong, at least some of his questions are right.

The writer believes that the ultimate goal of electronic spectroscopy is photochemistry. When we know the structure of the excited molecule and the way of reaching it we want to know how it will react. For this we need all the excited states in a given region of the spectrum. The weak, the hidden bands are very important. Unfortunately, optical spectroscopy is ill-prepared to uncover these. Therefore all forms of spectroscopy that give hope to magnify weak bands, be these spin, parity, symmetry or angular momentum forbidden, are bound to gain increasing importance in the future. The various types of electron impact spectroscopies, natural and magnetic circular dichroism, two- and multiphoton spectroscopies are all in this category. These are beyond the limited scope of this chapter, however.

This chapter is not a review. There is no need for one at the present stage in the writer's opinion. Rather, it is meant to provoke discussions about a few problems which he believes are in the line of progress.

I Rydberg and Valence-Shell Transitions

Preliminaries

The far ultraviolet spectra of molecules are characterized by the coexistence of bands due to valence-shell (intravalency) and Rydberg transitions. Chemists who received their training in electronic spectroscopy in connection with the π electron systems of aromatic molecules are usually surprised when they are told about molecular Rydberg transitions. Yet, beyond 200 nm and sometimes even at longer wavelengths the existence of Rydberg bands is a fact of life. No spectrum can be interpreted without them.

For atoms Rydberg transitions are simply those in which the principal quantum number (n) increases. Since the size of atomic orbitals rapidly increases with n the orbital of the excited electron is of a much larger size than those of the electrons remaining in the core. Rydberg bands can be ordered into series converging to an ionization potential related to a given state of the positive ion:

$$\tilde{\nu} = Ip - \frac{R}{(n - \Delta)^2} \tag{1}$$

where $\tilde{\nu}$ is the wavenumber of the given band, R is the Rydberg constant and n is the principal quantum number of the excited electron. For hydrogenoids Δ, the "quantum defect" is zero. For atomic species with more than one electron it takes account of the fact that the energy of an electron in a given orbital depends on both the principal and the azimuthal quantum number, ℓ. The value of Δ for s, p, d, . . . type orbitals reflects the degree of "penetration" of the Rydberg orbital. In a mathematical sense penetration is a consequence of the requirement that orbitals must be mutually orthogonal. In particular, orbitals with a given principal quantum number must be orthogonal to orbitals with the same ℓ, but lower n value. Since their angular parts are the same orthogonality must be achieved through their radial parts. For example, 3s orbitals must be orthogonal to the 2s and 1s orbitals (their "precursors" in the core) and this introduces two local maxima and two radial nodes in the electron density function close to the nucleus. Physically, this amounts to less efficient screening of the nucleus from the electron in the Rydberg orbital exposing the latter to somewhat stronger attraction. This is translated by substracting Δ from n in (1) making the effective principal quantum number of the Rydberg orbital smaller. (It is to be noted that the atomic number Z is absent from (1). This implies that if Δ was equal to zero the Rydberg electron would be attracted by only one positive charge because of perfect screening and that it is hoped that whatever adjustment is needed can be done through Δ.) np, nd, . . . orbitals are of larger size than s orbitals with the same principal quantum number. All, except 2p, 3d, 4f, . . . have radial nodes but these are more remote from the nucleus. Penetration is not the only factor contributing to Δ. The core-Rydberg exchange integral is positive for triplets and negative for singlets causing a difference. Mutual polarization between the core and the Rydberg orbital also has some influence. Both quantum defects and terms $(R/(n - \Delta)^2)$ have characteristic values for ns, np, nd . . . . type orbitals of different atoms and for molecules containing them.

93

The existence of molecular Rydberg transitions seems to have been first predicted by Niels Bohr. The field was pionnered by Price and his Coworkers from the spectroscopic side (see [1] and [2] for relatively recent reviews) and by Mulliken from the theoretical side[3-5]. The striking fact is that the Rydberg formula (1) is valid for molecules with quantum defects similar to those which are encountered for the related atoms. To some extent in its Rydberg states the molecule behaves like an atom and this trend increases with increasing size of the Rydberg orbital that is, with increasing n values in a given series.

Atoms as well as molecules have electronic transitions that are not of the Rydberg type. For atoms the famous D-lines of sodium: (3s,3p) are an example. For molecules all the familiar (π, π^*) and (n, π^*) transitions of olefins and aromatic molecules are examples of non-Rydberg, valence-shell (or intravalency) type transitions. For typical valence-shell transitions the orbital of the excited electron is not much larger than the molecular core. Bands due to such transitions cannot be ordered into series. The orbital of the excited electron is usually antibonding in one or more bonds while Rydberg orbitals because of their large size are, in most cases, essentially non-bonding.

Unfortunately the distinction is not clearcut. We have to seek answers to the following questions:
a) are the lowest excited states of molecules Rydberg or valence-shell? ;
b) can the dinstinction be maintained in the energy range where both valence-shell and Rydberg states can exist? ;
c) how many Rydberg orbitals (and states) of a given type can be expected?
We are going to elaborate on these to some extent. The Reader who is interested in gaining more knowledge could consult references 3 to 10.

The first problem (a) can be illustrated on the example of the hydrogen molecule. Calling the atoms A and B one can form molecular orbitals (MO) from atomic orbitals (AO) $1s_A$ and $1s_B$. Ommitting normalization factors we obtain the two well known MOs $\sigma_g 1s \equiv 1s_A + 1s_B$ and $\sigma_u 1s \equiv 1s_A - 1s_B$. In the ground state both electrons are assigned to $\sigma_g 1s$. A transition is possible in which an electron jumps from $\sigma_g 1s$ to $\sigma_u 1s$. This we might consider as a valence-shell transition since both $\sigma_g 1s$ and $\sigma_u 1s$ are built from the same (atomically) unexcited AOs. If only valence-shell type excited orbitals were possible the spectrum of H_2 would consist of only two bands: a singlet and a triplet corresponding to the above transition. In actual fact it contains many other bands.

For the H_2 molecule $1s_A$ and $1s_B$ constitute the minimal set of AOs. Naturally, we can include higher AOs. For the sake of discussion we shall take only $2s_A$ and $2s_B$ although it would be more correct to include the σ type 2p AOs at the same time. Then we obtain MOs of the form:

$$\varphi_i = C_{1A} 1s_A + C_{1B} 1s_B + C_{2A} 2s_A + C_{2B} 2s_B \tag{2}$$

where the C are numerical coefficients. Since 1s and 2s mix little we obtain, in addition to $\sigma_g 1s$ and $\sigma_u 1s$:

$$\sigma_g 2s \equiv 2s_A + 2s_B$$

$$\text{and } \sigma_u 2s \equiv 2s_A - 2s_B \tag{3}$$

in a fair approximation.

Now, the question is which one of $\sigma_u 1s$ and $\sigma_g 2s$ has the lower energy? $\sigma_u 1s$ is antibonding but it is made from atomically unexcited AOs; $\sigma_g 2s$ has the (+) sign and it is bonding or non-bonding (because of its large size) but it is built from excited AOs. The general answer is that the order of these MOs varies according to particular molecules. As we shall see later in some molecules the first excited MO is of the valence-shell type, in others it is of the Rydberg type.

The second problem (b) appears clearly on the example of molecules which possess a central atom, like methane. Assuming T_d symmetry the ground state configuration is $(1s_c)^2(2a_1)^2(1t_2)^6$ and the first excited states are issued from the configuration $(1s_c)^2(2a_1)^2(1t_2)^5 3a_1$. The $3a_1$ MO is of the form $(\lambda 2s_c - 1s_a - 1s_b - 1s_c - 1s_d)$. At the united atom (UA) limit it becomes 3s (Rydbergization) so that according to Mulliken[4] it will be of the intermediate type:

$$C' 3s_{UA} + C'' (\lambda 2s_c - 1s_a - 1s_b - 1s_c - 1s_d) \tag{4}$$

where in the united atom (neon) $C' = 1$ and $C'' = 0$, but at actual internuclear distances both C' and C'' will be non-zero. (4) merely expresses the intermediate Rydberg – valence-shell nature of the orbital and its dependence on the internuclear distance. The 3s therein is an orbital of the united atom not of carbon or the hydrogens. Thus (4) is not meant to be minimized to yield two different MOs.

One might, however, extend the basis set to include C3s and other higher AOs in addition to C1s, C2s and the H1s. Then one of the new MOs that are obtained would be essentially the C3s Rydberg orbital. It is, of course, very similar to (4): it has the same symmetry and the same number of nodes. According to Mulliken they are actually the same that is, one of them is redundant.

In the case of methane (and also molecules like NH_3, H_2O, HF) the problem is fortunately an academic one. The lowest excited MO is of intermediate type and no evidence for a second level of the same symmetry and nodal properties has been found. In other types of molecules, however, the problem might become real.

Instead of Rydbergization the problem can be discussed in terms of mixing between Rydberg and valence-shell configurations (Cf. the recent review by Peyerimhoff[6]) from which two states might result.

A number of additional questions come into one's mind when the molecule does not contain just one central atom. Let us think about neopentane or hexadecane. There is no a priori reason to take the excited electron from the field of one given carbon atom rather than from another. (For the sake of simplicity we omit the hydrogen orbitals). Then, if we look for the lowest (3s) Rydberg MO, for example, we might form linear combinations of the five (or sixteen) C3s AOs one for each of the carbon atoms of the respective molecule. This would yield five or sixteen Rydberg MOs, the one of lowest energy being the all plus combination. The latter could be considered the 3s Rydberg MO. Some of the MOs obtained in this way could, at short internuclear distances, become similar to an MO obtained from the minimal basis set calculation. Anyway, one might wonder how many of the five or sixteen Rydberg MOs "exist" that is, how many of them can be used to build the wave functions of states. Sinanoglu pointed out[10] that problems of linear dependence arise for large Rydberg AOs with mutual overlap integrals close to unity leading to the disappearance of the negative combinations (in the two center case). Mulliken[4] has been led to the same conclusion by examining the behavior of the latter at zero internuclear distance. According to accepted ideas there can be only one ns type Rydberg MO-originating with each MO belonging to the ground state; there can be one np and one nd manifold from each with the possibility of splitting into three or five components when degeneracy is lifted. Then how many of the five or sixteen MOs which we obtained from the C3s in the above example are "real"? The expected answer is one ns, three np and five nd type Rydberg MOs from each originating level. The others should reduce to zero if we could make all the nuclei coalesce. A formal proof would be needed.

It follows from the above considerations that for larger molecules we cannot think about Rydberg MOs – at least not the lower ones – as being much larger than the molecular core. It would be probably more correct to say that they usually cover the fields of more than one atom or bond.

Now let us consider a molecule containing atoms belonging to different rows of the periodic system, say carbontetrabromide. Students who are getting familiarized with the Rydberg formula

$$\tilde{\nu} = I_P - \frac{R}{(n - \Delta)^2}$$

and are told that n is a principal quantum number invariably ask: the principal quantum number of what? For an originating MO mainly concentrated around the carbon it seems to be reasonable to take n = 3 for the lowest Rydberg orbitals. On the other hand, if the originating MO has a high density around the bromines n = 5 would have to be taken. But how if it has significant amplitudes on both the carbon and the bromines? All one can do then is to take fractional numbers and then increase them by 1, 2, . . . to obtain the higher members of the given series.

Using the term values instead of n and Δ has definite advantages (see the next section).

II The Routine; an Example

Somewhat superficially the excited states that a molecule is expected to have can be predicted as described below. As to the valence-shell MOs one can perform (or find) a good quantum chemical calculation carried out without Rydberg AOs. After that the Rydberg levels can be treated "atomically".

Let us illustrate conditions on the example of benzene, a part of whose spectrum is well known. Benzene has forty two electrons but as it is often done we shall only take the six π electrons. With these six carbon $2p_z$ AOs we carry out a molecular orbital calculation and obtain six MOs. Due to the high symmetry of benzene they are actually determined by symmetry. Under D_{6h} symmetry two of the MOs are non-degenerate, the two others doubly degenerate.

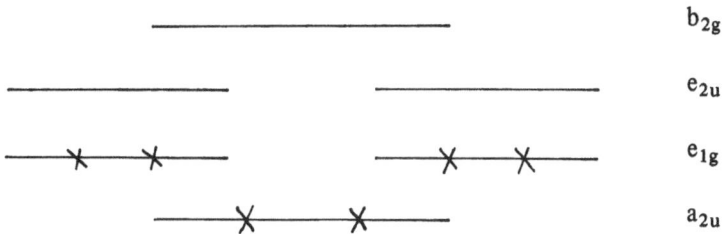

In the ground state a_{2u} and e_{1g} are filled. The transitions from these two MOs to the empty e_{2u} and b_{2g} MOs can be considered as valence-shell transitions (π, π^*) since all these MOs, including those which are empty in the ground state have been obtained from a basis set containing only atomically unexcited AOs. (This might seem to be simple. In actual fact only a part of the bands which could result from these transitions had been identified).

But then where are the Rydberg transitions? To obtain these the basis set must be extended and this will be mentioned again in the next section. However, if we accept to renounce knowing the MO form of the Rydberg orbitals we can do simply this: we assume that the molecule must have 3s, 3p, 3d, 4s, 4p, 4d, 4f, ... levels then determine their species under D_{6h} symmetry from the (group theoretical) correlation tables and find the selection rules from the character table. The correlation tables are given in Appendice IV of Herberg's Volume III[7]. We obtain that

s becomes a_{1g}
p becomes $a_{2u} + e_{1u}$
d becomes $a_{1g} + e_{1g} + e_{2g}$
f becomes $a_{2u} + b_{1u} + b_{2u} + e_{1u} + e_{2u}$

For an one electron function s and d are "g" and p and f are "u". Then the following states are obtained.

The (π, π^*) singlet, valence-shell states of benzene are:
a) From transition $e_{2u} \leftarrow e_{1g}$:

$$^1B_{2u} \leftarrow {}^1A_{1g} \qquad \text{forbidden}$$
$$^1B_{1u} \leftarrow {}^1A_{1g} \qquad \text{forbidden}$$
$$^1E_{1u} \leftarrow {}^1A_{1g} \qquad \text{allowed X, Y}$$

These three bands are, of course, well known. The two forbidden ones are made allowed vibronically. (The X and Y axes are in the molecular plane).
b) From transition $b_{2g} \leftarrow e_{1g}$:

$$^1E_{2g} \leftarrow {}^1A_{1g} \qquad \text{forbidden}$$

c) From transition $e_{2u} \leftarrow a_{2u}$:

$$^1E_{2g} \leftarrow {}^1A_{1g} \qquad \text{forbidden}$$

C. Sandorfy

d) From transition $b_{2g} \leftarrow a_{2u}$:

$$^1B_{1u} \leftarrow {}^1A_{1g} \qquad \text{forbidden}$$

While tentative assignments have been made, b), c) and d) have not been identified with certainty.

For the (singlet) Rydberg states one obtains:

e) From transition $ns(a_{1g}) \leftarrow e_{1g}$:

$$^1E_{1g} \leftarrow {}^1A_{1g} \qquad\qquad\qquad \text{forbidden}$$

f) From transition $np(a_{2u} + e_{1u}) \leftarrow e_{1g}$:

$$^1E_{1u} \leftarrow {}^1A_{1g} \qquad\qquad \text{allowed X, Y}$$
$$^1A_{1u} + {}^1A_{2u} + {}^1E_{2u} \leftarrow {}^1A_{1g} \qquad A_{2u} \text{ allowed, Z}$$

g) From transition $nd(a_{1g} + e_{1g} + e_{2g}) \leftarrow e_{1g}$:

$$^1E_{1g} \leftarrow {}^1A_{1g} \qquad\qquad \text{all forbidden}$$
$$^1A_{1g} + {}^1A_{2g} + {}^1E_{2g} \leftarrow {}^1A_{1g}$$
$$^1B_{1g} + {}^1B_{2g} + {}^1E_{1g} \leftarrow {}^1A_{1g}$$

and so on. There is, of course, a triplet for every singlet.

The 3s band has not been identified in absorption but Allen and Schnepp[11] found a band at the right place in the circular dichroism spectrum. An np series is well known. There are three other series with a near zero quantum defect which can be components of the $g \leftrightarrow g$ forbidden nd manifold or the nf manifold. The latter corresponds atomically to transitions with $\Delta L = 2$ and would be weak too. (For reviews see[7], [9] and [12].

Still other Rydberg bands can originate with the a_{2u} MO of the ground state.

h) From transition $ns(a_{1g}) \leftarrow a_{2u}$:

$$^1A_{2u} \leftarrow {}^1A_{1g} \qquad \text{allowed, Z}$$

i) From transition $np(a_{2u} + e_{1u}) \leftarrow a_{2u}$:

$$^1A_{1g} \leftarrow {}^1A_{1g} \qquad \text{forbidden}$$
$$^1E_{1g} \leftarrow {}^1A_{1g} \qquad \text{forbidden}$$

j) From transition $nd(a_{1g} + e_{1g} + e_{2g}) \leftarrow a_{2u}$:

$$^1A_{2u} \leftarrow {}^1A_{1g} \qquad \text{allowed, Z}$$
$$^1E_{1u} \leftarrow {}^1A_{1g} \qquad \text{allowed, X, Y}$$
$$^1E_{2u} \leftarrow {}^1A_{1g} \qquad \text{forbidden}$$

and so on.

Benzene has a relatively low ionization potential (9.247 eV or 74580 cm^{-1}) so that all transitions from a_{2u} and many from e_{1g} will lie beyond it and lead to super-excited states[13, 14]. The second 1P was shown to belong to the highest filled σ orbital[15].

It is not our intention to treat the spectrum of benzene in detail; we merely used it as an example for the benefit of the non-initiated Reader as a simple MO + Ryd-berg scheme without energies, vibronic interactions, σ electrons, and so on. Since the lowest Rydberg states of benzene do not have the same symmetry as its lowest valence-shell states some of the problems mentioned in the previous section are avoided. Mixing might occur for the E_{1u} states, however. While caution is needed the scheme is helpful in locating the bands that a given molecule should possess.

III Lines of Recent Progress

During the last ten years major advances have been made in what can be called chemical spectroscopy in the vacuum ultraviolet. The expression "chemical spectro-scopy" has, of course, no exact meaning. In practice, however, it does have one. It covers spectra of larger molecules measured usually under moderate resolution, an obligation imposed by the size and physico-chemical properties of the compounds. It follows that we have to renounce the whole volume of information stemming from the study of molecular rotation and even vibrational fine structure will be only partly resolved and that only in some cases. Bands are very often broad, because of the high probability of dissociation or predissociation in the excited states. As a partial compensation we can use chemical knowledge and analogies to correlate spectra and to seek understanding in global terms and in function of molecular structure. In what follows a few contributions which in the writer's opinion have been essential for progress in the spectroscopy of larger molecules will be pinpointed.

a) A clear understanding of the Rydberg — valence-shell distinction is a pre-requisite for the interpretation of electronic spectra, especially in the far ultraviolet where both types of transition generally occur. Mention of this problem has already been made in Sect. I. For different aspects of the problem the reader might consult Refs.[3] to[10].

b) Ultraviolet photoelectron (PE) spectroscopy has become an indispensible companion of electronic spectroscopy. Since the break achieved by Turner[16, 17] these spectra can be obtained with sufficient accuracy to be used jointly with ab-sorption spectra and the results of advanced quantum chemical calculations.

Ionization potentials can be, naturally, obtained from the absorption spectra themselves. Indeed, until this day, the most accurately known IPs are obtained from the convergence limits of Rydberg series. This is, however, limited to well resolved spectra of small molecules and radicals. In most cases such high quality electronic spectra cannot be obtained for the larger molecules which are of prime interest to the chemist. Even if instrumental conditions would make it possible the bands usually remain broad for physical reasons (dissociation, predissociation) and the coincidence and mutual overlap of bands. Therefore an independent source of ioniza-tion potentials was badly needed. It is enough to have another look at the Rydberg formula to understand this:

$$\tilde{\nu} = I_p - \frac{R}{(n - \Delta)^2}$$

The IP is needed to assign the bands together with quantum defects or term values. Rydberg bands "follow" the IP. With the knowledge of the ranges for ns, np, nd, ...

Δ and term values we can locate the corresponding bands. Very high term values often indicate bands due to valence-shell transitions. The IPs are needed for all this, however.

In addition to the IP itself PE spectra also yield the energies of a part of the excited states of the positive ion. Supposing the validity of Koopmans' theorem[18] these can be assigned to ionization of an electron from successive MOs of the ground state. This helps in assigning Rydberg series converging to these excited states of the positive ion. If handled with caution they can be related to MOs obtained through quantum chemical calculations.

Chemical spectroscopy in the far UV is not conceivable anymore without the knowledge of photoelectron spectra. In addition to Turner's book[17], a number of treatises on PE spectroscopy are now available[19-23].

c) Robin's recognition of the diagnostic value of Rydberg terms and the extensive use he made of them[9].

As already mentioned in the previous section, the terms $R/(n - \Delta)^2$ have certain advantages over the quantum defects. They have characteristic values for s, p, d, ... type Rydberg states for given atoms. They have similar values even if the atom is incorporated into a molecule provided the transition originates with an orbital having a large amplitude around that atom. The term values depend on penetration, exchange and polarization contributions. These might all undergo modifications in the molecule. Penetration, in particular, becomes an even more complicated phenomenon. Since the Rydberg orbital must be orthogonal to all MOs which are in the core and since the core contains two or more nuclei orthogonality in the molecule can be achieved by penetrating to various extents into two or more atomic cores. The extent of penetration around any given atom will then generally decrease and the term value with it. This effect is quite significant for s type Rydberg orbitals which are strongly penetrating; it is much less significant for the p and negligible for the d. This is a major topic in Robin's "Higher excited states of polyatomic molecules" and the Reader is strongly advised to have another look at Chap. 1 in this book[9]. It is not possible to reproduce the details of his treatment here but the following examples may illustrate conditions for some simple organic molecules (Table 1).

For some of these molecules the 3p manifolds are widely split. The range for 3p bands is usually $22000-18000$ cm^{-1}, for the 3d about $14000-12000$ cm^{-1}. Valence-shell transitions have, in general, term values too high to be either s, p or d.

The following comment seems to be in order. High resolution spectroscopists are accustumed to long Rydberg series from which the IP and Δ are determined by fitting the observed frequencies as closely as possible. The first member of a given series (with the smallest Rydberg orbital) usually deviates significantly from the value it should have according to the Rydberg formula; it still "sees" the core. This deviation can amount to a couple of thousand reciprocal centimeters. Interactions between various Rydberg states and Rydberg and valence-shell states made possible by symmetry conditions can also be examined under those conditions. In "chemical" spectroscopy it quite often happens that instead of a series we see only the first member of the series or one or two members. The bands are in most cases broad and since the intensity of the bands decreases rapidly with increasing n the higher

Table 1.

Molecule	3s term value (cm^{-1})	3p term value (cm^{-1})
CH_4	31500	–
C_2H_6	29500	21700
C_4H_{10}	25200	21000
H_2O	41800	21150
		19740
CH_3OH	34100	26150
		21300
CH_3OCH_3	25700	18470
CH_3F	33000	21300
CF_4	30000	20800
C_4F_{10}	33270	23070
NH_2CH_3	31200	20700
$NH(CH_3)_2$	28800	18800
$N(CH_3)_3$	24400	18560
$N(C_2H_5)_3$	23700	18500

3s and 3p term values for some simple molecules.

members are easily lost in the background. Even then, thanks to the photoelectron IPs and the term values, we can make Rydberg assignments.

This is abhorred by high resolution spectroscopists. They are quick to point to the many perturbations that may affect the term values and question their reliability for making assignments. In chemical spectroscopy we are less demanding, however. Even a deviation of 1000 cm^{-1} may not affect a given electronic assignment and that assignment might be just what we need. Term values together with photoelectron and quantum chemical input lead to a consistent general picture of the spectra of organic molecules.

d) Until recently, quantum chemistry remained in an exploratory stage as far as excited states of polyatomic molecules are concerned. While the Pariser-Parr-Pople method made it possible to interpret the (π, π^*) valence-shell transitions of aromatic molecules in a global manner in other areas spectroscopists continued to relie on correlation diagrams. Theoretical support was especially wanting in the far ultraviolet where Rydberg transitions account for many of the bands. Minimal basis set calculations do not suffice. This situation has changed due mainly to the works of Peyerimhoff and Buenker. A summary of their procedure is given in their review which appeared in 1975 in "Advances in Quantum Chemistry"[24]. It consists of using Gaussian-type double-zeta quality basis sets and Gaussian lobe functions introduced by Whitten[25] completed with polarization functions. The SCF calculation is then followed by a large configuration mixing calculation where the less important con-

101

figurations are taken care of by an extrapolation technique[26]. More characteristic from the spectroscopic point of view is the manner in which they provide for Rydberg states. In their own words[24]: " . . . it is obvious that addition of long-range united atom type functions not generally present in basis sets employed for ground state calculations is essential in order to allow for a valid description of typical excited Rydberg (diffuse) species". This is necessary even for states of the intermediate type (cf. the previous chapter). For Rydberg states single primitive Gaussians placed at either the midpoint of bonds or at one or more atoms (other than hydrogen) was found to be adequate provided the orbital exponents were optimized. Otherwise, the location of these long-range functions is not particularly critical.

With this procedure they obtained sound agreement with experiment for a number of typical polyatomic molecules and radicals and their results can actually serve as a guide to experimentalists in many cases.

Another important contribution of Buenker and Peyerimhoff[27] has been the duplication of the Mulliken-Walsh orbital correlation diagrams[28, 29] by ab initio calculations (see Chemical Reviews, 1974) and examination of their relation to Koopmans' theorem.

e) The aim of spectroscopy is gaining knowledge on the structure of excited molecules and on the way to attain them. What follows excitation is photophysics and photochemistry. In particular, the primary processes in photolysis are closely connected with the properties of the related excited state. Photolysis of organic molecules in the far ultraviolet has made significant progress in recent years. Earlier developments were reviewed by McNesby and Okabe[30]. An important step forward has been the introduction of rare gas resonance lamps as irradiation sources by Ausloos. The use of these lamps together with a combination of isotopic substitution and scavanger techniques enabled Ausloos and his co-workers[31–33] to investigate systematically the photolysis of various categories of organic molecules. Their results offer good possibilities for linking together photochemical and spectroscopic knowledge. They will be often cited in connection with individual molecules. As to reactions occurring in triplet excited states the powerful mercury sensitization method of Gunning and Strausz[24] should be cited.

Although not specifically developped for the vacuum ultraviolet we should like to cite in this context the Woodward-Hoffmann rules as applied to excited state reactions[35], the classification of photochemical reactions by Dauben, Salem and Turro[36] through the use of the electronic states of diradicals and the relationships between potential energy surfaces along the pathway of photochemical reactions[37–39] and references therein). Then we have to mention the concerted efforts of quantum chemists towards the calculation of potential surfaces (for a general view, see the recent review by Bader and Gangi[40]). Spectroscopists seeking to penetrate into photochemistry will find the book by Barltop and Coyle very helpful[41]. Problems in atmospheric chemistry are giving new impetus to both spectroscopic and photochemical research, including the vacuum ultraviolet (for a comprehensive treatise see Heicklen[42]).

f) A very important instrumental advance is the introduction of synchrotron radiation as source for far ultraviolet and photoelectron spectroscopy. This intense, continuous source makes it possible to obtain high quality spectra in wide ranges of electromagnetic spectrum[43]. Results obtained through the use of synchrotron radiation will be treated by Koch in a later volume.

IV Towards a Rydberg Chemistry

Much of our ideas on electronic excitation and ionization tacitly assume the validity of Koopmans' theorem[18]. The latter which is often used with SCF—MO calculations states that if $E_T(A)$ and $E_T(A^+)$ are the total energies of molecule A and its positive ion respectively then, approximately

$$E_T(A) = E_T(A^+) + \epsilon_D \qquad (5)$$

where ϵ_D is the energy of the MO from which the electron was taken. This neglects changes in the energy of all other electrons due to the removal of one electron from ϵ_D (rearrangement) and the changes it produces in correlation energies. Since these two often cancel each other in a fair approximation the negative of ϵ_D can be taken for the energy of ionization from that orbital. This is the approximation underlying comparisons between photoelectron spectra and molecular orbital—SCF calculations. Caution must be exercized however: the approximation loses its validity in several cases, in particular if ionization from close-lying orbitals is involved or when more than one state is issued from the same configuration. The order of photoelectron bands is then not necessarily the same as the on-paper order of the ground state molecular orbitals. This is, of course, well known.

The following reasoning of Buenker and Peyerimhoff is quite important[27]. Let us differentiate the previous equation:

$$\partial E_T(A) \,/\, \partial R = \partial E_T(A^+) \,/\, \partial R + \partial \epsilon_D \,/\, \partial R \qquad (6)$$

where R is a distance or other geometrical variable. Now, let ψ_D be the MO whose energy is ϵ_D. The last term, $\partial \epsilon_D \,/\, \partial R$ is the difference in the slopes of the total energy curves of A and A^+. If ψ_D is a bonding MO $\partial \epsilon_D \,/\, \partial R$ is positive for A for values of R larger than its equilibrium value. Thus, at the equilibrium geometry of the ion, A^+, where $\partial E_T(A^+) \,/\, \partial R = 0$, $\partial E_T(A) \,/\, \partial R$ is also positive. This implies that $R_o(A) < R_o(A^+)$. The conclusion is that in this approximation taking out an electron from an orbital which was bonding in a given bond will yield an ion in which the respective bond is weaker. The opposite follows for antibonding initiating orbitals while removal of an electron from a non-bonding orbital will have no great effect.

These conditions can be expected to have a bearing on the primary processes in the photolysis of ions. We are interested in Rydberg states. Therefore, still in the spirit of Koopmans' theorem, we shall inquire about the effect of the superposition of a Rydberg orbital on the ion. First, however, we will indulge in some general considerations concerning the possible photochemical importance of Rydberg states.

Since most known photochemical reactions occur at the lowest excited states of molecules a Rydberg state will be photochemically important if it is one of the lowest excited states of the given molecule. This will be especially true if it is actually the lowest of them. Now, on the basis of a great amount of experimental and theoretical knowledge accumulated during the last ten years we might sum up the situation as follows.

1) The lowest excited state of saturated hydrocarbons is best described as a 3s Rydberg state. (Having a 3s orbital for the excited electron). It is followed toward higher frequencies by 3p, 3d, . . . type Rydberg states. Typical valence-shell states have not been located with any degree of certainty for these molecules.

In cyclopropane there are both Rydberg and valence-shell type states among the lowest excited states, the lowest being Rydberg and the second lowest valence-shell[44].

2) In fluoroalkanes the lowest excited states are Rydberg[9, 45, 46]. In the fluoro-methanes 3s and 3p are followed by strong valence-shell transitions. In higher per-fluoro paraffins the 3s band seems to lie lower than the lowest valence-shell transition[9, 47].

3) In saturated heteroatomic molecules the lowest excited state is either of intermediate type like in alcohols or ethers[48] or of valence-shell type like for alkyl-halides and fluoroalkylhalides[51]. In every case the transition originates with the heteroatom lone pair AO. (For reviews see[9, 52, 53].

4) in monoolefins the well known (N,V) band (which, however, has some much argued about Rydberg admixture) coincides or is actually preceded by the 3s Rydberg band. In ethylene itself the latter is superimposed on the low frequency wing of the (N,V) band while in highly methylated or fluorinated olefins the band of lowest frequency is definitely the Rydberg 3s[54, 55] (sec. VII).

5) In butadiene and other diolefins valence-shell (N,V) and Rydberg bands still coincide near the onset of absorption[56, 9].

6) In all aromatic molecules the valence-shell $(\pi, \pi*)$ bands are the bands of lowest energy. We usually encounter two or three of them before reaching the lowest Rydberg band.

7) In heteroatomic, unsaturated molecules the first band is usually a valence-shell type (n, $\pi*$) band. In non-conjugated aldehydes and ketones it is followed toward higher frequencies by two or more Rydberg bands which have lower transition energies than the $(\pi, \pi*)$ band[57, 58]. In amides, carboxylic acids, esters and acyl halides (except, perhaps, fluorides) the (n, $\pi*$) band is at higher frequencies but it is usually still the band of lowest frequency. The next band is usually the Rydberg 3s, followed by bands due to valence-shell and higher Rydberg transitions[9].

8) In molecules containing atoms belonging to the second row of the periodic system the situation is in most cases similar to that found with molecules having the related precursor atom as the most characteristic atom. The lowest frequency bands of sulfides are usually of valence-shell or mixed type like those of oxides. In silane the first band is due to a transition to a relatively small 4s Rydberg orbital, in analogy to methane's 3s[9, 59].

Table 2 summarizes the facts mentioned in the above discussion.

Table 2. The lowest singlet excited states of different types of organic molecules

Paraffins	R[a]
Cyclopropane	1st R, 2nd VS[b]
Fluoroparaffins	1st R
Alkylhalides	1st VS, 2nd R
Alcohols	mixed
Monoolefins	1st R, 2nd VS
Diolefins	R, VS coincide
Higher olefins	VS
Aromatic molecules	VS
Ketones, aldehydes	1st VS, then R
Amides, esters	1st VS, 2nd R

[a] R = Rydberg.
[b] VS = Valence-shell.

Since the Rydberg state of lowest energy in organic molecules is reached by exciting an electron to a 3s orbital the latter is the most important one from the photochemical point of view. It is usually the least typically Rydberg among all Rydberg orbitals. In a loose manner its main characteristics could be summarized as follows.

a) While mainly non-bonding the 3s orbital is not extremely large. In a Na atom the main peak of the 3s wave function is at 3.4 bohrs from the nucleus with nodes near 1.0 and 2.0 bohrs (1 bohr = 0.529 Å). Although these values have only a remote relevance to organic molecules and will vary from molecule to molecule it is likely that in relatively small molecules like methane, for example, there will be a node between the carbon and the hydrogens giving the 3s orbital some antibonding character. This is likely to be so for ethane and other hydrocarbons as well where the center of the 3s orbital will be located near the center of a C–C bond.

This antibonding character will, to some extent, help with the dissociation of the related bond in the 3s Rydberg state. The main cause for dissociation will be, of course, the removal of an electron from a bonding ground state orbital.

b) Since an s orbital is necessarily totally symmetrical it will provide some bonding character between sites which are connected by symmetry operations. This could give some additional strength to bonds between atoms connected by symmetry. In particular, it could provide bonding between non-neighbor atoms which are connected by symmetry operations like the hydrogens in methane or ethane, for example.

The intuitive conclusion from these considerations is that excitation to the lowest s type Rydberg state can be expected to favor the breaking of bonds between atoms which are not symmetrically connected and the simultaneous elimination of smaller molecules from larger ones by providing bonding interactions between non-neighbor atoms. (The photolysis of all n-paraffins yields predominantly H_2 not 2H; see the next section). Contrary to this in excitations from bonding to antibonding orbitals which occur in typical valence-shell transitions the antibonding character is expected to prevail leading to simple bond cleavages rather than to molecular elimination.

Rydberg orbitals of the p type are less penetrating than the s but they may still have some bonding or antibonding character. The p manifold has three components one of the σ and two of the π type. They will remain triply degenerate in cubic molecules only and that only if the Jahn-Teller effect can be disregarded. In a molecule like methane since the angular node is on the central atom and the radial node is well beyond the hydrogens a 3p orbital is unlikely to have much influence on photolysis. The two π components remain often degenerate in molecules of lower symmetry. In a molecule like ethane, for example, which has one threefold axis the angular nodes would make the σ component slightly antibonding and the π component slightly bonding. The radial node could provide some antibonding character but it would fall beyond the hydrogens. It might have some importance in larger molecules. The π component could give some remote bonding between rather farlying atoms (in the Na atom the radial node is at about 4.5 bohrs from the nucleus).

This section while admittedly speculative has been an attempt to show that penetrating Rydberg molecular orbitals might contribute to bonding or antibonding to some extent. The role that the radial nodes might play in causing antibonding properties is one point that, in our opinion, would deserve further consideration. The possibility of creating bonding interactions between atoms which are not linked together by chemical bonds could be important at the stage of dissociation when some of the covalent links become weak. Penetrating Rydberg orbitals should be capable of forming what one might perhaps call remote one electron bonds. It follows that the Rydberg — valence-shell distinction should have a definite bearing on the photolytic primary process. Other conditions being equal, in mainly anti-bonding valence-shell excited states simple bond cleavages should be favored while in the mainly non-bonding Rydberg excited states molecular elimination should be of more frequent occurrence. In subsequent sections where individual molecules are treated we shall systematically include the known photochemical primary steps in order to see if the above predictions are born out[60, 61].

It should be emphasized at this point that the above considerations mean no more than that the initial Rydberg or valence-shell character of the orbital of the excited electron might be *one* of the factors determining the primary process. The character of the orbital corresponding to the vertical transitions might be altered along the reaction coordinate. The potential energy curves of Rydberg and valence-shell states often cross each other. For an understanding of the course of a photochemical reaction the knowledge of the potential surface is, of course, needed. In the construction of the latter, the existence of Rydberg states should not, however, be forgotten.

At this stage one might ask if there is any connection between the Rydberg or valence-shell character of the orbital of the excited electron and Woodward and Hoffman's[35] and Salem's scheme of interpretation of photochemical reactions[36–38]. In the words of Dauben, Salem and Turro[36] the latter is based on the assumption that " . . . all photochemical processes are controlled by generation of primary products which have the characteristics of diradicals". The states of the diradicals which they consider are the typical diradical states ^3D and ^1D with the electrons located one on each of the atomic orbitals involved plus two states of ionic character, Z_1 and Z_2, in which the two electrons are located in the same AO at one of the two atoms. The procedure consists in predicting the number and location of the "radical sites" and in the construction of correlation diagrams between the reactant excited states and the D and Z states of the primary photochemical product. The interpretation of such diagrams then entails symmetry, spin and energy considerations. "A reactant excited state which leads to a highly excited state of primary product can generally be ruled out as the photoreactive state, while a state correlating directly with the primary product is a strong candidate to be the photoreactive state".

Many interesting examples are found in Dauben, Salem and Turro's article. The reactive states they considered are (n, π^*) or (π, π^*) and the correlation diagrams obtained through these lead to a rather complete classification of a great variety of photochemical reactions. There are, however, molecules having Rydberg states among their lowest excited states and one might wonder if their inclusion would not change the picture at least in some cases.

A typical example is their diagrams related to photochemical coplanar hydrogen abstraction reactions (Fig. 3 in Ref. [36)]. The abstracting molecule might be a ketone or an aldehyde. The lowest excited states are the triplet and singlet (n, π^*) states. These are followed by two or three Rydberg states originating with the n orbital with 3s, 3p and 3d excited orbitals. The π^* state is at high energies beyond these Rydberg states. Thus between the ^{1}(n, π^*) and ^{1}(π, π^*) states there is a gap of about 4 or 5 eV in which the lowest Rydberg states are located. With which state of the final primary product these Rydberg states would correlate is not an easy question to answer. In the case of proton abstraction reactions by aldehydes or ketones the approach of the proton would be facilitated because of the large size of the Rydberg AO. On the other hand it might produce the primary product in a relatively high state of excitation. The (n, 3s) state wave function is symmetrical with respect to the reaction plane while the ^{1}D and ^{3}D states are antisymmetrical so that they cannot correlate with each other. Correlation with the Z states would be compatible with symmetry. Diradical states would become possible with one electron in an excited AO. This would seem to represent a symmetric state energetically more favorable than the Z states. Thus while upon (n, π^*) excitation the reaction is likely to proceed as predicted by Dauben, Salem and Turro at higher photon energies (higher than about 6 eV) at least a part of the molecules might choose a different path. Rydberg diradical states should then be added to the D and Z states as well as Rydberg states to the reactant states.

Solvent effects would also have a different influence on reactions occurring in Rydberg states than in (π, π^*) excited states. The former are generally broadened and shifted to higher frequencies in solution while the latter are stabilized in polar solvents of the proper polarity favoring the Z states of diradicals. Rydberg orbitals are highly exposed to desactivation by intermolecular interactions and this might increase the importance of three body collisions.

V σ-Electron Spectra

Paraffins

The lowest excited states of paraffins are Rydberg or of intermediate type but closer to the Rydberg than to the valence-shell type. This has been realized as long ago as 1935 by Mulliken[62].

The UV spectra of all paraffins (except cyclopropane and cyclobutane) belong to one of two fairly well defined types: they resemble either the spectrum of methane or the spectrum of ethane (Fig. 1). That the two exhibit significant differences can be understood on symmetry ground. The initiating MO of methane is a triply degenerate t_2 orbital of C—H bonding character, the ground state configuration being $[1a_1]^2[2a_1]^2[1t_2]^6$. The 3s Rydberg orbital is, of course, a_1. Thus the transition is $^1(A_1, T_2)$, the only allowed transition under T_d symmetry (x, y, z). Consequently, the spectrum begins with a strong band at the low frequency end. The 3p bands belong to transition $^1(1t_2, 3p(2t_2))$ or $^1(A_1, A_1 + E + T_1 + T_2)$ only T_2 being allowed. The very broad region of absorption which after a relative minimum follows the 3s band to higher wave numbers (maximum near 106,000 cm^{-1}) does not make assignments easy. It must contain the 4s band (near 93,500 cm^{-1}) and the members of the 3p and 3d manifolds.

The 3s band is a diffuse, Jahn-Teller split band, extending from about 78000 to about 84000 cm^{-1}. It resembles the first photoelectron band (vertical IP about 110000 cm^{-1}) which is also Jahn-Teller split. Interestingly, the PE band exhibits vibrational fine structure[63] while the UV band does not[64-66]. This might indicate that CH_4^+ dissociates more slowly than the neutral molecule in the 3s state, due, perhaps, to the antibonding property of the 3s orbital. According to Arents and Allen[67] the most probable structures resulting from the Jahn-Teller deformation are tetragonal (D_{2d}) and trigonal (C_{3v}) methane. Robin[9] has reviewed the existing

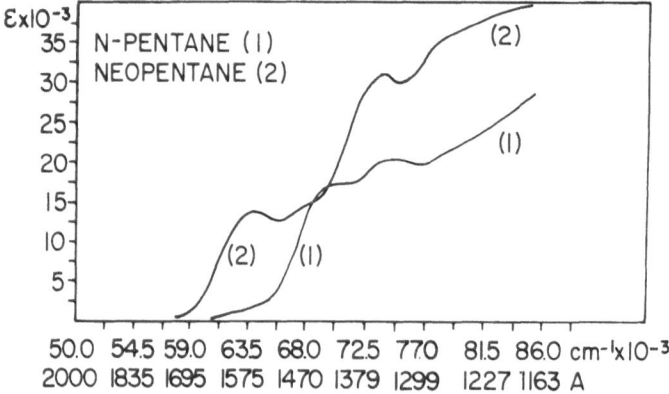

Fig. 1. The far UV absorption spectrum of n-pentane (1) and of neopentane (2).
[Reproduced from Lombos, B. A., Sauvageau, P., Sandorfy, C.: Chem. Phys. Letters 1, 221 (1967), with permission from the North-Holland Publishing Company]

knowledge on the spectrum of methane. Due to its diffuse character only the 3s band can be seriously discussed. A large scale quantum chemical calculation would be helpful: it could enable one to locate the forbidden bands belonging to the 3p manifold which could account for the shoulders that some authors observed. The extent of mixing between the T_2 states belonging to the 3s, 3p and possibly higher transitions would be also useful to know.

The photolysis of methane in the 3s band can be interpreted in terms of the two well known primary steps[33, 68]:

$$CH_4 + h\nu \rightarrow CH_2 + H_2 \tag{7}$$

$$CH_4 + h\nu \rightarrow CH_3 + H \tag{8}$$

the first one being about twenty times more frequent than the second one. The initiating orbital is CH bonding. In D_{2d} methane all C–H distances are the same and even if the structure was closer to C_{2v} there would be two pairs of equal C–H distances. This would favor step (7). In trigonal methane one C–H distance is much longer than the other three and this would favor step (8). Then it would seem that many more CH_4 molecules deform themselves into D_{2d} (or C_{2v}) than into C_{3v}. From our point of view the interesting observation is that the most important primary step is (7) and not

$$CH_4 + h\nu \rightarrow CH_2 + 2H \tag{9}$$

This is in line with the proposal made above that the 3s orbital provides positive overlap between the symmetrically related H atoms favoring the elimination of H_2 molecules rather than the release of H atoms. In trigonal methane the weakly bound hydrogen is not related to the others by symmetry and its H 1s must not have the same sign in the 3s MO than the others. Were the MO of the excited electron of the valence-shell type (and smaller) it would do little to bring the hydrogens together to produce H_2 molecules.

The ground state configuration of ethane is, under D_{3d} symmetry.

$$[1a_{1g}]^2[1a_{2u}]^2[2a_{1g}]^2[2a_{2u}]^2[1e_u]^4[3a_{1g}]^2[1e_g]^4 \, {}^1A_1$$

The t_2 orbital of methane had to split into $e + a_1$ for each CH_3 radical and then the interaction between the two "sides" makes them split into $1e_g$ and $1e_u$ and $2a_{1g}$ and $2a_{2u}$ respectively. The UV band of lowest wave number is due to transition to the $3s(a_{1g})$ Rydberg level and it is $g \leftrightarrow g$ and forbidden whether the transition originates with $1e_g$ or $3a_{1g}$. Thus the spectrum[69–72] starts with a weak band followed by the much more intense, mostly allowed 3p bands giving it an aspect very different from that of methane. Another characteristic of ethane is the closeness of the two frontier orbitals $1e_g$ and $3a_{1g}$. The photoelectron spectrum exhibits three bands belonging to ionization from these levels (vertical values 12.0, 12.7 and 13.4 eV) which belong to the Jahn-Teller split ${}^2E_g$ ion and to the ${}^2A_{1g}$ ion. The $1e_g$ orbital is C–H bonding and C–C antibonding while the $3a_{1g}$ orbital is strongly C–C bonding and weakly C–H bonding. Thus the respective states of the ion as well as the excited states of

109

C. Sandorfy

the neutral molecule which originate with $1e_g$ or $3a_{1g}$ are expected to have very different properties and a different photochemistry. The two orbitals are so close, however, that the order of the bands in the PE spectrum might not reflect their order in the neutral molecule. Furthermore, for the same reason, there are many, nearly coinciding bands in the UV spectrum[73].

The spectrum of ethane played an important role in the evolution of chemical far UV spectroscopy. It has become known as late as 1967 and was shown to fit into the series of paraffin spectra. Its fine structured bands constituted the first known example of a system of bonding σ-electrons having stable excited states[74]. The great resemblance of the fine structure of the most conspicuous component of the 3p manifold to that seen in the PE spectrum[17] strongly indicates the Rydberg character of the lowest excited states. The problems relating to this complex spectrum have been reviewed[73] and it seems to be sufficient to state the present position. The following singlet vertical excitation energies have been calculated by Buenker and Peyerimhoff[75] who used an enlarged basis set allowing for Rydberg orbitals and large scale configuration mixing. These results are in very satisfactory agreement with the observed spectrum. They indicate that

1) the ground state $1e_g$ (C–H) and $3a_{1g}$ (C–C) orbitals are nearly degenerate;
2) so are the two 3s states which account for the weak tale at the onset of the spectrum (about 67000 cm^{-1});
3) there are five allowed transitions to 3p states all within 0.14 eV.;
4) about half of the intensity belongs to the parallel (C–C) polarized ($3a_{1g}$,

Table 3.

Excitation		State	Polarization	Energy of excitation (eV)	Oscillator strength
Ground state		$^1A_{1g}$		0.0	
3s	← $1e_g$	$^1E_{1g}$	Forbidden	9.16	
3s	← $3a_{1g}$	$^1A_{1g}$	Forbidden	9.21	
3pσ	← $1e_g$	^1E_u	x, y	9.91	2 x (0.02–0.03)
3pσ	← $3a_{1g}$	$^1A_{2u}$	z	9.86	0.14
3pπ	← $1e_g$	^1E_u	x, y	9.99	0.00
3pπ	← $1e_g$	$^1A_{2u}$	z	9.99	0.02
3pπ	← $1e_g$	$^1A_{1u}$	Forbidden	10.04	
3pπ	← $3a_{1g}$	^1E_u	x, y	10.00	2 x (0.02–0.03)
∞	← $3a_{1g}$	$^2A_{1g}$		12.22	
∞	← $1e_g$	^2E_g		12.25	

The calculated excited states of ethane according to Buenker, R. J., Peyerimhoff, S. D.: Chem. Phys. 8, 56 (1975), Reprinted with permission from the Authors and the North-Holland Publishing Company.

$3p\sigma(a_{2u})$) transition which accounts for the broad absorption region underlying the structured bands[76];

5) the other half of the intensity is contributed by perpendicular bands which account for the fine structured bands superimposed on the broad band (70000 to 90000 cm^{-1}). At higher frequencies 3d and 4s bands follow.

Subsequently, Richartz, Bruna, Buenker, and Peyerimhoff[77, 78] calculated potential surfaces for ethane and the three lowest-lying states of its positive ion. The Jahn-Teller effect splits the ^2E_g state into a ^2B_g and a ^2A_g component (under C_{2h} symmetry). The first, well structured PE band (12.0 eV) corresponds to the ^2B_g ionic state. The observed spacing, 1170 cm^{-1} is due to both the ν_{11} asymmetric HCH bending and ν_3 C–C stretching vibrations. The ^2A_g component of the Jahn-Teller split ^2E_g state interacts with the other ^2A_g ($3a_{1g}$) state of the ion causing the complicated band structure of the PE bands at 12.7 and 13.4 eV. This strongly suggests that the initiating orbital of the structured 3p Rydberg band is also the $1e_g$ orbital. Furthermore, the photochemistry of ethane indicates that most (but not all) of its photolysis takes place at the ($1e_g$, 3s) state.

The photolysis of ethane yields mostly hydrogen and mostly in the form of molecular hydrogen. It has been explained in terms of four primary steps[79]:

$$C_2H_6 + h\nu \rightarrow CH_4 + CH_2 \qquad (10)$$

$$C_2H_6 + h\nu \rightarrow C_2H_5 + H \qquad (11)$$

$$C_2H_6 + h\nu \rightarrow C_2H_4 + H_2 \qquad (12)$$

$$C_2H_6 + h\nu \rightarrow CH_3 + CH_3 \qquad (13)$$

Their relative importance depends of the energy of the exciting photons but (11) and (12) are always predominant. That the C–H bonds and not the C–C bond is broken is well in line with the assumption that photolysis takes place in a state obtained through excitation from the C–H bonding $1e_g$ orbital and not from the C–C bonding $3A_{1g}$. Furthermore, it is important that at 8.4 eV (67750 cm^{-1}, the xenon resonance line) 85% of the decomposition occurs at (12), with only a minor contribution from (11). Since at 8.4 eV only the 3s Rydberg levels can be excited this shows again that the relatively large 3s orbital favors the extraction of molecular hydrogen. All six hydrogen atoms being connected by symmetry both 1,1 and 1,2 H_2 extractions could be expected with the shorter H–H distance favoring 1,1. Actually the latter process was found to be preponderant[79].

The spectra of other paraffins resemble either that of methane or that of ethane. Molecules closer to the tetrahedral type, like neopentane, like isobutane have strong 3s bands followed by a more or less pronounced gap due to the weakness of most of the 3p bands whereas more elongated molecules like propane and the other n-paraffins have 3s bands less well separated from the 3p (Fig. 1). As an hommage to those who in the nineteen forties developped the theory of the (π, π^*) spectra the writer would like to call these two types of spectra "round field" and "long field" respectively[80].

Characteristic of the spectrum of ethane were the close-lying initiating orbitals $1e_g$ and $3a_{1g}$. Then in order to understand the spectra of the higher normal paraffins we have to ascertain if such close-lying top occupied orbitals "exist" in the latter molecules as well. A very recent, large scale ab initio calculation by Richartz, Buenker and Peyerimhoff on propane has provided this information[81]. They have actually found that its three highest occupied MOs, $6a_1$ $4b_2$ and $2b_1$ are of very nearly equal energy. The many coinciding bands of each of the 3s, 3p, 3d, ... types which originate with these three MOs are certainly one of the causes of the diffuseness of its spectrum. The regions for the 3s,3p and 3d bands are near 71000, 77000 and 83000 cm^{-1}. Robin[9] assigned one of the 3s bands to a weak shoulder at about 66000 cm^{-1}. Whether this shoulder belongs to a separate electronic transition is debatable, however, and Richartz et al. assigned all three transitions to the 3s Rydberg orbital to the broad peak near 71000 cm^{-1}. It is much more intense than the 3s region in ethane itself but we have to remember that these bands are forbidden for ethane but not for the other n-paraffins. The spectra of the latter still "remember" their ethane ascendency, however.

For the higher normal paraffins production of H_2 molecules is still the most important primary step at 8.4 eV. According to recent results of Ausloos und his co-workers at the absorption threshold of all n-alkanes up to at least C_{12} the quantum yield for the H_2 elimination process is greater than o.8. At higher energies simple cleavage processes (both C–H and C–C) increase in importance. This may be partly occurring at states where the electron was promoted from a MO highly populated in C–C bonds. Also the shorter lifetime of the higher excited states might favor the simpler cleavage processes. Branched alkanes tend to eliminate lower alkanes at the absorption onset[83]. This could be expected since the center of the 3s orbitals is likely to be close to the branching carbon so that the radial nodes could help breaking both C–C and C–H bonds.

The case of cyclopropane has been discussed in detail by Robin[9]. Quantum chemical calculations show that the frontier orbital is $3e'\sigma$, "a sigma orbital composed of 2p atomic orbitals of the carbon atoms oriented tangentially to the internuclear triangle, and having maximum overlap symmetrically in-plane, outside each edge". The lowest energy photoelectron band which corresponds to ionization from this orbital exhibits a large split (about 8000 cm^{-1}) due to the Jahn-Teller effect.

The UV spectrum of cyclopropane[84] is quite different from those of the alkanes we have examined so far. Whereas acyclic alkanes and cyclic alkanes other than cyclopropane and cyclobutane have only Rydberg bands in the lower frequency (and photochemically important) part of the spectrum, for cyclopropane the lowest Rydberg bands intermingle with bands due to valence-shell transitions. Two structured bands at 63000 and 78000 cm^{-1} (159 and 128 nm) have been assigned to the $(3e'\sigma, 3p)$ and $(3e'\sigma, 4p)$ transitions on the grounds of the similarity of their vibrational fine structure with that of the respective photoelectron bands and their term values. Two other bands, near 70000 cm^{-1} (143 nm) and 83000 cm^{-1} (120 nm) (the latter very strong) must be valence-shell. While the Rydberg bands "disappear" in a solid film, the valence-shell bands remain; this is a criterion introduced by Robin[85] for distinguishing between Rydberg and valence-shell type bands. The main contributor to the intensity of the 70000 cm^{-1} band is probably the $^1(A_1', E')$

component of the $(3e'4e')$ transition while the 83000 cm^{-1} band receives most of its strength from the $(3e', a_2')$ transition (also $^1(A_1', E')$). Several other bands might contribute, however, and the situation is not altogether clear.

Since all these transitions originate with the $3e'\sigma$ orbital they would considerably weaken a C–C bond and consequently it is natural to except C–C cleavage. Actually, this is what is observed[82]:

$$cyclo-C_3H_6 + h\nu \rightarrow C_2H_4 + CH_2 \tag{14}$$

As to the identity of the state in which most cyclopropane molecules photolyse the most likely candidate seems to be the 3p Rydberg state (or rather states since under D_{3h} symmetry 3p splits to $a_2'' + e'$ and $(a_2'' + e') \times e' = E'' + A_1' + A_2' + E'$; transitions to E' and A_2'' states being allowed). However, it is important to note that the 3s bands have not been so far identified. They should be near 56 or 57000 cm^{-1} with normal term values. They must be photochemically important even though they are very weak.

Fluorocarbons

While fluorine atoms do, of course, have lone pairs of electrons they are so tightly bound that excitation or ionization from the MOs formed from them produce only high energy bands in both PE and UV spectra. This is also true for the MOs mainly populated in the C–F bonds. Thus in saturated molecules containing only C, H and F (except for CF_4) the bands of lower energy are due to bonding (C–C and C–H) σ electrons just as in paraffins.

Let us compare CH_4 and CF_4. As we have seen the t_2 orbital in methane gives two bands in the PE spectrum due to splitting through the Jahn-Teller effect. The two bands are centered at 109700 and 117100 cm^{-1} (or 13.6 and 14.3 eV). The first PE band of CF_4 is at 129900 cm^{-1} (or 16.1 eV)[9, 17]. It is, indeed, much more difficult to take away an electron from CF_4 than from CH_4. Whether the initiating orbital is of C–F bonding character or if it is formed essentially from the fluorine lone pairs ($\bar{F}$) has been a controversial subject. Robin[9] has given arguments in favor of a t_1 lone pair assignment. In fact a $^1(t_2,3s)$ transition would yield a strong lowest Rydberg band while the $^1(t_1,3s)$ transition would be forbidden. The band is centered at 100700 cm^{-1} (about 20000 cm^{-1} higher than for methane) and it is weak, much weaker than in methane.

This large hypochromic shift is less evident for the partly fluorinated molecules. In CH_3F, CH_2F_2 and CHF_3 the triply degenerate t_2 orbital of CH_4 splits and the MO which the lowest energy transitions originate with is doubly degenerate in CH_3F (e) and non-degenerate in CH_2F_2 (b_2) and CHF_3 (a_1). In all three it is mainly populated in the C–H bonds. Because of the splittings that occur the top filled MO is actually higher and the IP and the lowest absorption bands somewhat lower for CH_3F and CH_2F_2 than for CH_4 itself. For CHF_3 the trend to higher energies prevails[53].

As mentioned above the lowest (vertical) IP of ethane is at 96800 cm^{-1} (12.0 eV) considerably lower than the IP of methane. If we substitute fluorine atoms into ethane the lowest IP moves gradually to higher energies and with at least two fluorines at the same carbon the frontier orbital is of predominantly C—C charac-ter[86]. The UV spectra follow the trend in the IPs,C_2F_6 has a high IP, 117800 cm^{-1} (14.6 eV). It can only correspond to an orbital mainly populated in the C—C bond and it reflects the stabilizing effect of the fluorines. The UV spectrum also starts at very high frequencies: the band of lowest wavenumber which we found[86] is at 97600 cm^{-1}; its term value indicates a 3p type Rydberg orbital. The forbidden 3s band should be near 87000 cm^{-1} but so far it has not been reported. The absence of the $1e_g$ C—H orbital of ethane makes the lower part of the spectra of C_2F_6 much simpler[86].

Ionization connected with F and C—F type MOs begins, in general, near 125000 cm^{-1} (15.5 eV). The related UV bands appear at correspondingly high wavenumbers.

A much more thorough discussion of fluorocarbon spectra has been given by Robin (Ref.[9], Chap. III.B.1 and Addendum). He pointed out that the first two bands in all these spectra are readily assigned to 3s and 3p Rydberg transitions with characteristic term values of about 31000 and 22000 cm^{-1} respectively. Quite generally the presence of the highly electronegative fluorine atoms increases the term values, especially those of the strongly penetrating 3s orbitals. Many of the bands which are found at higher wavenumbers can be accounted for as higher Rydberg transitions. (3d, 4s, ... from the frontier orbital and transitions converging to the second, third, ... PE bands). Clearly, however, valence-shell transitions must also be present. These might constitute the diffuse background starting above 105000 cm^{-1} underlying the Rydberg bands and be mainly responsible for the strong bands found at higher wavenumbers. The valence-shell transitions can be of the (σ, σ^*) and of the $(\overline{F}, \sigma^*)$ types. The latter are well known for the higher halogens (where they become the first bands of the spectra) and must exist for the fluorocompounds as well.

The UV spectra of the series of perfluoroalkanes have been measured by Bélanger, Sauvageau and Sandorfy[47]. Robin and Kuebler[9] measured the photo-electron spectra. The IP goes from 130700 cm^{-1} in CF_4 to 101200 in n-C_8F_{20}. The interesting point is that in the UV spectra the strong bands found in the low wave-number part of the spectrum move much more rapidly to lower wavenumbers than the IP and the shifting does not seem to come to a halt up to C_8 at least. This is very different from what the related hydrocarbons do. They converge to about 65000 cm^{-1} for C_6. For this reason Robin suggests that these bands are actually valence-shell bands while the 3s Rydberg bands are in the low wavenumber shoulder of these bands. A similar situation exists in polysilanes[87, 88]. The vacuum UV photolysis of ethyl fluoride at 8.4 eV was studied by Chan, Inel and Tschuikow-Roux[89]. The principal primary processes were found to involve the elimination of HF and H_2 and, to a lesser extent C—F and C—C bond fission. Since the only excited state that can be reached at 8.4 eV is the 3s Rydberg state the predominance of the molecular elimination processes was to be expected according to the ideas put for-ward in Sec. IV.

Silanes

The interpretation of the UV absorption spectra of silicon compounds involves several delicate problems. Among these are the Rydberg vs valence-shell distinction, the order of Rydberg states in silicon compounds and the participation of 3d atomic orbitals in the wave functions of their ground and excited states.

Both the PE[90–92] and UV absorption spectra[93, 94] of silanes have been known for several years. The first PE band corresponds to the $^2T_2$ ion and it is Jahn-Teller split with peaks at 99200 and 103200 cm^{-1} (12.3–12.8 eV). All the known electronic transitions of silane also originate with the $^2t_2$ orbital. Harada et al.[94] found a weak band at 64100 cm^{-1} followed by intense bands at 72500 and 78000 and a very strong band at 87000 cm^{-1}. The Rydberg term values which are to be expected have been thoroughly discussed by Robin[9]. The order of the lowest Rydberg orbitals for the Si atom is 4s < 3d < 4p but for the united atom argon it is 4s < 4p < 3d. The 4s term value is expected to be intermediate between those of Si (27000 cm^{-1}) and Ar (34000 cm^{-1}). The 4p and 3d terms are expected to be similar around 17000 cm^{-1}. Valence-shell transitions, as in methane, could be $^1(2t_2,4a_1)$, $^1(A_1,T_2)$ and $^1(2t_2,3t_2)$; $^1(A_1, A_1 + E + T_1 + T_2)$.

Recent calculations by Schwarz[59] who studied the continuous change from valence to Rydberg type states in the series HCl, H_2S, PH_3, SiH_4 by *ab initio* SCF techniques and the (Z + 1) core-analogy model have shown that in the lowest excited state of SiH_4 the orbital of the excited electron is a relatively small sized Rydberg orbital. This result is corroborated by analogy with methane itself. The question is, which of the observed bands corresponds to the $^1(2t_2,4s)$ transition? (Fig. 2). If it is the weak band at 64100 cm^{-1} which appears in the spectrum of Harada et al., then the term value would be about 37000 cm^{-1}. This is certainly too high for a 4s Rydberg orbital in a silicon compound. The weakness of this band (an allowed transition $^1(A_1,T_2)$) would also be hard to understand. It is more natural then to assign the $^1(2t_2,4s)$ transition to the strong shoulder at 72500 cm^{-1} with a term value of 28700 cm^{-1}. The more intense band at 78000 cm^{-1} can contain the 4p and 3d

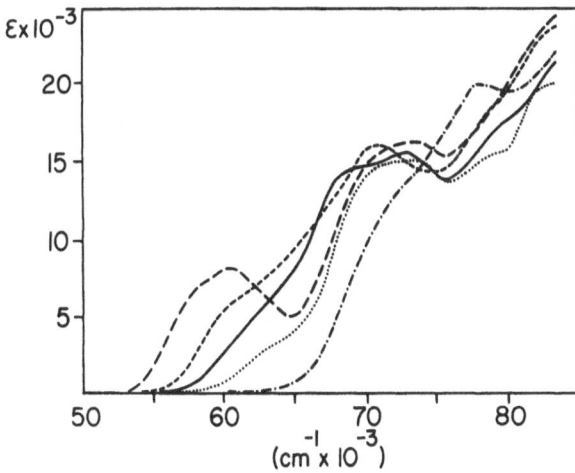

Fig. 2. The far UV absorption spectra of SiH_4 – . – . – . – . – . – .,, CH_3SiH_3 . . . . , $(CH_3)_2SiH_2$ ——— , $(CH_3)_3SiH$ - - - - - - - - - - - - , and $(CH_3)_4Si$ – – – – – –. [Reproduced from Roberge, R., Sandorfy, C., Matthews, J. J., Strausz, O. P.: J. Chem. Phys. *69*, 5105 (1978), with permission from the American Institute of Physics]

Rydberg bands. According to Schwarz these have valence-shell character. Actually, the term value is quite high, 23200 cm^{-1}. Just as for methane, the orbital of the excited electron in these states is probably of an intermediate type. These assignments are in basic agreement with those of Robin[9] except that he assigns the weak band at 64100 cm^{-1} to an additional valence-shell transition. In our opinion, however, the existence of this band is questionable.

Some weak bands should, of course, exist such as those corresponding to the forbidden components of the $^1(2t_2,4p)$ transition (A_1, E, T_1) or to valence-shell states of the same symmetry. Whether or not there is a band at 64100 cm^{-1} due to a separate electronic transition is a problem of some importance. Further experimental work, in particular electron impact measurements, would be needed as well as advanced quantum chemical calculations to determine the extent of interaction between the two T_2 states arising from the $^1(2t_2,4s)$ and $^1(2t_2,4p)$ transitions, and to locate the forbidden bands. The main point at this stage is, however, that the 64100 cm^{-1} band cannot belong to the $^1(2t_2,4s)$ transition.

The PE spectrum of methylsilane has been published by Price[96]. It bears a pronounced resemblance to that of ethane. This seems to show that in methylsilane there are still two close-lying frontier orbitals, similar to the $1e_g$ and $3a_{1g}$ orbitals of ethane. The UV spectrum too is closer to the ethane type.

The highly diffuse character of the UV absorption spectra of silane derivatives precludes the possibility of gaining knowledge on the structure and geometry of the excited states. By means of photoelectron spectra, Rydberg term values and by analogies with the spectra of corresponding paraffins it is still possible however, to interpret the spectra of silane derivatives in a general manner. An extensive work aiming at this has been recently carried out by Roberge, Sandorfy, Matthews and Strausz[95].

The spectra resemble those of paraffins and can be interpreted in Rydberg terms. The 4s and 4p regions can be identified in all the spectra. According to their relative intensities and the more or less pronounced gap which is observed between them, the spectra are either of the methane ("round field") or of the ethane ("long field") type (cf. the section on paraffins). The former have relatively strong 4s bands as is usual for molecules of tetrahedral symmetry, whereas the latter such as ethane have weak 4s bands. (In ethane such bands are g↔g and forbidden). Therefore, the 4s–4p gap is characteristic of "round" molecules. Replacing methyl groups by ethyl groups (or presumably longer alkyls) tends to transform the "round field" type spectra to the "long field" type.

Fluorine substitution causes hypsochromic shifts similar to those which were found in the spectra of fluoroparaffins. It is small when only one or two fluorines are present but becomes large for the trifluoro methanes and silanes. In ethane type molecules where there are two nearly degenerate frontier orbitals, the fluoroeffect has another interesting manifestation. It influences these two frontier orbitals to different degrees. The consequence of this is a sharpening of the lowest energy PE band which becomes a single band, and a deepening of the gap between the 4s and 4p regions in the UV spectrum because the bands due to transitions from the more affected frontier orbital have been shifted to higher frequencies.

The main primary process in the photolysis of silane is, at 8.4 eV:

$$SiH_4 + h\nu \rightarrow SiH_2 + H_2 \tag{15}$$

(φ = 0.58) This was to be expected on the ground of the intermediate Rydberg –
valence-shell character of the $(2t_2,4s)$ excited state. The 4s Rydberg orbital should
help in bringing the hydrogen atoms together. The $H_2/2H$ ratio is somewhat smaller
than for methane; this might indicate that in silane 4s is closer to the valence-shell
type than 3s in methane.

Methylsilane also photolyses mainly through molecular elimination of hydrogen[97]. This is in keeping with photolysis in the (e,4s) state, the initiating e orbital
being C–H bonding and the 4s orbital again helping to bring the hydrogen atoms
together. However, the relative proportion of Si–C cleavage to H_2 elimination is
higher than in ethane, so that presumably a higher percentage of CH_3SiH_3 photolyses in the $(a_1,4s)$ state. Since the two states are close to one other this might
keenly depend on the exciting wavelength.

VI Lone Pair Spectra

Halides

In molecules like organic halides other than fluorides, alcohols, ethers, amines, etc.,
the ionization potential and the near-totality of the known electronic transitions
originate with MOs associated with the lone pairs. While these orbitals are often
much less localized around a given heteroatom than was believed for some time, the
lone pair concept is still useful.

In all alkyl halides except fluorides the lowest ionization potential and all electronic transitions up to about 90000 cm^{-1} originate with the lone pair electrons of
the halogens. The related band is at 11.3 eV (91100 cm^{-1}) in the photoelectron
spectrum of $CH_3Cl(17)$. The second photoelectron band (14.4 eV or 116200 cm^{-1})
corresponds to ionization from an orbital strongly populated in the C–Cl bond, the
third one to an orbital mainly populated in the CH_3 group (15.4 eV or
124200 cm^{-1}). The first UV band of CH_3Cl is centered around 59000 cm^{-1}
(169 nm)[49, 50]. It is not a Rydberg band, its term value is too high for either s, or p,
or d. It is generally assigned to a valence-shell transition of the (n, σ^*) type in which
one of the electrons of the chlorine lone pair is raised to an orbital antibonding in
the C–Cl bond. We shall call it $(C–Cl)^* \leftarrow \overline{Cl}$. This is a quite general phenomenon:
saturated molecules containing lone pairs often do have transitions to σ^* orbitals
which occur before the Rydberg orbitals. Since the transition is a valence-shell transition and since the ground state orbital is essentially non-bonding, the photochemical behavior in the excited state is largely determined by the anti-bonding
character of the $(C–Cl)^*$ orbital and we expect C–Cl cleavage. Indeed, the primary
step in the photodecomposition of alkyl halides is[98]:

$$CH_3X + h\nu \rightarrow CH_3 + X \tag{16}$$

117

All alkyl chlorides possess a weak absorption band between 50000 and 59000 cm^{-1} (170–200 nm) and they all behave in the same way.

The $(C–Cl)* \leftarrow \overline{Cl}$ band is followed at higher frequencies by Rydberg bands, 4s, then 4p. They belong to the lowest ionization potential. Beyond 90000 cm^{-1} we find Rydberg bands related to higher ionization potentials. If we irradiate into the Rydberg bands we might expect a more complicated pattern of photodecomposition. The initiating orbital is still the same lone pair orbital but the excited orbital is no more antibonding, the molecule will resemble the ion. The hole on the chlorine is likely to be at least partly filled by an influx of electrons from the adjacent bonds. Chlorine atoms will probably be liberated but they could well pull out a hydrogen atom to form HCl with the help of the large Rydberg orbital leading to a more complicated pattern involving molecular elimination processes. This was actually observed in the photolysis of ethylchloride[99].

All this applies to chlorofluorocarbons as well[50, 51, 53]. These compounds have recently been the object of a great deal of controversy. In the stratosphere where certain chlorofluorocarbons are known to accumulate their respective photosensitivity will depend on the exact location of their absorption bands. Compounds containing only one chlorine like CF_3Cl, CF_2HCl, C_2F_5Cl, have their absorption maxima near 66000 cm^{-1} (152 nm) and their absorptions are weaker than for dichloro or trichloro fluorocarbons. These compounds are unlikely to undergo photodecomposition in the stratosphere. Chlorofluorocarbons containing two or more chlorines absorb at lower frequencies and more strongly. CF_2Cl_2 (Freon–12) and $CFCl_3$ (Freon–11) have their first absorption maxima near 56000 cm^{-1} (170 nm) and 54000 cm^{-1} (185 nm) respectively. At those wavelengths the photon flux emanating from the Sun is sufficiently intense in the stratosphere to cause photodecomposition of these molecules[100]. Bromine containing fluorocarbons which absorb at even lower frequencies should be even more sensitive. Photochemically CH_3Cl, CH_2Cl_2, . . ., would be more vulnerable than the related fluorocarbons but the former are also chemically vulnerable and might be destroyed before reaching the stratosphere. Since, as stated above, in fluorocarbons containing higher halogens the band of lowest wavenumber is of valence-shell $(C–X)* \leftarrow \overline{X}$ type the antibonding character of the excited orbital leads to the cleavage of the carbon-halogen bond with the production of atomic halogens. Rowland and Molina[100] drew attention to the possibility that chlorine atoms might enter into a chain reaction with ozone which in the upper atmosphere forms a layer protecting life against the Sun's UV radiation. This might lead to partial depletion of the ozone layer. The extent to which this might occur is not a spectroscopic but a photochemical problem, however. It received a great deal of attention in recent years[98, 101].

For our present purposes the interesting point is that all this so-called fluorocarbon controversy and the danger to the ozone layer is linked to the valence-shell and antibonding character of the $(C–Cl)*$ orbital. Were this state a Rydberg state elimination of small molecules (HCl, FCl, Cl_2, . . ., according to cases) would be expected and the photochemical consequences would be different.

Tables 4 and 5 taken from a review paper by the writer[53] contain some illustrative data on the lower transitions of such molecules and do not require further comment.

A fundamental discussion of the PE spectra of mono-, di-, tri- and tetramethyl-halides has been given by Turner[17]. As to the UV spectra reference should be made to Russell[49] and to Robin[9].

Just as in alkylhalides in the spectra of chlorosilanes the highest occupied MOs in the ground state are formed from chlorine lone pair orbitals ($\overline{Cl}$) and all known UV transitions originate with these.

The simplest molecule of this kind is chlorosilane, SiH_3Cl. Its PE spectrum has been measured by several Authors[102, 103] who assigned the first PE band 93600 cm^{-1} (11.61 eV, vertical) to ionization from the doubly degenerate (e, C_{3v})

Table 4. The lowest ionization potential and approximate maxima of the UV absorption bands of lowest frequency of chloro- and bromo-fluoromethanes. [Reproduced from Sandorfy, C.: Atmospheric Environment _10_, 343 (1976), with permission of Pergamon Press Ltd.]

	Lowest vertical IP		Valence-shell		Rydberg
	eV	cm^{-1}	cm^{-1} (nm)		cm^{-1} (nm)
CH_3Cl	11.3	91100	59000 (169)		61400 (158)
CF_3Cl	13.0	104900	71500 (140)		78100 (128)
CF_2HCl	12.6	101600	66200 (151)		74500 (134)
CFH_2Cl	11.7	94400	62500 (160)		65300 (153)
CH_2Cl_2	11.4	92000	58000 (172)		67000 (149)
CF_2Cl_2	12.3	99200	56500 (177)	65400 (153)	74000 (135)
$CFHCl_2$	12.0	96800	57500 (174)	60600 (165)	70000 (143)
$CHCl_3$	11.5	92800	58000 (172)		66200 (151)
$CFCl_3$	11.9	96000	54000 (185)	60500 (165)	70000 (143)
CCl_4	11.7	94400	57500 (174)		69000 (145)
CF_3Br	12.0	96800	48850 (205)		70500 (142)
CF_2Br_2	11.2	90300	44000 (227)	52000 (192)	66200 (151)
CF_2ClBr	11.8 (Br) 12.9 (Cl)	95200 104100	48800 (205)	61000 (164)	66750 (150)

Table 5. The lowest ionization potential and approximate maxima of the UV absorption bands of lowest frequency of some chlorofluoromethanes. [Reproduced from Sandorfy, C.: Atmospheric Environment *10*, 343 (1976), with permission of Pergamon Press Ltd.]

	Lowest vertical IP		Approximate maxima of the lowest UV absorption bands Valence-shell		Rydberg	
	eV	cm^{-1}	cm^{-1}	nm	cm^{-1}	nm
C_2F_5Cl	13.0	104900	66700	150	78300	128
CH_3CF_2Cl	12.5	100800	63900	156	73900	135
CF_2Cl-CF_2Cl	12.85	103600	64500	155	77200	130
$CF_2Cl-CFHCl$	12.0	96800	57500	174	70000	143
$CFCl_2-CF_2Cl$	12.05	97200	62500	160	73000	137

$\overline{Cl}$ orbital, the second band to the mainly C–Cl (a_1) bonding orbital 108100 cm^{-1} (13.4 eV) and the third one 110500 cm^{-1} 13.7 eV) to the degenerate (e) CH$_3$ bonding orbital. The corresponding values for CH$_3$Cl are 11.28, 14.4 and 15.5 eV, respectively. It is to be noted that the IP of SiH$_3$Cl is higher than the IP of CH$_3$Cl whereas GeH$_3$Cl's IP is 11.30 eV, like for CH$_3$Cl. According to the ab initio calculations of Howell and Van Wazer[104] the 3d AOs have an important stabilizing effect on chlorosilane. These Authors included into the basis set first, 3d AOs on Si only, then on Cl only, and finally on both Si and Cl. The stabilization of the *total* energy amounted to about 2.5, 0.6 and 2.9 eV respectively demonstrating the importance of the 3d of the silicon atom. At the same time these calculations have shown that in order to have a sizeable effect, the substituent (Cl) must have AOs (3pπ) with energies sufficiently close to the Si 3d. It appears that this is not the case for alkyl or fluorosilanes or for chlorogermanes, among others. The frontier orbital in chlorosilane is stabilized by about 0.25 eV only. Supposing the validity of Koopmans' theorem this is in good agreement with experiment. Well before the PE spectra became known, Bell and Walsh[105] on the bases for their UV absorption measurements concluded to this increase in IP: for each silyl compound studied, the spectrum was shifted to higher frequencies relative to that of the corresponding methyl compound. They attributed this to the partial loss of lone pair character and the gain in bonding character of the originating orbital on passing from the carbon to the silicon compounds. Just as Howell and Van Wazer they attributed this to the intervention of the Si 3d.

The band of lowest frequency in alkyl chlorides (discussed above) is due to transitions of the (C–X)* ← $\overline{X}$ type where n is essentially the pπ lone pair AO of the halogen. On the basis of the above argument in SiH$_3$Cl this band is expected to move to higher frequencies. Actually, it has not been observed. It must exist, however, and it is likely to be hidden under the envelope of the more intense Rydberg bands. The $^1(\overline{Cl}, 4s)$ and $^1(\overline{Cl}, 4p)$ bands have been located at 67000 and 74900 cm^{-1} respectively[9, 104]. The term values are 26600 and 18700 cm^{-1}, somewhat lower than those for the related carbon compounds.

Roberge[106] measured the spectra of two other monochlorosilanes ($(CH_3)_3SiCl$ and $(CH_3)_2SiHCl$) in the hope of locating the "missing" $(Si–Cl)^* \leftarrow \overline{Cl}$ band but could not locate it with certainty. The same applies to the dichlorosilanes whose spectra were measured by Causley and Russell[107]. Roberge found a strong indication for this band in the UV spectrum of $SiHCl_3$. In the PE spectrum which was reported by Frost et al.[103] there are three well defined peaks and one shoulder (11.99, 12.45, 12.60 (sh) and 13.08 eV) which are readily assigned to the $\overline{Cl}$ levels. ($a_1 + a_2 + e + e$; C_{3v}). For chloroform the corresponding values are 11.4, 11.8, 12.1 and 12.9 eV. In the UV spectrum the $^1(\overline{Cl}, 4s)$ and $^1(\overline{Cl}, 4p)$ bands can be assigned to the well pronounced shoulders at 71000 and 74700 cm^{-1} (term values 25700 and 22000 cm^{-1} respectively). In addition at the low frequency end of the spectrum there is a long descending tail, centered at 66000–67000 cm^{-1}. This would have a term value of about 30000 cm^{-1} which is too high for this type of molecule and since the 4s Rydberg bands could unambiguously located at higher frequencies it can be assigned to the $(Si–Cl)^* \leftarrow \overline{Cl}$ transition. It is about 10000 cm^{-1} *higher* than in chloroform. This lends credibility to the suggestion that in the mono- and dichlorosilanes the $(Si–Cl)^* \leftarrow \overline{Cl}$ band shifted to higher frequencies relative to the corresponding carbon compounds and "disappeared" underneath the stronger bands that are located there.

The UV spectrum of $SiCl_4$ has been discussed by Causley and Russell[108] (Fig. 3). Their most striking observation is the huge hypsochromic shift which the

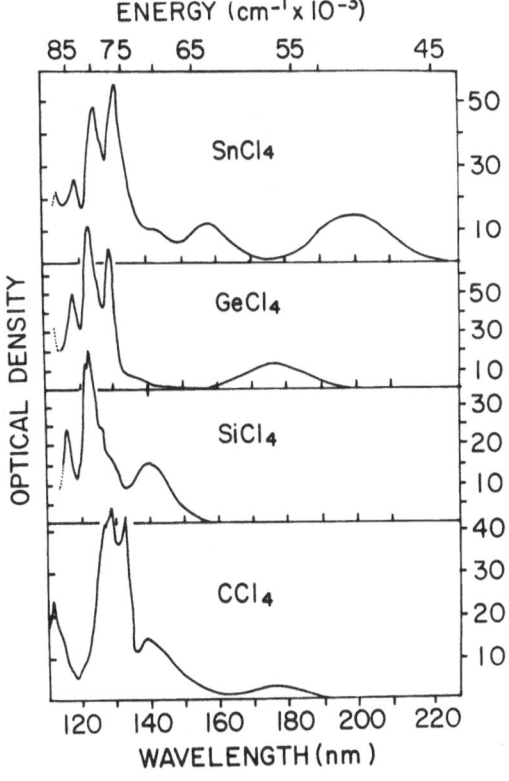

Fig. 3. The far UV absorption spectra of CCl_4, $SiCl_4$, $GeCl_4$ and $SnCl_4$. [Reproduced from Causley, G. C., Russell, B. R.: J. Electron Spectrosc. *11*, 383 (1977), with permission from the Authors and the Elsevier Scientific Publishing Company]

C. Sandorfy

band of lowest frequency undergoes in passing from CCl_4 to $SiCl_4$, from 57000 to 72000 cm^{-1}, while it is at 57000 in $GeCl_4$ and at 50000 cm^{-1} in $SnCl_4$! They assign these bands to $(Si-Cl)* \leftarrow \overline{Cl}$, etc., excitations. This observation, together with the general blue shift in chlorosilanes observed by Bell and Walsh[105] and the theoretical study of SiH_3Cl by Howell and Van Wazer[104] all strongly indicate $d\pi - p\pi$ mixing in chlorosilanes. This has been confirmed by Frost et al.[103] who have shown that after taking all other factors into account that might contribute to the blue shift the need for involving $d\pi - p\pi$ interaction still remains.

Oxo-compounds

Water, alcohols and ethers all contain oxygen lone pairs ($\overline{O}$) and their spectra are dominated by transitions from $\overline{O}$ orbitals. The spectrum of water is discussed by Herzberg[7], Claydon[109] and Hudson[110] and it is, of course, essential to the understanding of the spectra of alcohols and ethers. Robin[9] gave a well thought synthesis on these spectra to which the present writer has little to add. A brief comment will suffice.

The ground state configuration of water is

$$[1a_1]^2[2a_1]^2[1b_2]^2[3a_1]^2[1b_1]^2$$

The $1b_1$ orbital is essentially an oxygen 2p lone pair with its axis perpendicular to the molecular plane. $3a_1$ is the oxygen 2p lone pair orbital lying along the symmetry axis; it is mixed with the H1s to some extent. Available for the construction of excited valence-shell MOs are $4a_1$ and $2b_2$ which are the symmetric and antisymmetric combinations of the O–H antibonding σ orbitals. The first band of water is a broad band centered at 60000 cm^{-1}. It originates with $\overline{O}$ ($2b_1$). The orbital of the excited electron is probably best described as being of intermediate Rydberg (3s) — valence-shell ($4a_1$) type with significant O–H antibonding character. There is another broad band centered around 79500 cm^{-1} which can be interpreted as a transition from $3a_1$, to the same intermediate 3s ($4a_1$) orbital. The corresponding term values are

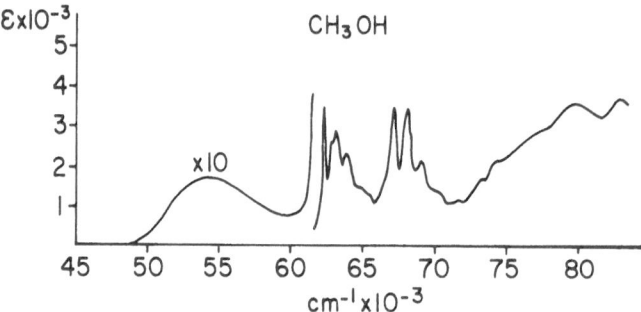

Fig. 4. The far UV absorption spectrum of methanol.
[Reproduced from Salahub, D. R., Sandorfy, C.: Chem. Phys. Letters *8*, 71 (1971), with permission from the North-Holland Publishing Company]

122

quite high: 41800 and 39500 cm^{-1} respectively. Superimposed to the second diffuse band are several sharp bands constituting the fine structure of the $(1b_1, 3p)$ Rydberg states. The three members of the manifold are $^1A_1$, $^1B_1$ and $^1B_2$. The two first are allowed with vertical frequencies 80624 and 82038 cm^{-1}. At higher frequencies 4s and 3d are in near coincidence.

Alcohols remember all this. The first band of methanol is a diffuse weak band centered at about 54000 cm^{-1} [48]. It closely resembles the 60000 cm^{-1} band of water. The first and second vertical IPs of water are at 101780 $(1b_1)$ and 119000 cm^{-1} $(3a_1)$ differing by 17200 whereas the same bands for methanol are at 88420 $(2a'')$ and 102400 $(1a'')$ differing by 14000 cm^{-1}. This shows that both $\overline{O}$ orbitals are destabilized by mixing with the π orbitals of the methyl group. Understandably, the first band was variously assigned as valence-shell σ^* or 3s. To our present knowledge it is quite safe to describe it as a 3s MO of intermediate type. Then in analogy with water the structured bands at about 62000 and 67000 cm^{-1} can be assigned to two members of the $(2a'', 3p)$ manifold (See Ref.[9] for a discussion of this split). Robin observed that for higher alcohols the 3s band remains near 55000 cm^{-1} and does not follow the IP. Despite this the 3s Rydberg assignment is valid. Indeed, this is an example of the force of Robin's term value method. As the alkyl group increases in size the $(\overline{O}/3s)$ orbital becomes more and more delocalized over the alkyl group and the term value decreases from the 41800 cm^{-1} of water to 27200 cm^{-1} for tertbutanol. Thus the location of the band remains nearly constant. A similar decrease in the value of the 3s term is observed for the paraffins themselves (Sect. V). The 3s band originating from the $a'(\sigma)$ orbital does not seem to have been observed but several higher Rydberg bands are known.

The most important primary steps in the photolysis of methanol when excited in the 3s band are

$$CH_3OH + h\nu \rightarrow CH_2O + H_2 \tag{16}$$

$$CH_3OH + h\nu \rightarrow CH_3O + H \tag{17}$$

The simple bond fission (17) is the more important one but both (16) and (17) occur in significant proportions. This seems to show that the 3s orbital is of the intermediate type with strong O–H antibonding character but that it can still be instrumental in bringing together the two hydrogens. This is a complicated case. A thorough assessment of the photochemistry of alcohols has been given recently by von Sonntag and Schuchmann[111].

The spectra of ethers are also essentially $\overline{O}$ spectra and resemble those of water and alcohols. The main difference is that the bands have usually well developed vibrational fine structure, even the 3s bands. The short progressions that are found show that the $(2b_1)$ lone pair orbital is more purely lone pair in ethers than in water or in alcohols. It also follows that the 3s orbital is more purely Rydberg with less antibonding character than in water or alcohols. This is evident from the ease of the photochemical cleavage of the O–H bond wherever it is present. The observed fine structure shows that the 3s excited state is stable in ethers while it is likely to be repulsive in alcohols and water.

C–O bonds are, of course, still broken during the photolysis of ethers and this might seem to contradict the purely lone pair character of the $2b_1$ orbital. It is believed that in order to understand this we have to invoke electronic rearrangement: when an electron is removed from the field of a strongly electronegative atom a rapid flow of electrons occurs in order to compensate for the loss. This then has the consequence of weakening the neighboring bonds. (We will have to invoke this principle in connection with ketones as well, for example). The weakening of bonds between an electronegative atom and hydrogen is bound to be sizeable in view of the weak electron attracting power of the proton.

Dimethylether has its IP at 80330 cm^{-1} (9.96 eV). The 3s band has its origin at 53140 cm^{-1}, the 3p band seems to have two origins at 58820 and 61390 cm^{-1} [112, 113] and a 3d type band is found at 68120 cm^{-1}. As for alcohols the 3s band moves little with increasing alkyl size while the IP does move; this can again be explained by the accompanying decrease in term value as the length of the alkyl groups increases.

The two most important primary steps in the photolysis of dimethylether at 192 nm (in the 3s band) are

$$CH_3OCH_3 + h\nu \rightarrow CH_3O + CH_3 \qquad (18)$$

$$CH_3OCH_3 + h\nu \rightarrow CH_2O + CH_4 \qquad (19)$$

with (18) being more important[111, 114].

Cyclic ethers have similar spectra. The 3s bands are usually highly structured exhibiting the low frequencies of pseudorotation[115].

The spectra of thiols and thioethers (sulfides) "remember" the spectrum of H_2S just as those of alcohols and ethers remember the spectrum of water (Fig. 5). All the known bands of these molecules are due to transitions from the sulfur lone pair orbital $(\overline{S})$. It is $2b_1$ in H_2S the axis of the π orbital being perpendicular to the molecular plane. The first IP of H_2S is at 84420 cm^{-1} (10.46 eV) more than 2 eV lower than in water[116]. The UV spectrum of H_2S has been known for many years from

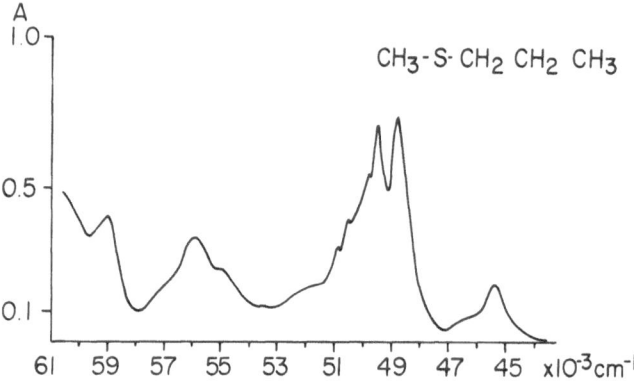

Fig. 5. The far UV absorption spectrum of methyl-n-propyl-sulfide.
[Reproduced from Olivato, P. R, Viertler, H., Wladislaw, B., Sauvageau, P., Sandorfy, C.: J. Chem. Phys. 70, 1677 (1979), with permission from the American Institute of Physics]

Price's original work[116, 117]. A recent, thorough analysis of the spectra of sulfur compounds has been given by Robin[9] on the basis of term values. His conclusions are in good agreement with large scale quantum mechanical calculations by Shih, Peyerimhoff and Buenker[118] and by Guest and Rodwell[119] on H_2S. The band of lowest frequency (40000–60000 cm^{-1} for H_2S is due to the $^1(\bar{S}, 4s)$ Rydberg transition $^1(2b_1, 4s)$; $^1(A_1, B_1)$ which is in near-coincidence with a valence-shell transition $^1(2b_1, 3b_2)$; $^1(A_1, A_2)$ which according to Shih et al.[119] has 3d admixture. That this (near UV) region of the spectrum cannot be explained with only one electronic transition has been recognized earlier by Barrett and Hitch[120] on spectroscopic ground. It is essential for the interpretation of the spectra of molecules containing divalent sulfur. At higher energies the absorption regions due to members of the $(2b_1, 4p)$ and $(2b_1, 3d)$ manifolds are readily recognized. They contain many sharp bands.

In the mercaptans the two bands which coincided for H_2S become well separated with maxima at about 44000 and 49000 cm^{-1}. The spectra of a number of mercaptans and sulfides have been measured by Clark and Simpson[121].

A typical sulfide spectrum that of methyl-n-propyl sulfide is shown in Fig. 5[122]. The $\bar{S}$ $(2b_1)$ IP is 68450 cm^{-1} (8.49 eV). At the low frequency end we find a well defined maximum at 45400 cm^{-1} and there are shoulders at about 44000 and 46300 cm^{-1}. The peak is likely to correspond to the $4s \leftarrow 2b_1$ Rydberg transition which is superimposed to a mainly valence-shell type band. ($2b_1, 3b_2$ in H_2S). The maximum gives a term value of 23050 cm^{-1}, acceptable for a 4s band of a sulfur containing molecule of this size.

In the 48000–52000 cm^{-1} region thioethers have two bands that are members of the $(\bar{S}, 4p)$ manifold. In the $(\bar{S}, 3d)$ region two bands can be distinguished, at 55000 and 55900 cm^{-1} with some additional shoulders. The band at 59000 cm^{-1} can reasonably be assigned to the 4s transition from the second highest occupied MO which is $5a_1$ in H_2S.

Thiols photolyze 90% by S–H cleavage[123, 124] at 254 nm a much higher proportion than for O–H cleavage in alcohols. This is very likely linked to the existence of the low lying valence-shell state. The same applies to the homolytic C–S fission in sulfides.

$CH_3SCH_2COCH_3$ has an interesting spectrum[125]. The lowest bands can be assigned as follows:

33250 cm^{-1}	$\pi^* \leftarrow \bar{O}$	band of the carbonyl
41200 cm^{-1}	$\sigma^* \leftarrow \bar{S}$	valence-shell
45500 cm^{-1}	$4s \leftarrow \bar{S}$	Rydberg
52400 cm^{-1}	$\begin{cases} 4p \leftarrow \bar{S} \\ 4s \leftarrow \bar{O} \end{cases}$	

Photochemists might like to attempt to photolyze this molecule in the various excited states, the three first, in particular. It might give rather different results on each.

VII π-Electron Spectra

Ethylene

Few molecules received more attention than ethylene. The analysis given in 1968–1969 by Merer and Mulliken[126] and by Merer and Schoonveld[127, 128] is still essentially valid. Subsequently McDiarmid[129–131] made an important contribution in showing that the O–O band of the V ← N transition is near 48000 and not near 38000 cm^{-1} that is, the extremely long tail once observed in liquid ethylene was due to a trace of oxygen[131]. In both V and T states (singlet and triplet (π,π^*)) the (adiabatic) excited state geometry is perpendicular but Mulliken has shown that the C–C distance does not need to be extremely long, a strong hyperconjugative effect reduces it to perhaps 1.44 Å. As to the valence-shell or Rydberg character of the V state Mulliken made the remark several years ago that the π^* MO reduces to $3d\pi$ at the united atom limit and therefore it should be considered as being of the intermediate type[5]. According to Buenker and Peyerimhoff[132, 133], however, this is a consequence of mixing between the (π, π^*) and $(\pi, 3p\pi)$ states. An interesting sequel to this is, according to the latter Authors, that the transition moment depends on the twist angle. For the coplanar V state π^* would be significantly Rydbergized but it would become more compact and valence-shell like as the twist angle increases. Consequently, the vertical transition may not have the largest Franck-Condon factor and the vertical transition might be at higher wavenumbers than the observed maximum at 61700 cm^{-1}.

However, ethylene does have a truly Rydberg 3s state which does not have the same symmetry as the V state. The corresponding band is located at the longwavelength tail of the (N,V) band with its origin at 57340 cm^{-1}. Merer and Schoonveld[128] have shown that in the 3s state the molecule deviated from coplanarity by 25 °. The related term value is 27400 cm^{-1} the IP is at 84800 cm^{-1} (10.51 eV, adiabatic). Higher Rydberg bands have also been identified[134, 135], they all converge to the π IP.

The many problems involved with the spectrum and excited states of ethylene have been most thoroughly reviewed by Robin[9] up to 1975. The interested Reader will find ample information therein and in the other works referred to in this section. It is felt that there is no need for a more detailed discussion at this point.

As to the photolytical primary steps of ethylene the dominant one is H_2 release but H atoms are also obtained in significant proportions. According to Borrell, Cervenka and Turner[136] at the 184.9 nm mercury line there are three major primary processes:

$$CH_2{=}CH_2 + h\nu \rightarrow CH{\equiv}CH + H_2 \qquad\qquad \varphi = 0.68 \qquad\qquad (20)$$

$$CH{\equiv}CH + H + H \qquad\qquad \varphi = 0.20 \qquad\qquad (21)$$

$$CH_2{\equiv}CH + H \qquad\qquad \varphi = 0.16 \qquad\qquad (22)$$

On the bases of experiments on *trans*-1,2 and 1,1 dideuteroethylene it has been established that at 8.4 eV or higher energies terminal elimination of H_2 or D_2 is more probable than 1,2 elimination of $HD^{[137-139]}$.

Calvert and Pitts[68] compare the quantum yield of primary processes.

$$CH_2=CH_2 + h\nu \rightarrow CH_2C + H_2 \qquad\qquad (23) \ (\varphi_1)$$

$$CH_2C + H + H \qquad\qquad (24) \ (\varphi_2)$$

$$HC\equiv CH + H_2 \qquad\qquad (25) \ (\varphi_3)$$

$$HC\equiv CH + H + H \qquad\qquad (26) \ (\varphi_4)$$

$$CH_2=CH + H \qquad\qquad (27) \ (\varphi_5)$$

$\varphi_2 + \varphi_4/\varphi_1 + \varphi_3$ is about 1.0 at 147 nm and 1.4 at 123.6 nm in conformity with the general observation that the importance of simple cleavage processes increases with increasing photon energy. Process (27) is minor but increases in importance with increasing photon energy. All this is compatible with photolysis at the perpendicular V state. It is possible, however, that a part of the molecules photolyzes at the Rydberg state. The 3s orbital would help forming H_2.

As to triplet states, for Rydberg states they are very close to the corresponding singlet and they are seldom considered separately. For valence-shell transitions they, of course, require a separate study. For this, mercury sensitization has been successfully applied in many cases[34]. In the case of ethylene the main primary process is still H_2 formation, both 1,1 and 1,2.

In highly methylated ethylene the (N,V) and 3s bands separate and the Rydberg band is definitely at lower frequencies[140]. The same applies to most fluoroethylenes[55]. In the chloroethylenes conditions are likely to be similar although some problems remain with the band assignments[141, 142]. Now, in a recent paper Ausubel and Wijnen[143] have shown that the photolysis of both *cis* and *trans* 1,2-dichloroethylenes (between 200 and 240 nm) involve two excited states. For the trans isomer they obtained, from the lower excited state, the following primary steps:

$$t\text{-}C_2H_2Cl_2^* + h\nu \rightarrow C_2H_2 + Cl_2 \qquad\qquad (28)$$

$$t\text{-}C_2H_2Cl_2^* + h\nu \rightarrow C_2HCl + HCl \qquad\qquad (29)$$

$$t\text{-}C_2H_2Cl_2^* + h\nu \rightarrow C_2H_2Cl + Cl \qquad\qquad (30)$$

where the rate constants were in the proportion $8 : 3.8 : 1$ in the above order. In addition from the higher excited state they obtained:

$$t\text{-}C_2H_2Cl_2^{**} + h\nu \rightarrow C_2H_2Cl + Cl \qquad\qquad (31)$$

The *cis* isomer behaved similarly.

It is then rather tempting to argue that it is the higher excited state which yields to additional Cl atoms is the V valence-shell state, while the lower excited state which eliminates mainly H_2 and HC1 is the 3s Rydberg state. Berry and Pimentel[144] observed laser emission from HF and HC1 following UV photolysis of fluoro- and chloroethylenes. It is quite possible that HF and HC1 are produced mostly in the Rydberg state.

Butadiene

There is hardly a theoretical chemist who at his beginnings did not make a calcula- tion on 1,3-butadiene. Yet, some problems remain.

A simple MO treatment yields four π MOs for trans-butadiene (C_{2h}) which are, in order of increasing energy; a_u, b_g, a_u and b_g so that the originating orbital for all known electronic transitions is b_g. The IP (adiabatic)[145] is 73240 cm^{-1} (9.08 eV; lower than for benzene (!)). The first (singlet) excited state is ^1B_u. The corre- sponding transition is readily assigned to the intense band system centered at 47800 cm^{-1} (210 nm) with fairly well resolved vibrational structure. It has been known for a long time ($V_1 \leftarrow N$). Three more $V \leftarrow N$ transitions can be derived from this simple scheme, two A_g (V_2 and V_3) and another B_u (V_4). Their identification turns out to be a difficult problem. V_2 and V_3 are forbidden for trans-butadiene due to its center of symmetry. They might still appear, of course, with a weak intensity. The importance of these bands is connected with the possibility that V_2 and V_3 (the two A_g "states" obtained in first approximation) undergo configurational mixing and that one of the resulting states is pushed down to energies similar or even lower than V_1 with the photochemical consequences that this might entail. The (+) and (−) combinations of the two A_g wave functions yield states $^1A_g^+$ and $^1A_g^-$. The $^1A_g^-$ state is expected to be at lower energies[146, 147].

Considerable experimental and theoretical efforts have been deployed towards the understanding of the butadiene spectrum and, in particular, to locate the lower ^1A_g band. The most thorough investigation of the UV spectrum of *trans*-butadiene in the gas phase is due to McDiarmid[148]. In addition to the (N,V_1) band she identified four Rydberg series, two of the np type and two with low quantum defects which could be nd or nf. They all converge to the π IP. There is some indica- tion for a valence-shell type band near 58800 cm^{-1} but it is a weak, diffuse feature and could not be assigned with certainty. Starting at 69000 cm^{-1} there is a broad band (already observed by Bélanger (given in[52])) which probably receives its intensity from $V_4 \leftarrow N$ and a transition starting from the highest occupied σ orbital which is close to the lower occupied π MO. The most important point is that MCDiarmid[148] finds a weak, structured band on the high frequency shoulder of the $V_1 \leftarrow N$ band with origin at 50144 cm^{-1}. This *could* be the lower ^1A_g band. How- ever, we still have to locate the Rydberg band which is also g ↔ g and forbidden. It could well compete for the assignment of this band. Both bands *must* exist and the choice is, indeed, a delicate one.

Robin[9] who reviewed the problem cites the successful Pariser-Parr + configura- tion interaction calculations of Allinger and Miller[149] who obtained 56000 and 59000 cm^{-1} for $V_2 \leftarrow N$ and $V_3 \leftarrow N$ respectively. In cis-dienes the V_2 and V_3

bands are in the $60000-65000$ cm^{-1} region. Both McDiarmid and Robin lean towards the Rydberg assignment for the 50144 cm^{-1} band although the term value seems to be somewhat low for a 3s state. Buenker, Shih and Peyerimhoff[150] carried out an all-valence-electron configuration interaction treatment on the low-lying excited states of butadiene including a very great number (150000) of selected singly and doubly excited configurations. They also assign McDiarmid's band to the $^1(b_g,3s)$ transition and locate the lower ^1A_g state at 56950 cm^{-1} (7.06 eV).

The importance of this problem is not linked to butadiene, however, It is rather that if butadiene has such lowlying bands all polyenes should have some. Now, in visual pigments, in rhodopsin, the molecule that absorbs light contains as chromophore 11–*cis* retinal, a Schiff-base containing six conjugated double bonds, five C=C and one terminal C=N which is bound to the protein opsin[151]. (For recent reviews see[152]). If there were one (or more) low-lying forbidden (N,V) bands these and not the V_1 state would be the state in which the primary process of vision might occur. Therefore following the suggestion of Schulten and Karplus[147] several attempts have been made to ascertain the existence of such bands in the spectra of various polyenes experimentally. These included two-photon excitation studies, a method which is well suited for the search for g ↔ g transitions. (Hudson and Kohler[153], Swofford and McClain[154], Holton and McClain[155]). Twaroski and Kliger[156] examined the situation for 1,3,5-hexatriene and after assessing previous works on this problem had to conclude that the prediction that forbidden ^1A type states lie below the allowed V_1 state can neither be confirmed nor refuted. Quite recently Vaida, Turner, Casey and Colson[157] applied the new techniques of multiphoton ionization and thermal lensing spectroscopy in the gas phase. They did find the band and resolved a good amount of fine-structure. However, at high pressure the band broadened out considerably and in virtue of Robin's criterion[158, 85] they were led. to adopt the Rydberg 3s assignment.

We should like to comment that even if the order of states was definitely known for butadiene or hexatriene this could not be extrapolated to the corresponding Schiff bases or, even less, to the Schiff base of retinal. In order to do this we have to get rid of the 3s Rydberg bands and, in Schiff bases, of the (n, π*) transitions. In hexatriene the IP is at 66800 cm^{-1} (8.28 eV)[160] so that the 3s Rydberg band is expected to be around $43000-44000$ cm^{-1} just where the strong (N,V$_1$) band is. Conditions are expected to be more favorable for octatetraene and longer polyene chains[161]. The IP of octatetraene is not known to us but it is quite certainly higher than 7.5 eV so that the lowest Rydberg band will fall at energies higher than the (N,V$_1$) band which is at about 33000 cm^{-1} [162]. Similarly we need four conjugated double bonds in a Schiff base to ensure that the (π,π*) bands have overtaken the (n,π*) band. So what is needed is double photon work on a molecule containing at least C=C–C=C–C=C–C=N or, well, the Schiff base of retinal itself. The work by Birge, Bennett, Pierce and Thomas[159] on all-trans-retinol comes closest to this. They found a two photon excitation maximum approximately 1600 cm^{-1} to the red of the one-photon absorption maximum.

Benzene has been used as an example in the introductory section on Rydberg and valence-shell states. We renounce treating it here. After Herzberg[7] the Reader should consult the beautiful review written by Robin[9], not forgetting about his

Addendum. Benzene taught us about vibronically allowed, electronically forbidden transitions; about the role of symmetry in spectroscopy; about superexcited states; σ orbitals coming before some π orbitals; about donor-acceptor complexes; . . . Quantum chemistry and molecular spectroscopy would be pover without this molecule. But let us let Kekulé dream his eternal dream . . .

VIII C=C, C=N and C=O

Most organic molecules contain bonding σ, π and non-bonding lone pair electrons. From this immensity we shall pick out the imines, and ketones and aldehydes and compare them to simple olefins (The lowest n,π and σ IPs for the simplest representative molecules are shown in Table 6[163]). Formaldimine is not a stable molecule but Peel and Willett[164] extracted its spectrum from the spectra of methylamine pyrolysis products recorded at different temperatures. Comparison is rendered somewhat difficult by the fact that some of the data represent adiabatic, others vertical IPs but it is still instructive (advert $\equiv$ adiabatic and vertical). In ethylene the separation between the π and the highest σ level is almost 2 eV. In formaldimine the first IP was assigned to the nitrogen lone pair[164], it has nearly the same value as the π IP in ethylene. The π level of formaldimine went up to 12.43 eV while the highest σ level comes in at 15.1 eV. In formaldoxime the n and π levels seem to have changed place (165) with π moved to 10.62 and n to 11.13 eV. The second π level was assigned to the band at 14.3 eV. Formaldehyde has its oxygen lone pair band at 10.88 eV, and, most important, the π level underwent a spectacular shift to higher energies: 14.09 eV. (The data for ethylene and formaldehyde are from Turner[17]). In support of these assignments there are the ab initio calculations on ethylene[133] and on formaldehyde[25, 166] by Buenker and Peyerimhoff; 4–31G ab initio calculations on formaldimine were published by Macaulay, Burnelle and Sandorfy[167] and on formaldoxime by Liotard, Dargelos and Chaillet[168]. A large scale calculation on C=N is yet to be performed.

Table 6.

$\begin{matrix} H \\ \\ H \end{matrix} C=C \begin{matrix} H \\ \\ H \end{matrix}$	$\begin{matrix} H \\ \\ H \end{matrix} C=N-H$	$\begin{matrix} H \\ \\ H \end{matrix} C=N-OH$	$\begin{matrix} H \\ \\ H \end{matrix} C=O$
10.51 (π) advert.	10.52 (n)	10.62 (π) advert.	10.88 (n) advert.
12.38 (σ) ad.	12.43 (π) vert.	11.13 (n) vert.	14.09 (π) ad.
	15.1 (σ)	14.3 (π) vert.	15.85 (σ) ad.
		14.9 (σ)	

The lowest n, π and σ ionization potentials of ethylene, formaldimine, formaldoxime and formaldehyde.

Alkyl substitution has a large effect on all these spectra. Robin's Tables IV.A.III and IV.C.I[9] illustrate this for C=C and C=O respectively.

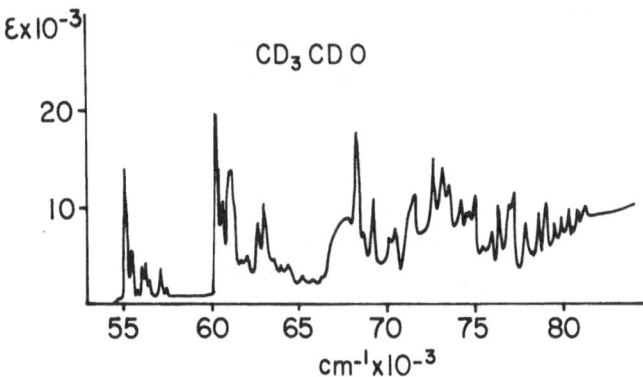

Fig. 6. The far UV absorption spectrum of acetaldehyde.
[Reproduced from Lucazeau, G., Sandorfy, C.: J. Mol. Spectrosc. *35*, 214 (1970), with permission from Academic Press Inc.]

As discussed above the UV spectrum of ethylene (and other monoolefins) is a π electron spectrum where the bands of lowest frequency are the (N,V_1) (π,π^*) band and the $(\pi,3s)$ Rydberg bands. Formaldehyde (and aliphatic aldehydes and ketones) have their well known (n,π^*) band in the 33000–36000 cm^{-1} region. After that, towards higher energies (Fig. 6) we find Rydberg bands, which originate with the lone pair, *not* with the π orbital. For formaldehyde the ns, np and nd series start at 57300, 64300 and 71600 cm^{-1} respectively, they all converge to the $\bar{O}$ IP at 87750 cm^{-1} (10.88 eV). The most intriguing thing about formaldehyde is the difficulty to find the (π,π^*) band. Whitten and Hackmeyer[169] and Buenker and Peyerimhoff[166] made extensive calculations on this problem. Both groups place the (π,π^*) band at very high frequencies (about 90000 and 92000 cm^{-1}) making it actually a superexcited band. This is in line with the high energy of the second (the π) PE band at 113600 cm^{-1} (14.09 eV)[170]. An important recent review on the spectrum of formaldehyde has been published by Moule and Walsh[171] and the UV spectrum has been reexamined by Lessard and Moule[172].

In acetaldehyde (and other aldehydes and ketones) the n–π gap is reduced. The $\bar{O}$ band is at 82500 cm^{-1} and the π band at 106100 cm^{-1}. Walsh[173] and Lucazeau and Sandorfy[174] studied the spectrum thoroughly. The (n,π^*) band is easily located at 34000 cm^{-1} followed 20000 cm^{-1} higher by the 3s (55050 cm^{-1}), 3p (60000 and 63500 cm^{-1}) and 3d (68000 cm^{-1}) bands. The (π,π^*) band is again elusive. We assigned it to a broader band underlying the 3d bands but Robin[9] puts it as high as 73000 cm^{-1}. There seems to be an $\sigma^* \leftarrow \bar{O}$ valence-shell band in coincidence with the 3s bands around 55050 cm^{-1}. As has been pointed out in Sect. IV the large gap between the (n,π^*) and (π,π^*) bands filled by Rydberg bands must be photochemically important.

Acetone received considerable attention in the last five years. Careful experimental studies were carried out by energy loss spectroscopy[175], low energy electron impact[176, 177] and optical absorption spectroscopy[178] and a large scale ab initio calculation has been performed[179]. The well known $^1(n,\pi^*)$ band has its maximum at 35650 cm^{-1}, the corresponding $^3(n,\pi^*)$ is at 33470 cm^{-1}. The maximum of the

C. Sandorfy

$^3(\pi,\pi^*)$ triplet is around 46200 cm^{-1}. Towards higher energies three peaks are
found at 51200, 52350 and 53500 cm^{-1} which belong to the fine structured ^{1}(n,3s)
band. The ^{1}(n,4s) member of the series has been located at 65200 cm^{-1}. The
^{1}(n,3s) region is centered at 59700 cm^{-1} and a band at 63200 cm^{-1} has been
assigned to the lowest member of the ^{1}(n,3d) manifold. The elusive $^1(\pi,\pi^*)$ band is
broad, dissociative, its maximum is very probably somewhat higher than
67000 cm^{-1}, and it contributes to the apparent intensity of the 3p and 3d bands.
The (π,π^*) states have actually mixed Rydberg-valence-shell character. On the whole
spectra of acetaldehyde and acetone are quite similar.

C=N molecules received by far less attention than C=C or C=O molecules. The
reason for this is probably their instability. Schiff bases are no less important than
olefins or ketones. Spectra of C=N compounds are expected to be intermediary between those of C=C and C=O compounds but they are also expected to be more
complicated since the $\bar{N}$ and π levels are close to one other (See Table 6 for the IPs)
(Fig. 7). All electronic transitions in the 30000–75000 cm^{-1} region must obviously
originate with these two orbitals. The (n,π^*) bands are found between 41000 and
43000 cm^{-1} for various non-conjugated imines (230–245 nm) with ϵ_{max} values of
the order of 100 or 200 (Fig. 8). The much more intense (π,π^*) band is also readily
located in the 58000–60000 cm^{-1} region with ϵ_{max} values above 10000. The
(n,π^*) band is at frequencies higher than for ketones while the $\bar{O}$ and $\bar{N}$ IPs are in the
opposite order. The (π,π^*) bands are at frequencies higher than for olefins and lower
than for ketones or aldehydes, in the same order as the π IPs.

In contrast to the spectra of simple olefin and carbonyl spectra the imine spectra
are diffuse. There are probably several reasons for this, among others the possibility
of syn-anti isomerism and the closeness of the bands originating with the n and π
orbitals which might offer a variety of paths for predissociation. Valence-shell transi

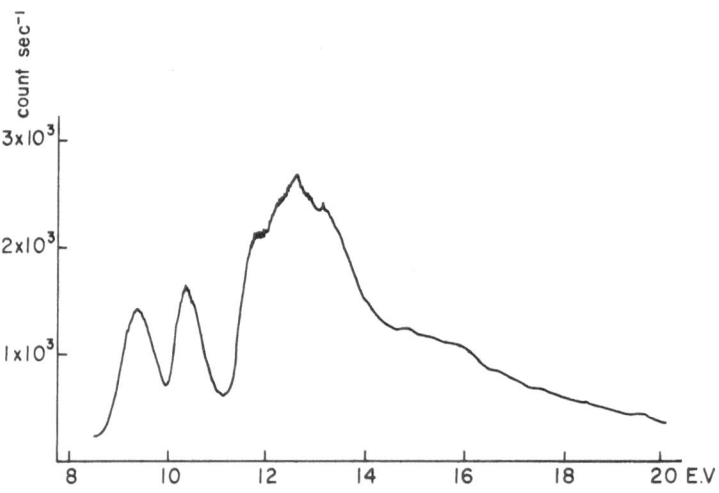

Fig. 7. The HeI photoelectron spectrum of C$_2$H$_5$–CH=N–C$_2$H$_5$.
[Reproduced from Vocelle, D., Dargelos, A., Pottier, R., Sandorfy, C.: J. Chem. Phys. 66, 2860
(1977), with permission from the American Institute of Physics]

132

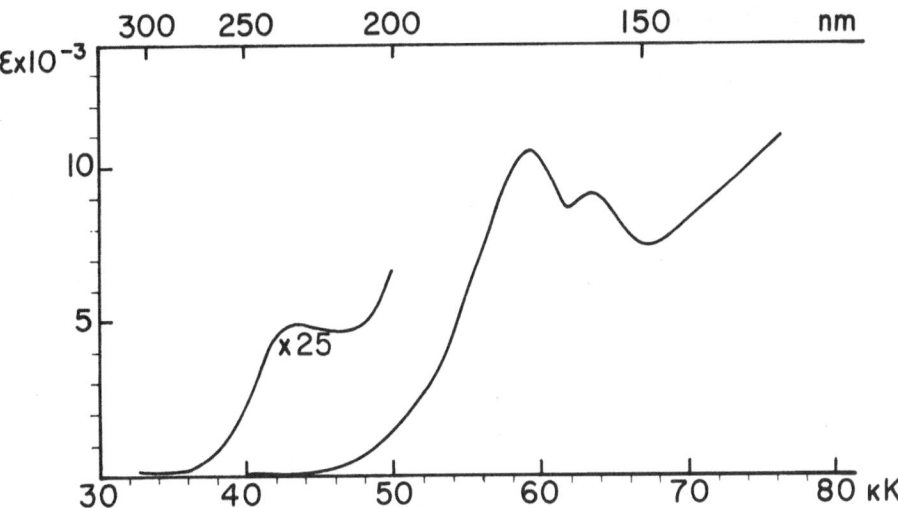

Fig. 8. The far UV absorption spectrum of $C_2H_5-CH=N-C_2H_5$.
[Reproduced from Vocelle, D., Dargelos, A., Pottier, R., Sandorfy, C.: J. Chem. Phys. **66**, 2860 (1977), with permission from the American Institute of Physics]

tions of the $\sigma^* \leftarrow n$ type might also occur. The diffuseness of the spectra makes the Rydberg bands difficult to locate. The (n, 3s) band should be between 48000 and 50000 cm^{-1}, the (n, 3p) and (π, 3s) between 55000 and 58000 cm^{-1}, the (π, 3p) and (n, 3d) around 63000–65000 cm^{-1}. In all simple imines there is a broad band near 64000 cm^{-1} which can be assigned to the (π, 3p) and (n, 3d) transitions. The other Rydberg bands are ill defined shoulders. The (n, 3s) band is likely to be responsible for a part of the broad and highly asymmetrical low wave-number shoulder of the (π, π^*) band.

Comparing C=C, C=N and C=O we can summarize the facts as follows. The lone pair IP is stabilized by a few tenths of an eV if nitrogen is replaced by oxygen. The π IP is about the same for C=C and C=N (depending on the alkyl substituents) and is then stabilized by 3 or 4 eV for C=O. The transition energies of the (n,π^*) transitions in C=N and C=O are in the reverse order of their lone pair IPs showing the importance of electronic rearrangement in (n,π^*) transitions. The location of the (π,π^*) band is not widely different for C=C and C=N. The jump from C=N to C=O is about 10000 cm^{-1} while the π IPs differ by as much as about 30000 cm^{-1}. The lowest Rydberg bands, of course, follow the IPs. They are usually on the shoulders of the (π,π^*) band for C=C and C=N but they are well separated and lie between the (n,π^*) and (π,π^*) bands for the carbonyl compounds.

The photolysis of aldehydes and ketones in the (n, π^*) band has been widely investigated. (For summaries see[32] and[41]). The dominant primary steps involve cleavage of the C—C bonds next to the carbonyl:

$$CH_3CHO + h\nu \rightarrow CH_3 + HCO \tag{32}$$

$$CH_3CHO + h\nu \rightarrow CH_4 + CO \tag{33}$$

133

C. Sandorfy

for the acetaldehyde and

$$CH_3COCH_3 + h\nu \rightarrow CH_3CO + CH_3 \tag{34}$$

$$CH_3COCH_3 + h\nu \rightarrow 2CH_3 + CO \tag{35}$$

for acetone. They might go through the triplet state. The carbonyl bond never breaks, it is a C–C bond adjacent to it that does. This is probably linked to the flow of electronic charge towards the oxygen following excitation of an electron from the area of this strongly electronegative atom. Assuming this the observed bond cleavages are in keeping with the valence-shell character of the (n,π^*) excited state.

What can one expect from a photolysis at higher frequencies? The next band in the UV spectrum is $(n,3s)$ *not* the (π,π^*) band. Thus at frequencies higher than 50000 cm^{-1}, primary steps involving molecular elimination are expected to become more important. This is born out by the results of Lin and Ausloos[180] which led to the following additional primary processes:

$$CH_3COCH_3 + h\nu \rightarrow H_2 + C_3H_4O \tag{36}$$

$$CH_3COCH_3 + h\nu \rightarrow CH_4 + CH_2CO \tag{37}$$

$$CH_3COCH_3 + h\nu \rightarrow C_3H_4 + H_2O \tag{38}$$

The photolysis of simple imines does not seem to have been studied extensively.

References

1. Price, W. C.: Advan. Spectrosc. *1*, 56 (1959)
2. Price, W. C. in: Chemical spectroscopy and photochemistry in the vacuum ultraviolet, pp. 1. Sandorfy, C., Ausloos, P., Robin, M. B. (eds.). Dordrecht, Holland: Reidel 1974
3. Mulliken, R. S.: J. Amer. Chem. Soc. *86*, 3183 (1964); *88*, 1849 (1966); *91*, 4615 (1969)
4. Mulliken, R. S.: Acc. Chem. Res. *9*, 7 (1976)
5. Mulliken, R. S.: J. Chem. Phys. *66*, 2448 (1977)
6. Peyerimhoff, S. D.: Gazz. Chim. Ital. *108*, 411 (1978)
7. Herzberg, G.: Molecular spectra and molecular structure, Vol. III. Electronic spectra and electronic structure of polyatomic molecules. Princeton: Van Nostrand 1966
8. Duncan, A. B. F.: Rydberg series in atoms and molecules. New York: Academic Press 1971
9. Robin, M. B.: Higher Excited States of polyatomic molecules, Vols. I and II. New York: Academic Press 1974–1975
10. Sinanoglu, O., in: Chemical spectroscopy and photochemistry in the vacuum ultraviolet, pp. 337. Sandorfy, C., Ausloos, P., Robin, M. B. (eds.). Dordrecht, Holland: Reidel 1974
11. Allen, S., Schnepp, O.: J. Chem. Phys. *59*, 4547 (1973)
12. Dunn, T. M., in: Studies on chemical structure and reactivity. Ridd, J. H. (ed.). London: Methuen 1966
13. Koch, E. E., Otto, A.: Chem. Phys. Lett. *12*, 476 (1972)
14. El-Sayed, M. F. A., Kasha, M., Tanaka, Y.: J. Chem. Phys. *34*, 334 (1961)
15. Asbrink, L., Edquist, O., Lindholm, E., Selin, L. E.: Chem. Phys. Lett. *5*, 192 (1970)
16. Al-Joboury, M. I., Turner, D. W.: J. Chem. Soc., 5141 (1963)

17. Turner, D. W., Baker, C., Baker, A. D., Brundle, C. R.: Molecular photoelectron spectroscopy. New York: Wiley-Interscience 1970
18. Koopmans, T.: Physica *1*, 104 (1934)
19. Eland, J. H. D.: Photoelectron spectroscopy. London: Butterworths 1974
20. Baker, A. D., Betteridge, D.: Photoelectron spectroscopy. London: Pergamon Press 1972
21. Rabalais, J. W.: Principles of ultraviolet photoelectron spectroscopy. New York: Wiley-Interscience 1977
22. Brundle, C. R., Baker, A. D.: Electron spectroscopy: Theory, techniques and applications. New York: Academic Press 1977
23. Carlson, T. A.: Photoelectron and Auger Spectroscopy. New York: Plenum Press 1975
24. Peyerimhoff, S. D., Buenker, R. J.: Advances in quantum chemistry *9*, 69 (1975)
25. Whitten, J. L.: J. Chem. Phys. *44*, 359 (1966)
26. Buenker, R. J., Peyerimhoff, S. D.: Theoret. Chim. Acta *39*, 217 (1975)
27. Buenker, R. J., Peyerimhoff, S. D.: Chem. Revs. *74*, 127 (1974)
28. Mulliken, R. S.: Rev. Mod. Phys. *14*, 204 (1942)
29. Walsh, A. D.: J. Chem. Soc., 2260 (1953) and subsequent papers
30. McNesby, J. R., Okabe, H.: Advances in Photochemistry *3*, 157 (1964)
31. Ausloos, P. J., Lias, S. G.: Ann. Rev. Phys. Chem. *22*, 185 (1971)
32. Ausloos, P. J., Lias, S. G., in: Chemical spectroscopy and photochemistry in the vacuum ultraviolet. Sandorfy, C., Ausloos, P. J., Robin, M. B. (eds.). Dordrecht, Holland: Reidel 1974
33. Ausloos, P. J.: Mol. Photochem. *4*, 39 (1972)
34. Gunning, H. E., Strausz, O. P.: Advances in photochemistry *1*, 209 (1963)
35. Woodward, R. B., Hoffman, R.: The conservation of orbital symmetry. Weinheim: Verlag Chemie 1970
36. Dauben, W. G., Salem, L., Turro, N. J.: Acc. Chem. Res. *8*, 41 (1975)
37. Salem, L., Leforestier, C., Segal, G., Wetmore, R.: J. Amer. Chem. Soc. *97*, 479 (1975)
38. Salem, L.: J. Amer. Chem. Soc. *96*, 3486 (1974)
39. Michl, J.: J. Mol. Photochem. *4*, 243 (1972)
40. Bader, R. F. W., Gangi, R. A.: Chapter 1 in Theoretical Chemistry, Vol. 2. A Specialist Periodical Report. The Chemical Society of London, 1975
41. Barltrop, J. A., Coyle, J. D.: Excited states in organic chemistry. New York: Wiley-Interscience 1975
42. Heicklen, J.: Atmospheric chemistry. New York: Academic Press 1976
43. Koch, E. E., Otto, A.: Int. J. Radiat. Phys. Chem. *8*, 113 (1976)
44. Basch, H., Robin, M. B., Kuebler, N. A., Baker, C., Turner, D. W.: J. Chem. Phys. *51*, 52 (1969)
45. Sauvageau, P., Gilbert, R., Berlow, P. P., Sandorfy, C.: J. Chem. Phys. *59*, 762 (1973)
46. Harshbarger, W. R., Robin, M. B., Lassettre, E. N.: J. Electron. Spectrosc. *1*, 319 (1973)
47. Bélanger, G., Sauvageau, P., Sandorfy, C.: Chem. Phys. Lett. *3*, 649 (1969)
48. Salahub, D. R., Sandorfy, C.: Chem. Phys. Lett. *8*, 71 (1971)
49. Russell, B. R., Edwards, L. O., Raymonda, J. W.: J. Amer. Chem. Soc. *95*, 2129 (1973)
50. Zobel, C. R., Duncan, A. B. F.: J. Amer. Chem. Soc. *77*, 2611 (1955)
51. Doucet, J., Sauvageau, P., Sandorfy, C.: J. Chem. Phys. *62*, 355 (1975)
52. Sandorfy, C.: J. Mol. Structure *19*, 183 (1973)
53. Sandorfy, C.: Atm. Env. *10*, 343 (1976)
54. Robin, M. B., Hart, R. R., Kuebler, N. A.: J. Chem. Phys. *44*, 1803 (1966)
55. Bélanger, G., Sandorfy, C.: J. Chem. Phys. *55*, 2055 (1971)
56. McDiarmid, R.: J. Chem. Phys. *64*, 514 (1976)
57. Ito, H., Nogata, Y., Matsuzaki, S., Kuboyama, A.: Bull. Chem. Soc. Japan *42*, 2453 (1969)
58. Lucazeau, G., Sandorfy, C.: J. Mol. Spectrosc. *35*, 214 (1970)
59. Schwarz, W. H. E.: Chem. Phys. *9*, 157; *11*, 217 (1975)
60. Sandorfy, C., in: Applications of MO theory in organic chemistry. Csizmadia, I. G. (ed.). Holland: Elsevier 1977
61. Sandorfy, C.: Z. Phys. Chem. *101*, 307 (1976)

62. Mulliken, R. S.: J. Chem. Phys. *3*, 517 (1935)
63. Rabalais, J. W., Bergmark, T., Werme, L. O., Karlsson, L., Siegbahn, K.: Phys. Scripta *3*, 13 (1971)
64. Duncan, A. B. F., Howe, J. P.: J. Chem. Phys. *2*, 851 (1934)
65. Ditchburn, R. W.: Proc. Roy. Soc. *A229*, 44 (1955)
66. Sun, H., Weissler, G. L.: J. Chem. Phys. *23*, 1372 (1955)
67. Arents, J., Allen, L. C.: J. Chem. Phys. *53*, 73 (1970)
68. Calvert, J. G., Pitts Jr., J. N.: Photochemistry. New York: Wiley-Interscience 1966
69. Lombos, B. A., Sauvageau, P., Sandorfy, C.: Chem. Phys. Lett. *1*, 42 (1967)
70. Raymonda, J. W., Simpson, W. T.: J. Chem. Phys. *47*, 430 (1967)
71. Lombos, B. A., Sauvageau, P., Sandorfy, C.: J. Mol. Spectrosc. *24*, 253 (1967)
72. Lassettre, E. N., Skerbele, A., Dillon, M. A.: J. Chem. Phys. *49*, 2382 (1968)
73. Sandorfy, C., in: Chemical spectroscopy and photochemistry in the vacuum ultraviolet. Sandorfy, C., Ausloos, P. J., Robin, M. B. (eds.). Dordrecht, Holland, Reidel 1974
74. Pearson, E. F., Innes, K. K.: J. Mol. Spectrosc. *30*, 232 (1969)
75. Buenker, R. J., Peyerimhoff, S. D.: Chem. Phys. *8*, 56 (1975)
76. Custer, E. M., Simpson, W. T.: J. Chem. Phys. *60*, 2012 (1974)
77. Richartz, A., Buenker, R. J., Bruna, P. J., Peyerimhoff, S. D.: Mol. Phys. *33*, 1345 (1977)
78. Richartz, A., Buenker, R. J., Peyerimhoff, S. D.: Chem. Phys. *28*, 305 (1978)
79. Lias, S. G., Collin, G. J., Rebbert, R. E., Ausloos, P. J.: J. Chem. Phys. *52*, 1841 (1970)
80. Klevens, H. B., Platt, J. R.: J. Chem. Phys. *17*, 470 (1949)
81. Richartz, A., Buenker, R. J., Peyerimhoff, S. D.: Chem. Phys. *31*, 187 (1978)
82. Scala, A. A., Ausloos, P. J.: J. Chem. Phys. *49*, 2282 (1968)
83. Rebbert, R. E., Lias, S. G., Ausloos, P. J.: J. Photochem. *4*, 121 (1975)
84. Basch, H., Robin, M. B., Kuebler, N. A., Baker, C., Turner, D. W.: J. Chem. Phys. *51*, 52 (1969)
85. Robin, M. B., Kuebler, N. A.: J. Mol. Spectrosc. *33*, 274 (1970)
86. Sauvageau, P., Doucet, J., Gilbert, R., Sandorfy, C.: J. Chem. Phys. *61*, 391 (1974)
87. Gilman, H., Atwell, W. H., Schwebke, G. L.: J. Organomet. Chem. *2*, 369 (1964)
88. Kumada, M., Tamao, K., in: Advances in Organometallic chemistry, Vol. 6. Stone, F. G. A., West, R. (eds.). New York: Academic Press 1968
89. Chan, S. C., Inel, Y., Tschuikow-Roux, E.: Can. J. Chem. *50*, 1443 (1972)
90. Pullen, B. P., Carlson, T. A., Moddeman, W. E., Schweitzer, G. K., Bull, W. E., Grimm, F. A.: J. Chem. Phys. *53*, 768 (1970)
91. Cradock, S.: J. Chem. Phys. *55*, 980 (1971)
92. Potts, A. W., Price, W. C.: Proc. Roy. Soc. London *A326*, 165 (1972)
93. Alexander, A. G., Strausz, O. P., Pottier, R., Semeluk, G. P.: Chem. Phys. Lett. *13*, 608 (1972)
94. Harada, Y., Murrell, J. N., Sheena, H. H.: Chem. Phys. Lett. *1*, 595 (1968)
95. Roberge, R., Sandorfy, C., Matthews, J. I., Strausz, O. P.: J. Chem. Phys. *69*, 5105 (1978)
96. Price, W. C., in: Molecular spectroscopy, Vol. 4. Hepple, P. W. (ed.). London: The Institute of Petroleum 1969
97. Strausz, O. P.: private communication
98. Rebbert, R. E., Ausloos, P. J.: J. Photochem. *4*, 419 (1975)
99. Ichimura, T., Kirk, A. W., Kramer, G., Tschuikow-Roux, E.: J. Photochem. *6*, 771 (1976/1977)
100. Rowland, F. S., Molina, M. J.: Rev. Geophys. Space Phys. *13*, 1 (1975)
101. Heicklen, J.: J. Photochem. *4*, 381 (1975)
102. Cradock, S., Whiteford, R. A.: Trans. Faraday. Soc. *67*, 3425 (1971)
103. Frost, D. C., Herring, F. G., Katrib, A., McLean, R. A. N., Drake, J. E., Westwood, N. P. C.: Chem. Phys. Lett. *10*, 347 (1971); Can. J. Chem. *49*, 4033 (1971)
104. Howell, J. M., Van Wazer, J. R.: J. Amer. Chem. Soc. *96*, 3064 (1974)
105. Bell, S., Walsh, A. D.: Trans Faraday Soc. *62*, 3005 (1966)
106. Roberge, R.: Ph. D. Thesis, Université de Montréal, 1978
107. Causley, G. C., Russell, B. R.: J. Electron Spectrosc. *8*, 71 (1976)
108. Causley, G. C., Russell, B. R.: J. Electron. Spectrosc. *11*, 383 (1977)

109. Claydon, C. R., Segal, G. A., Taylor, H. S.: J. Chem. Phys. *54*, 3799 (1971)
110. Hudson, R. D.: Rev. Geophys. Space Phys. *9*, 305 (1971)
111. Von Sonntag, C., Schuchmann, H. P., in: Advances in Photochemistry, Vol. 10. New York: Wiley-Interscience 1977
112. Hernandez, G. J.: J. Chem. Phys. *38*, 1644 (1963)
113. Harrison, A. J., Price, D. R. W.: J. Chem. Phys. *30*, 357 (1959)
114. Von Sonntag, C., Schuchmann, H. P., Schomburg, G.: Tetrahedron *28*, 4333 (1972)
115. Hernandez, G. J.: J. Chem. Phys. *38*, 2233 (1963)
116. Price, W. C.: J. Chem. Phys. *4*, 147 (1936)
117. Price, W. C., Teegan, J. P., Walsh, A. D.: Proc. Roy. Soc. London *A201*, 600 (1950)
118. Shih, S. K., Peyerimhoff, S. D., Buenker, R. J.: Chem. Phys. *17*, 391 (1976)
119. Guest, M. F., Rodwell, W. R.: Mol. Phys. *32*, 1075 (1976)
120. Barrett, J., Hitch, M. J.: Spectrochim. Acta *24A*, 265 (1968)
121. Clark, L. B., Simpson, W. T.: J. Chem. Phys. *43*, 3666 (1965)
122. Olivato, P. R., Viertler, H., Wladislaw, B., Sauvageau, P., Sandorfy, C.: J. Chem. Phys. *70*, 1677 (1979)
123. Bridges, L., Hemphill, G. L., White, J. M.: J. Phys. Chem. *76*, 2668 (1972)
124. Callear, A. B., Dickson, D. R.: Trans. Faraday Soc. *66*, 1987 (1970)
125. Olivato, P. R., Viertler, H., Wladislaw, B., Cole, K. C., Sandorfy, C.: Can. J. Chem. *54*, 3026 (1976)
126. Merer, A. J., Mulliken, R. S.: Chem. Rev. *69*, 639 (1969)
127. Merer, A. J., Schoonveld, L.: J. Chem. Phys. *48*, 522 (1968)
128. Merer, A. J., Schoonveld, L.: Can. J. Phys. *47*, 1731 (1969)
129. McDiarmid, R., Charney, E.: J. Chem. Phys. *47*, 1517 (1967)
130. McDiarmid, R.: J. Chem. Phys. *50*, 1794 (1969)
131. McDiarmid, R.: J. Chem. Phys. *55*, 4669 (1971)
132. Buenker, R. J., Peyerimhoff, S. D.: Chem. Phys. Lett. *36*, 415 (1975)
133. Buenker, R. J., Peyerimhoff, S. D.: Chem. Phys. *9*, 75 (1975)
134. Wilkinson, P. G.: Can. J. Phys. *34*, 596, 643 (1956)
135. Wilkinson, P. G., Mulliken, R. S.: J. Chem. Phys. *23*, 1895 (1955)
136. Borrell, P., Cervenka, A., Turner, J. W.: J. Chem. Soc. B, 2293 (1971)
137. Okabe, H., McNesby, J. R.: J. Chem. Phys. *36*, 601 (1962)
138. Gorden, R., Ausloos, P. J.: NBS J. Res. *75A*, 141 (1971)
139. Ausloos, P., Lias, G., in: Chemical spectroscopy and photochemistry in the vacuum ultraviolet. Sandorfy, C., Ausloos, P. J., Robin, M. B. (eds.). Dordrecht, Holland: Reidel 1974
140. Robin, M. B., Hart, R. R., Kuebler, N. A.: J. Chem. Phys. *44*, 2664 (1966)
141. Walsh, A. D., Warsop, P. A.: Trans. Faraday Soc. *64*, 1418, 1425 (1968)
142. Walsh, A. D., Warsop, P. A., Whiteside, J. A. B.: Trans. Faraday Soc. *64*, 1432 (1968)
143. Ausubel, R., Wijnen, M. H. J.: J. Photochem. *4*, 241 (1975)
144. Berry, M. J., Pimentel, G. C.: J. Chem. Phys. *51*, 2274 (1969)
145. Brundle, C. R., Robin, M. B.: J. Am. Chem. Soc. *92*, 5550 (1970)
146. Buenker, R. J., Whitten, J. L.: J. Chem. Phys. *49*, 5381 (1968)
147. Schulten, K., Karplus, M.: Chem. Phys. Lett. *14*, 305 (1972)
148. McDiarmid, R.: J. Chem. Phys. *64*, 514 (1976)
149. Allinger, N. L., Miller, M. A.: J. Amer. Chem. Soc. *86*, 2811 (1964)
150. Buenker, R. J., Shih, S. K., Peyerimhoff, S. D.: Chem. Phys. Lett. *44*, 385 (1976)
151. Wald, G.: Science *162*, 230 (1968)
152. Kliger, D. S., Menger, E. L.: Acc. Chem. Res. *8*, 81 (1975) and following papers
153. Hudson, B. S., Kohler, B. E.: Chem. Phys. Lett. *14*, 299 (1972)
154. Swofford, R. L., McClain, W. M.: J. Chem. Phys. *59*, 5740 (1973)
155. Holton, G. R., McClain, W.: Chem. Phys. Lett. *44*, 436 (1976)
156. Tworowski, A. J., Kliger, D. S.: Chem. Phys. Lett. *50*, 36 (1977)
157. Vaida, V., Turner, R. E., Casey, J. L., Colson, S. D.: Chem. Phys. Lett. *54*, 25 (1978)
158. Robin, M. B., Hart, R. R., Kuebler, N. A.: J. Chem. Phys. *44*, 1803 (1966)
159. Birge, R. R., Bennett, J. A., Pierce, B. M., Thomas, T. M.: J. Amer. Chem. Soc. *100*, 1533 (1978)

C. Sandorfy

160. Price, W. C., Walsh, A. D.: Proc. Roy. Soc. London *A185*, 182 (1945)
161. Favrot, J., Leclercq, J. M., Roberge, R., Sandorfy, C., Vocelle, D.: Photochem. Photobiol., *29*, 99 (1979)
162. Woods, G. F., Schwartzman, L. H.: J. Amer. Chem. Soc. *71*, 1396 (1949)
163. Vocelle, D., Dargelos, A., Pottier, R., Sandorfy, C.: J. Chem. Phys. *66*, 2860 (1977)
164. Peel, J. B., Willett, G. D.: J. Chem. Soc. Faraday Trans. II, *71*, (11), 1799 (1975)
165. Dargelos, A., Sandorfy, C.: J. Chem. Phys. *67*, 3011 (1977)
166. Peyerimhoff, S. D., Buenker, R. J., Kammer, W. E., Hsu, H. L.: Chem. Phys. Lett. *8*, 129 (1971)
167. Macaulay, R., Burnelle, L. A., Sandorfy, C.: Theoret. Chim. Acta *29*, 1 (1973)
168. Liotard, D., Dargelos, A., Chaillet, M.: Theoret. Chim. Acta *31*, 325 (1973)
169. Whitten, J. L., Hackmeyer, M.: J. Chem. Phys. *51*, 5584 (1969)
170. Baker, A. D., Baker, C., Brundle, C. R., Turner, D. W.: Int. J. Mass Spectr. Ion Phys. *1*, 285 (1968)
171. Moule, D. C., Walsh, A. D.: Chem. Rev. *75*, 67 (1975)
172. Lessard, C. R., Moule, D. C.: J. Mol. Spectrosc. *60*, 343 (1976)
173. Walsh, A. D.: Proc. Roy. Soc. London *A185*, 176 (1945)
174. Lucazeau, G., Sandorfy, C.: J. Mol. Spectrosc. *35*, 214 (1970)
175. Huebner, R. H., Celotta, R. J., Mielczarek, S. R., Kuyatt, C. E.: J. Chem. Phys. *59*, 5434 (1973)
176. St-John, W. M., Estler, R. C., Doering, J. P.: J. Chem. Phys. *61*, 763 (1974)
177. Van Veen, E. H., Van Dijk, W. L., Brongersma, H. H.: Chem. Phys. *16*, 337 (1976)
178. Scott, J. D., Russell, B. R.: J. Chem. Phys. *63*, 3243 (1975)
179. Hess, B., Bruna, P. J., Buenker, R. J., Peyerimhoff, S. D.: Chem. Phys. *18*, 267 (1976)
180. Lin, L. J. T., Ausloos, P. J.: J. Photochem. *1*, 453 (1973)

Received October 26, 1978

Some Aspects of the Photoelectron Spectroscopy of Organic Sulfur Compounds

Rolf Gleiter and Jens Spanget-Larsen

Institut für Organische Chemie der Universität, D-6900 Heidelberg, Germany

Dedicated to Professor Dr. H. Hartmann on the occasion of his 65th birthday

Table of Contents

The chemistry of sulfur comprises a multitude of classes of compounds which vary widely in structure and bonding[1]. The present survey is essentially confined to organic molecules with sulfur bonded to one or two carbon atoms, e.g., sulfides and thiocarbonyls. The presentation is based on illustrative examples, rather than exhaustive coverage of the literature; references and photoelectron data for more than hundred sulfur compounds, many of which are not treated explicitly in the text, are collected in the table at the end of this article.

The principles of photoelectron spectroscopy have been discussed extensively in recent publications. Consequently, we give only a very brief introduction to the subject, with emphasis on aspects we consider particularly relevant in relation to organic sulfur compounds. For detailed treatments we refer to the classical book by Turner and his co-workers[2] and to the recent volume edited by Brundle and Baker[3]. The latter contains numerous references to relevant books, reports and review articles.

1 Introduction

Interaction of a molecule M with electromagnetic radiation with energy quanta in the approximate range 1–8 eV may lead to excitation of its valence electronic system. The molecular absorption spectrum can usually be measured with a conventional UV/VIS spectrometer, yielding information on the transitions from the ground state χ(M) to the various excited electronic states $\widetilde{A}$(M*), $\widetilde{B}$(M*), etc. (Fig. 1). Increasing the energy of the radiation beyond a certain threshold normally leads to emission of electrons, thereby generating the radical cation M‡ in its ground state χ(M‡) or in one of its excited states $\widetilde{A}$(M‡), $\widetilde{B}$(M‡), etc. (Fig. 1).

The various ionic states are conveniently studied by means of photoelectron spectroscopy. In a photoelectron spectrometer molecules of the sample are irradiated with monochromatic electromagnetic radiation and the ejected electrons are sorted in an electric field according to their kinetic energy. The number of electrons counted for a certain kinetic energy is recorded in counts per second (cps, c/s).

The irradiating photon source can be an X ray source (~10^3 eV)[3]. This kind of radiation ejects mainly inner shell electrons, and solid state as well as gas phase investigations are possible. The most common acronyms for this technique are XPS or ESCA (Fig. 2). The photon source most frequently used in the study of the valence electronic region is the He(Iα) photon energy line in the far UV (21.2 eV), produced by a Helium discharge lamp[2, 3]. In the following we shall be concerned with gas phase photoelectron (PE) spectra obtained with this photon source, or that operating with the He(IIα) photon energy line (40.8 eV).

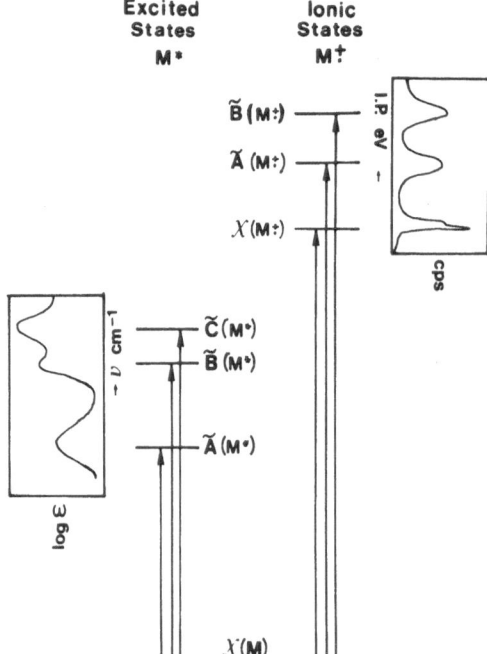

Fig. 1. Diagram illustrating the relation between electronic absorption spectroscopy (left) and photoelectron spectroscopy (right)

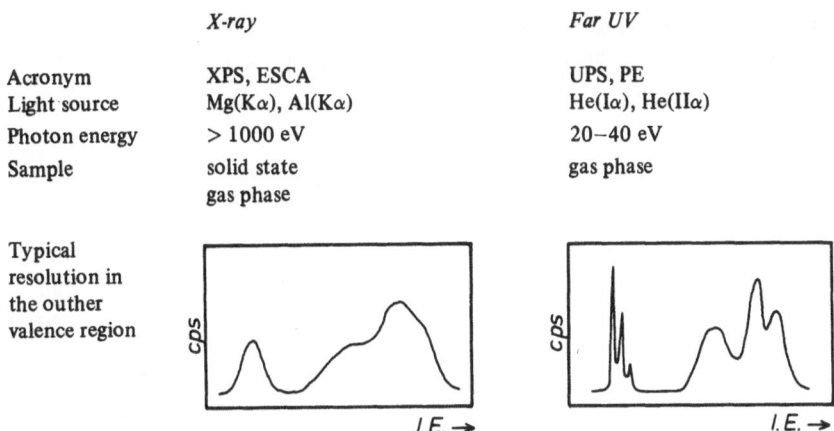

Fig. 2. Characteristic features of X-ray and far UV photoelectron spectroscopy

Using a monochromatic photon source, the ionization energies of M corresponding to the energy differences between the ground state and the different cation states of M (Fig. 1) can be obtained as

$$I = h\nu - E_{kin} \tag{1}$$

where $h\nu$ is the photon energy and E_{kin} is the kinetic energy of the emitted electron. Apart from its energy, a PE band is furthermore characterized by its intensity and its shape, features which are often helpful in the assignment of the individual ionization energies to predicted transitions.

Practically all theoretical predictions of molecular PE spectra are based on canonical molecular orbital (MO) theory, at least as a starting point. Consider the Slater determinant corresponding to the ground configuration of M; this is usually a good approximation to the wave function of the ground state $\chi(M)$. Removal of an electron from one of the occupied MOs generates the corresponding "Koopmans' configuration", which often leads to a satisfactory representation of one of the cation states, the so-called primary hole states (Fig. 3). Provided the MOs of *M* are not affected by the removal of an electron, the energy required to remove an electron

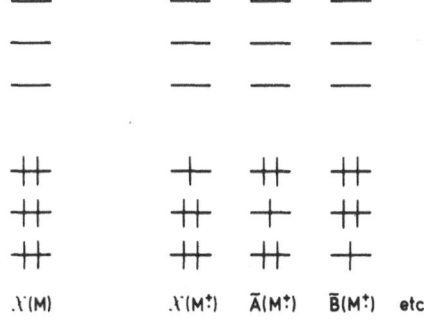

Fig. 3. Schematic representation of the ground configuration of M, corresponding to the ground state $\chi(M)$, and the "Koopmans' configurations" derived by formal removal of one electron from the ground configuration of M, corresponding to the primary hole states of $M^{\ddagger}$

141

from the J'th MO is equal to minus the MO energy, $-\epsilon_J$. As a first approximation, the J'th vertical ionization energy, $I_{v,J}$, can thus be estimated from the energy of the J'th MO:

$$I_{v,J} = -\epsilon_J \qquad (2)$$

This result is usually referred to as "Koopmans' theorem" [4].

The close relation between ionization energies and MO energies is largely responsible for the popularity of PE spectroscopy, which is probably the closest one can get to an experimental probe of the orbital structure of molecules. However, it must be emphasized that MOs are *not* observables; they merely serve as convenient building blocks in a description of molecular states, as indicated above. Furthermore, Koopmans' theorem (2) is a crude approximation which, e.g., neglects electron reorganization and correlation effects. Fortunately, reorganization and correlation effects tend to cancel and Koopmans' theorem is a fairly good approximation for many molecules [4]. As an example where Koopmans' theorem fails badly we shall consider the case of CS. This is the simplest compound with a carbon-sulfur bond, although it is not typical of the compounds we are going to discuss later on.

In Fig. 4 are indicated the calculated energies of the three highest occupied MOs of CS as well as the results of a treatment going beyond Koopmans' theorem [5, 6]. The highest occupied MO (HOMO) is the 2π orbital; reorganization and correlation effects are predicted to practically cancel. In the case of the second highest MO, 7σ, stabilization of the corresponding cation hole state due to electron reorganization (relaxation) is found to predominate, as it does in most cases, leading to the prediction of an ionization energy considerably lower than the one obtained from Eq. (2). As a result, Koopmans' theorem fails to predict the order of the two lowest states of CS^+.

An even more serious breakdown of Koopmans' theorem is found for the third MO, 6σ [6]. The corresponding Koopmans' configuration, $(6\sigma)^{-1}$, interacts strongly with the configuration $(7\sigma)^{-1}\,(2\pi)^{-1}\,(3\pi)$ which involves excitation of an electron

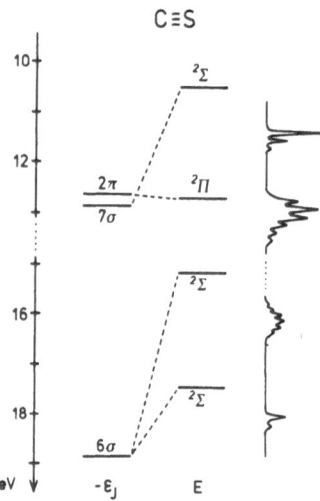

Fig. 4. Calculated MO energies ϵ_J and $M \rightarrow M^+ + e$ transition energies E for the diatomic species CS [5, 6]. The shape of the PE bands is indicated to the right [7−9]

into a virtual orbital. This leads to the contribution of $(6\sigma)^{-1}$ to two cation states with about equal weights and consequently to a breakdown of the orbital picture of ionization: there is no unique correspondence between PE peaks and MOs (Fig. 4). The most striking evidence for the invalidity of the one-particle approach is found for the innermost valence orbital, 5σ, not indicated in Fig. 4; the original line is smashed into numerous weak contributions extending from 23 eV to 33 eV.

With these results in mind, it is clear that general statements like "Chemists can see the orbital structure of even fairly large molecules and no longer have to rely on the predictions of theoreticians"[10] are somewhat over-optimistic. Fortunately, the orbital picture of ionization can be considered as a good approximation for the outermost valence levels, although Koopmans' theorem frequently fails to predict the correct ordering of the ionization energies (e.g., the azines[11, 12]). Limitation to the outer valence region is no severe restriction, since this is the region of immediate interest to the chemist. Anyway, the complexity of the PE spectra of organic compounds usually restricts the interpretation to the low energy region where individual bands may be detected (see below).

In the remainder of this chapter we briefly consider the application of band shape and intensity criteria in the assignment of PE bands to molecular states. The assignment of the spectra of small molecules is very often aided by a considerable fine structure as exemplified by the PE spectrum of CS in Fig. 4. Indeed, the vibrational structure of the PE spectra of small species like CS and H_2CS has been calculated[5, 13], yielding a rather striking match of the experimental spectra and thereby confirming the assignment based on the energy criterion. Also qualitative

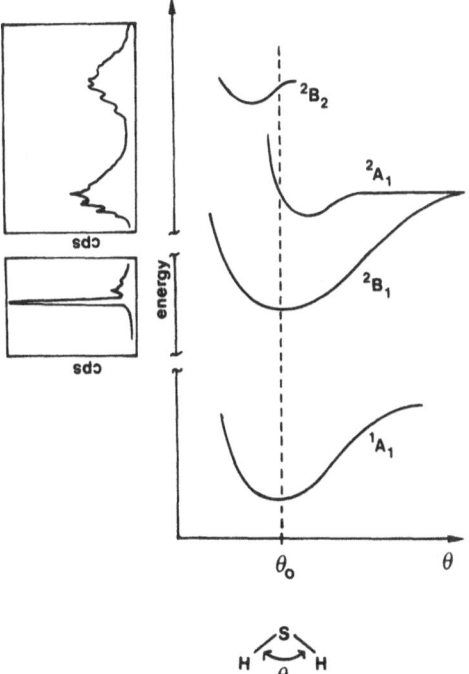

Fig. 5. Qualitative diagram indicating the approximate dependence of the potential energy on the bond angle θ for the ground state of H_2S ($^1A_1$) and the three lowest states of H_2S^+ ($^2B_1$, $^2A_1$, $^2B_2$)[15]. The shape of the PE bands is shown to the left[2]

143

band shape criteria in connection with the Franck-Condon principle can be used to test a proposed assignment. As an example we will consider the PE spectrum of H_2S[2, 14)] (Fig. 5). According to a calculation[15)] of the three lowest electronic states of $\underline{H_2S^+}$, the equilibrium $H-S-H$ angle θ is $97°$ in $\chi(^2B_1)$, $126°$ in $\widetilde{A}(^2A_1)$ and $56°$ in $\widetilde{B}(^2B_2)$. The angle is $92°$ for the ground state of neutral H_2S. Using the Frank-Condon principle as a guide[16)], this leads to the expectation that in the first PE band of H_2S the O−O transition is the most intense one, while the two next PE bands should exhibit more Gaussian-like shapes. This is indeed observed as can be seen in Fig. 5.

The intensity of a PE band depends on the photon wavelength. As a rule, He(I) photons favor the ejections of electrons from diffuse orbitals, while He(II) photons favor ejection from contracted orbitals[10)]. Atomic orbitals of third row atoms tend to be more diffuse than those of second row atoms, indicating, e.g., that MOs predominantly localized on sulfur should give rise to intense bands in He(I) and weaker bands in He(II) spectra, relative to MOs composed of carbon orbitals. As an application of these qualitative ideas, we consider the He(I) and He(II) PE spectra of carbon subsulfide[17)] shown in Fig. 6. Calculations[17)] indicate that the five highest occupied MOs all have large amplitudes on the sulfur atoms, except for the third highest which is essentially C−C bonding. One would thus expect that the PE band associated with ionization of the third MO should be relatively weak in the He(I) spectrum and relatively strong in the He(II) spectrum. Inspection of Fig. 6 shows that this is consistent with the behaviour exhibited by the third PE band at 12.9 eV, in accordance with the assignment based on other criteria[17)]. However, this qualitative intensity criterion neglects bond contributions and should probably be applied with some caution.

The application of band shape and intensity criteria is of course greatly facilitated when the individual PE bands are well separated and show fine structure. The

$$S=C=C=C=S$$

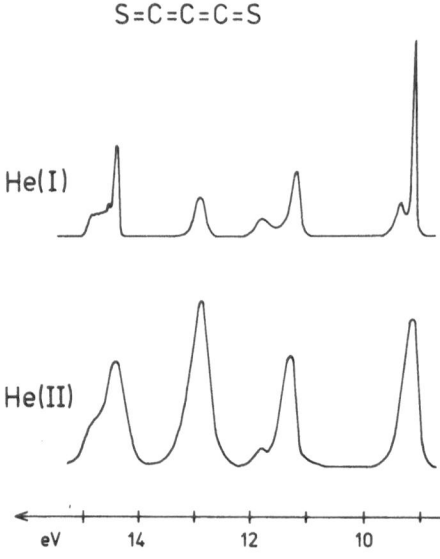

He(I)

He(II)

eV   14   12   10

Fig. 6. He(I) and He(II) PE spectra of carbonsubsulfide[17)]

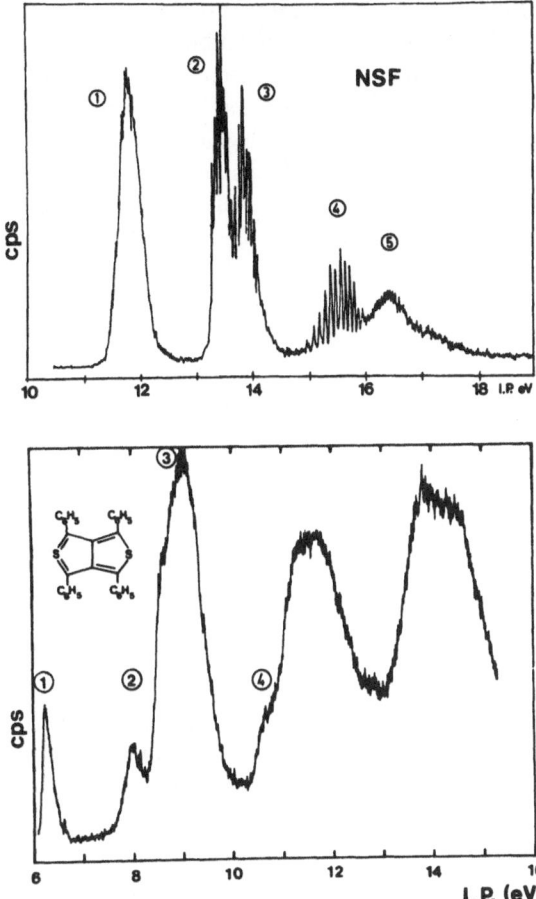

Fig. 7. He(I) PE spectra of NSF[18–20] and tetra-phenylthieno[3.4-c]thiophene[21]

comparison in Fig. 7 of the PE spectrum of a small inorganic compound with that of a large organic molecule indicates some of the difficulties in the application of PE spectroscopy in organic chemistry:

1) strongly overlapping bands
2) lack of vibrational fine structure

The strongly overlapping bands are due to the large number of close lying cation states within a limited range of energy, which often restricts the interpretation to the low energy region where individual peaks may be resolved. The frequent lack of vibrational fine structure is due to the circumstance that a large number of the 3N-6 vibrational modes are excited during ionization, rendering resolution impossible.

Further complication is encountered when the molecule adopts more than one conformation in the gas phase. PE spectroscopy occasionally provides a clue to the preferred geometry, e.g., in the case of sulfides and disulfides.

In view of the complications pointed out above together with the shortcomings of Koopmans' theorem, it would seem that the analysis of the PE spectrum of a large

145

organic molecule is extremely difficult. However, it is usually possible to measure the PE spectra of a series of closely related compounds. Correlation of PE bands in such a series provides rather detailed information on the character and sequence of the individual ionization processes. This information in combination with the results of MO models and perturbation theory may lead to an unambigous assignment of ionization energies to MOs or ionic states. For a detailed discussion of this technique we refer the reader to the recent treatise by Heilbronner and Maier[22]; illustrative examples from the field of sulfur organic chemistry are given in the following sections.

II Saturated Sulfides

1 Hydrogen Sulfides

The simplest sulfide is hydrogen sulfide, H_2S. The first three PE bands at 10.5, 13.4 and 15.5 eV and their assignment to molecular states are indicated in Fig. 5. These bands are associated with ejection of electrons from the three highest occupied MOs $2b_1$, $5a_1$ and $2b_2$, respectively[2, 14, 23, 24]; complete breakdown of the MO picture of ionization is found for the inner valence level $4a_1$ [24].

The first two vertical ionization energies of the hydrides of the 6th main group of elements are indicated in Fig. 8[25]. As anticipated, the shifts are relatively large when passing from H_2O to H_2S. Note the rather small decrease of the binding energies when going from H_2S to H_2Se, indicating that replacement of sulfur by selenium can be considered as a minor perturbation of the valence electronic system.

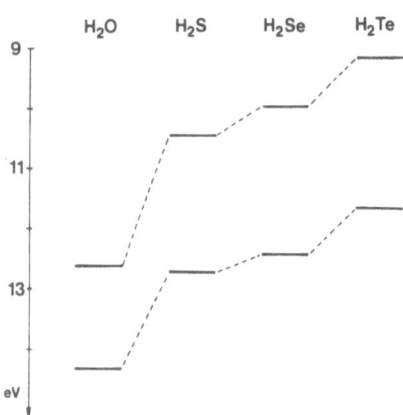

$2b_1(n_S)$

The HOMO of H_2S, $2b_1$, can be represented by a non-bonding S3p orbital, oriented perpendicular to the molecular plane. The non-bonding character indicates

Fig. 8. Correlation of the lowest two ionization energies in a series of hydrides of the 6th group of elements[25]

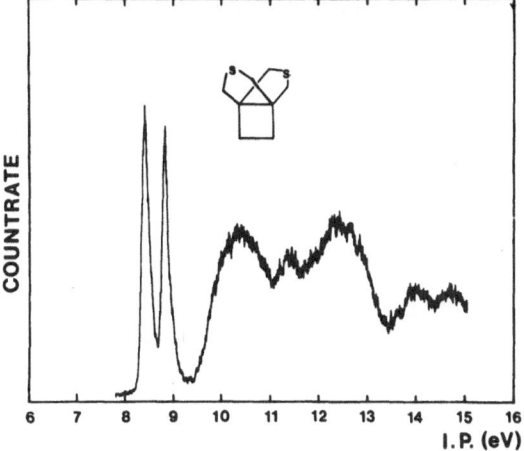

Fig. 9. He(I) PE spectrum of 3,7-dithiabicyclo[3.3.2]decane[31]

that the molecular geometry is insignificantly affected by ionization, giving rise to a strongly Franck-Condon allowed transition and a sharp narrow peak in the PE spectrum (Fig. 5). The HOMO of H_2S is the precursor of closely related n_S orbitals associated with the sulfide functions in more complicated sulfides; these are the most important orbitals of these compounds in relation to PE spectroscopy.

2 Alkyl Sulfides

The PE spectra of alkyl sulfides are characterized by the presence of relatively sharp peaks at low energy (~ 8–10 eV), corresponding to the ionization from the n_S orbitals. These peaks are usually well separated from the often compact structures at higher energy, e.g. Fig. 9; this greatly facilitates the analysis of the PE data.

Correlation of the first ionization energy in a series of alkyl derivatives of H_2S is shown in Fig. 10[26–30]. The inductive and hyperconjugative effect of the alkyl group causes a gradual decrease of the binding energy with increasing chain length of the substituents; the decrease is particularly marked for the first few members of the series. After Et_2S the energy remains rather constant.

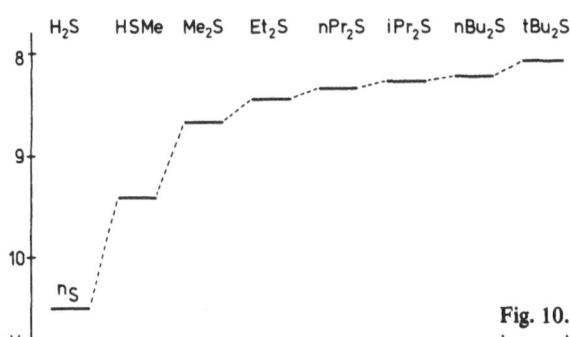

Fig. 10. Correlation of observed n_S levels in a series of alkyl sulfides[26–30]

In sulfides with more than one sulfur center, more n_S bands are observed (Fig. 9). In the propellanes $1-5$ two thiolane rings are sterically fixed. The distance between the sulfur atoms varies slightly, from 4.6 Å in 3 to 5 to 4.9 Å in 1. The PE data for 1 to 5 are summarized in Fig. 11 [31].

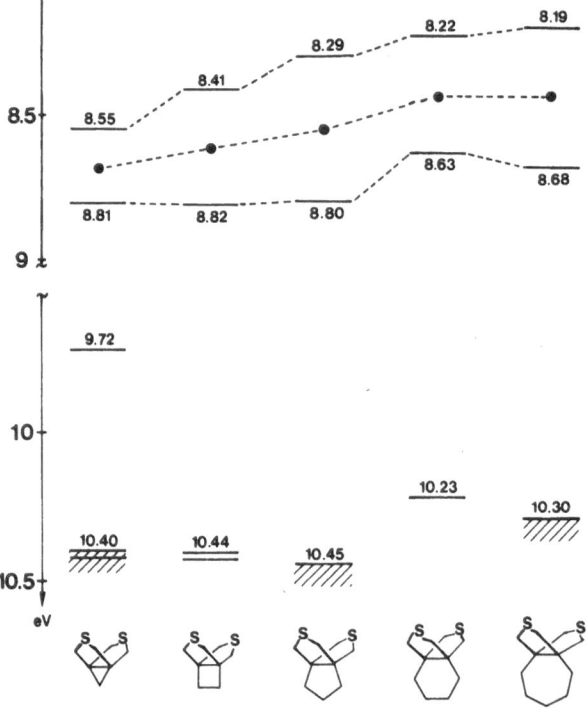

$n = 1$ to 5
1 to 5

$(CH_2)_n$

The interpretation of the results is straight forward. Due to the "through space" interaction of the two n_S orbitals, the symmetry adapted linear combinations $n_S^{\pm} = 1/\sqrt{2}\,(n_{S1} \pm n_{S2})$ are split. The observed energy difference is between 0.3 eV and 0.5 eV, corresponding to the quantity $2|\beta|$, where β is the "resonance integral" between the basis orbitals n_{S1} and n_{S2} in a simple Zero-Differential-Overlap (ZDO) model. A decrease of the center of gravity, indicated by a dot in Fig. 11, is observed with increasing size of the alkyl moiety.

6

7

The 3p character of the n_S orbitals is nicely demonstrated by comparison of the PE spectra of 1,3,5-trithiane (6)[32, 33] and tris(thiomethyl)methane (7)[34, 35].

Fig. 11. Correlation of n_S levels in the propellanes $1-5$[31]. The dots indicate the centers of gravity

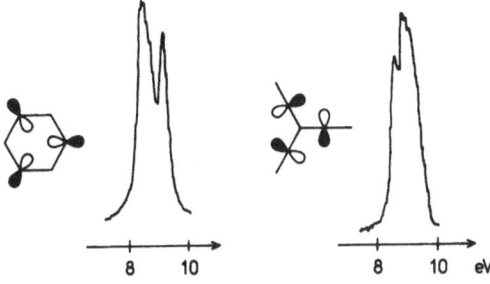

Fig. 12. n_S bands in the PE spectra of 1,3,5-trithiane 6 (left)[32, 33] and tris(thiomethyl)methane 7 (right)[34, 35]

Both compounds can be considered to contain a C_3 axis of symmetry. "Through space" interaction gives rise to n_S-combinations of e and a symmetry in the C_3 point group. The Jahn-Teller effect can be considered to be minimal for the essentially non-bonding $e(n_S)$ levels[32]. However, the splitting pattern is e above a in 6 and a above e in 7, as indicated by the relative intensities of the n_S bands in the PE spectra, see Fig. 12. This can be explained by the in-phase overlap in case of 6 and the out-of-phase overlap in case of 7 (Fig. 12), causing a change of sign of the effective β value and thereby a reversion of the splitting pattern when passing from 6 to 7[35].

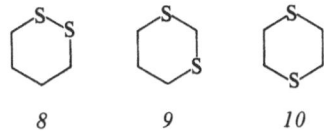

8   9   10

In case of the dithianes 8 to 10, the distance between the sulfur centers varies considerably. Nevertheless, the energy difference between the first two peaks in their PE spectra varies only slightly[30, 32], as shown in Fig. 13. The observed difference is 0.95 eV in 8 and 0.41 eV in 9, consistent with the decrease of the "through space" interaction. Perhaps unexpected is the relatively large split in the case of 10,

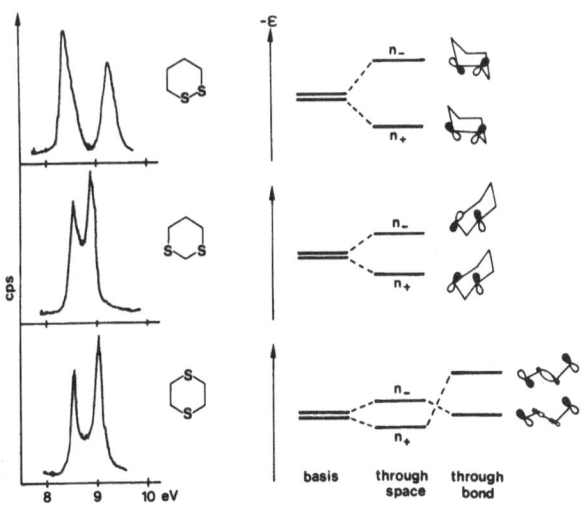

Fig. 13. n_S bands in the PE spectra of the dithianes 8, 9 and 10. "Through space" and "through bond" contributions to the splitting of the levels are indicated to the right[30, 32]

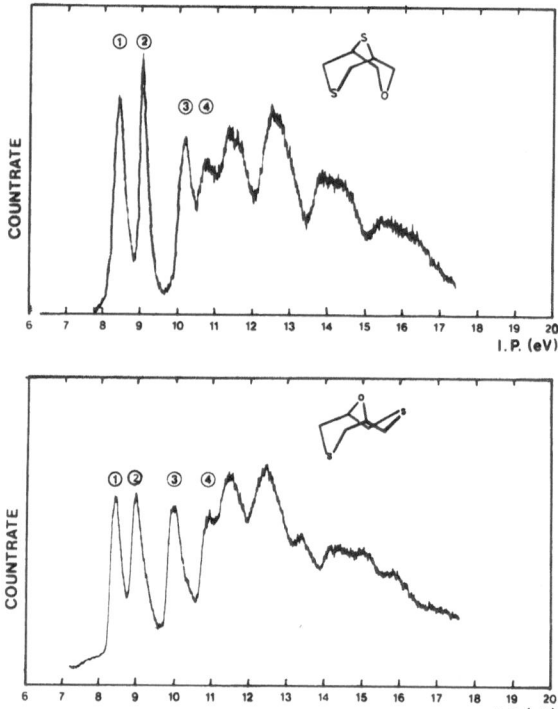

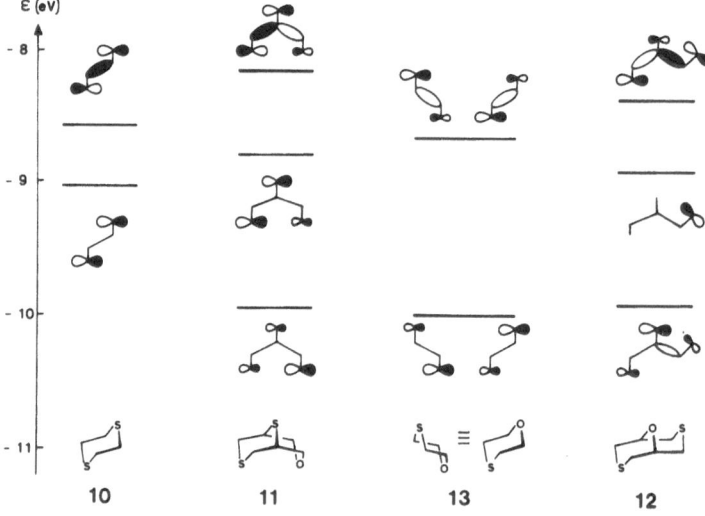

Fig. 14. PE spectra of the 3,7,9-heterobicyclo[3.3.1]-nonanes *11* and *12*[39])

Fig. 15. Observed ionization energies and calculated app. orbital shapes for 1,4-dithiane *10*[32]), 3,9-dithia-7-oxa[3.3.1]nonane *11*[38]), 1,4-thioxane *13*[32]) and 3,7-dithia-9-oxa[3.3.1]nonane *12*[38])

where the "through space" interaction is quite small. However, a strong "through bond" interaction[36, 37] via σ orbitals corresponding to C_2-C_3 and C_5-C_6 is dominating in the case of 10[32], leading to reversal of the ordering of n_S^+ and n_S^- and a relatively large energy separation, as indicated in Fig. 13.

In mixed oxygen and sulfur compounds such as the 3,7,9-heterobicyclo[3.3.1]-nonanes 11 and 12, competitve interactions between the n_O and n_S orbitals might

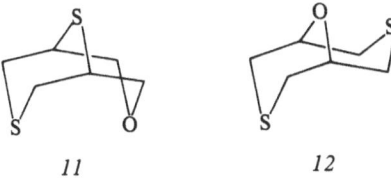

11 12

be present. In particular, a strong "through space" interaction between n_O and n_S has been postulated[38]. In the PE spectra of 11 and 12 (Fig. 14)[39] three sharp peaks are observed at low energy. A comparison with the results of PE spectroscopic investigations on 1,4-dithiane (10) and 1,4-thioxane (13)[32] suggests the interpretation indicated in Fig. 15. The interaction between n_S and n_O is relatively weak and dominated by the "through bond" contribution. The first two ionization energies can be assigned to MOs with mainly n_S character, and the third ionization energy corresponds to ejection from a predominantly n_O orbital at higher binding energy.

A somewhat similar situation applies to the interaction between the n_S orbital and the ethylenic π orbital in the thiapropellene 14. The PE spectrum of 14 shows

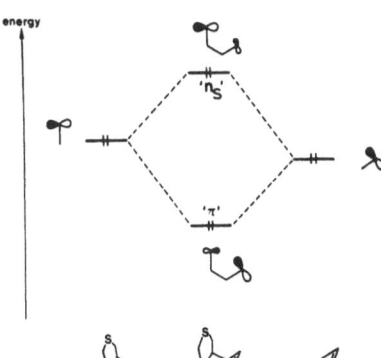

14

two bands at 8.07 eV and 9.06 eV[40]. The first band can be assigned to a predominantly n_S orbital, the second to a predominantly π orbital. The relatively low value of the first ionization energy indicates that "through space" interaction according to the diagram in Fig. 16 is significant.

energy

'n_S'

'π'

Fig. 16. Qualitative diagram indicating "through space" interaction in the thiapropellene 14[40]

3 Spiro Compounds

A particular kind of "through space" interaction is known as "spiro conjuga-tion"[41-43]. The He(I) and He(II) PE spectra of the spiro compounds *15* and *16* are shown in Fig. 17[44, 45]. The bands in the low energy region are relatively more

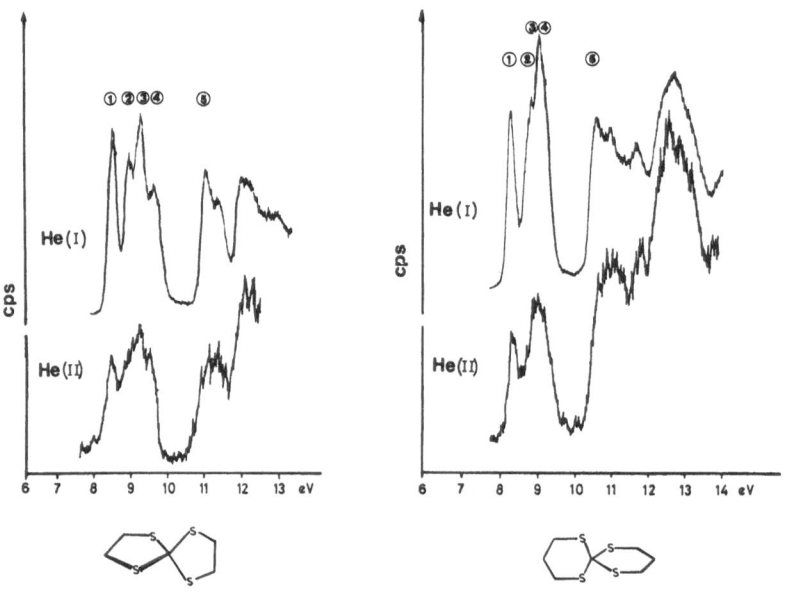

15   *16*

intense in the He(I) than in the He(II) spectrum, indicating that the bands corre-spond to ejection of electrons from MOs mainly localized on the sulfur atoms (cf. Sect. I), i.e. the n_S-type orbitals.

The first four bands in the PE spectra of *15* and *16* can be assigned to MOs derived in a stepwise ZDO procedure[45]. Consider first the case of an idealized D_{2d} geometry. As indicated in Fig. 18, "through space" interaction between n_S orbitals within the constituent rings gives rise to a_2 and b_1 fragment orbitals (local C_{2v} sym-metry). "Spiro conjugation" leads to a splitting of the a_2 fragment orbitals into an a_2 and a b_1 combination (D_{2d} symmetry), while the b_1 fragments form the basis for a doubly degenerate e level. However, the qualitative interaction diagram in

Fig. 17. He(I) and He(II) PE spectra of the spiro compounds *15* and *16*[44, 45]

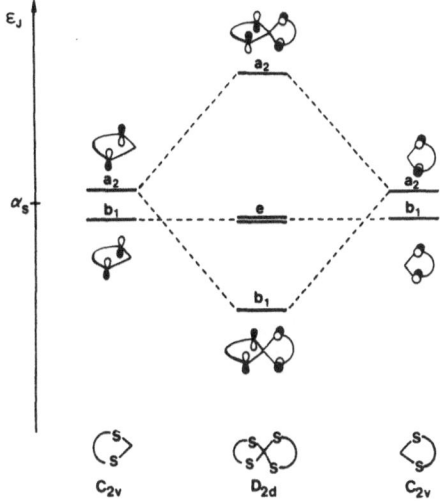

Fig. 18. Qualitative diagram indicating the spiro interaction between the n_S orbitals in *15* and *16* in the case of an idealized D_{2d} symmetry

Fig. 18 does not lead to a convincing agreement with the observed energy levels. The results of a refined model, taking into account the deviation from D_{2d} symmetry as well as strong "through bond" interaction with an occupied σ MO (e in D_{2d}), are indicated in Fig. 19. The agreement with the experimental splitting pattern is plausible.

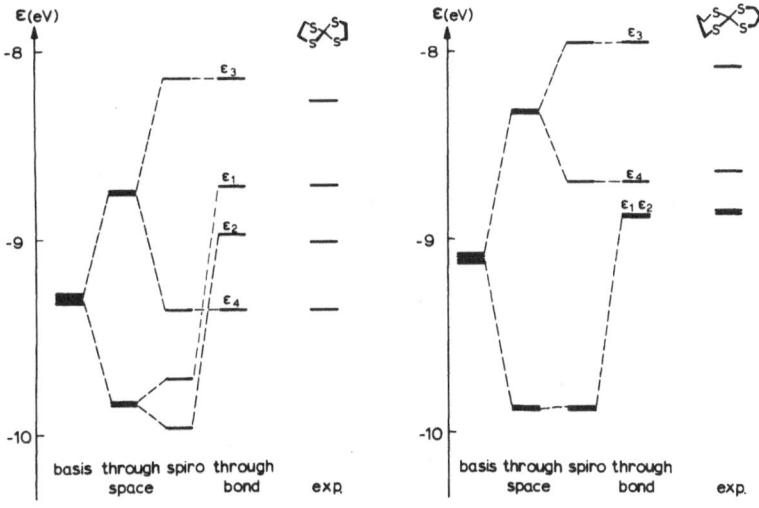

Fig. 19. n_S levels in *15* (left) and *16* (right) derived by a refined model[15], with indication of the experimental values (Fig. 17)

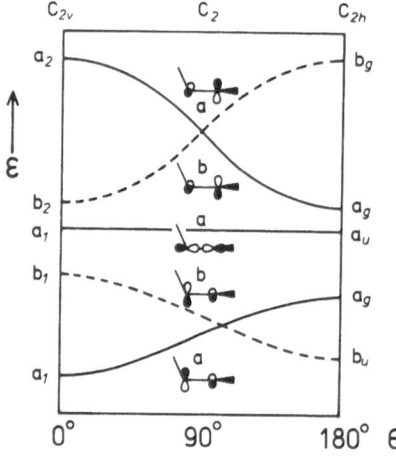

Fig. 20. Approximate correlation diagram indicating the dependence of the highest occupied MO energies of H_2S_2 on the dihedral angle θ. The shape of the MOs is indicated for θ close to 90°; the shape depends strongly on θ

4 Disulfides

In most of the sulfides considered so far, different sulfur centers have been separated by more than one bond. In this paragraph we consider the PE spectra of disulfides, where two sulfur atoms are linked together.

The conjugative interaction between the linked S3p orbitals depends on the dihedral angle θ, as shown in Fig. 20 which indicates the conformational dependence of the highest occupied MOs of hydrogen disulfide, H_2S_2. The dihedral angle in H_2S_2 and its simplest alkyl derivatives is close to 90° (90.6°[46] and 84.7°[47] have been reported for H_2S_2 and Me_2S_2, respectively). In this region, the two highest occupied MOs are near-degenerate n_S^- and n_S^+ combinations, as indicated in Fig. 20, giving rise to two closely spaced peaks in the PE spectrum.

The first two ionization energies in a series of alkyl disulfides are indicated in Fig. 21[30, 48−50]. The interpretation is straight forward: the splitting of the levels, ΔI, is very small in all compounds, except for the t-butyl derivative where the very large bulk of the alkyl groups causes an increase of the dihedral angle θ, leading to a considerable splitting of the n_S ($\Delta I = 0.65$ eV). The center of gravity, indicated by a dot in Fig. 21, increases monotonically with increasing size of the alkyl groups (cp., e.g., Fig. 10).

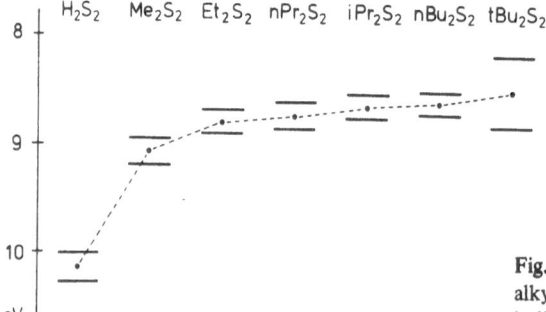

Fig. 21. Observed n_S levels in symmetry alkyl disulfides[30, 48−50]. The dots indicate the centers of gravity

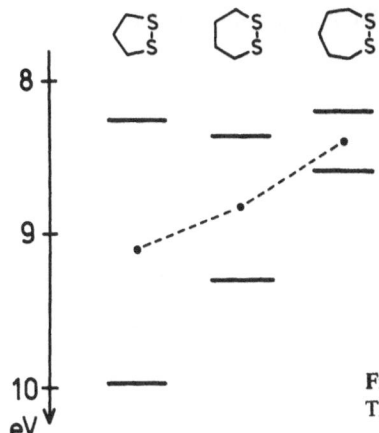

Fig. 22. Observed n_S levels in cyclic disulfides[29, 30, 35]. The dots indicate the centers of gravity

Larger variation of ΔI is found in the series of cyclic disulfides 1,2-dithiolane (17)[35], 1,2-dithiane (18)[30] and 1,2-dithiepine (19)[29], as shown in Fig. 22. The

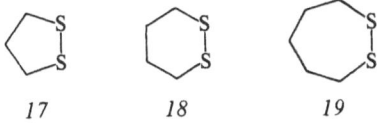

17   18   19

very large value in the case of 17, $\Delta I = 1.72$ eV, is consistent with the small θ value which can be estimated for this compound (35° has been reported for the 1,2-dithiolane derivative lipoic acid[51]). However, in this region the classification of the second highest occupied MO as n_S-type may become problematic[30].

According to simple ZDO MO theory, the splitting of the two n_S levels is equal to $2|\beta|$, where β is the "resonance integral" between the linked S3p atomic orbitals. β can be assumed to vary with the dihedral angle as $\beta = \beta° \cos\theta$, where $\beta°$ is the β-value for $\theta°$ corresponding to maximum conjugation. This leads to the following formula

$$\Delta I = 2|\beta°|\cos\theta \tag{3}$$

which indicates that the splitting of the n_S levels is proportional to $\cos\theta$. In Fig. 23 is shown the correlation between ΔI and $\cos\theta$ for Me_2S_2, 17, and 18 ($\theta \sim 85°$,

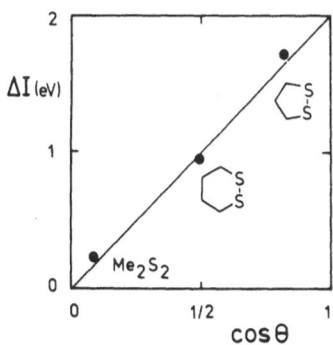

Fig. 23. Correlation between the observed splitting of the n_S levels, ΔI, and the cosine of the dihedral angle θ for three disulfides

$35°$ and $60°$, respectively [47, 51, 52]). The correlation is very convincing, leading to the approximate relation (ΔI in eV)

$$\Delta I \sim 2.0 \cos \theta \tag{4}$$

Eqs. (3) and (4) indicate that $|\beta°|$ is about 1.0 eV, a value which is entirely reasonable [53–55].

Equation (4) can be used to estimate dihedral angles for dialkyl disulfides from measured ΔI values. For 19 ($\Delta I = 0.39$ eV) one obtains $\theta \sim 79°$, and for $(t\text{-Bu})_2 S_2$ ($\Delta I = 0.65$ eV) one gets $\theta \sim 109°$, assuming $\theta > 90°$ in the latter case. These estimates are consistent with those based on other, more elaborate schemes [30, 35, 49, 50]. In particular, Baker et al. [49] applied Bergson's equation [56].

$$\Delta I = \frac{2\gamma |\cos \theta|}{1 - (0.129 \,|\cos \theta|)^2} \tag{5}$$

where γ is determined empirically from UV absorption. They obtained $\theta = 97.5 \pm 0.5°$ for $(t\text{-Bu})_2 S_2$.

The situation is more complicated in disulfides containing additional sulfide functions. Pfister-Guillouzo et al. [50, 57, 58] have investigated the PE spectra of 1,2,4-trithiolane (20) and the alkyl-tetrathianes 21–23. The PE spectrum of 20

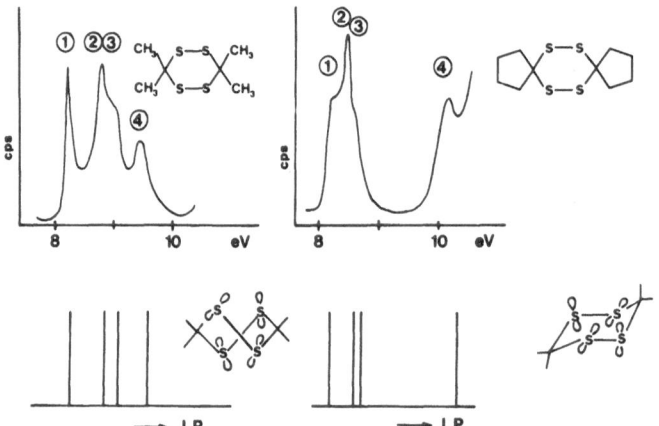

20	21	22	23

shows low energy peaks at 8.12, 8.48 and 9.78 eV. Assuming that the first and the third band can be assigned to n_S combinations of the disulfide group, a ΔI value of 1.66 eV is obtained, consistent with a dihedral angle deviating significantly from

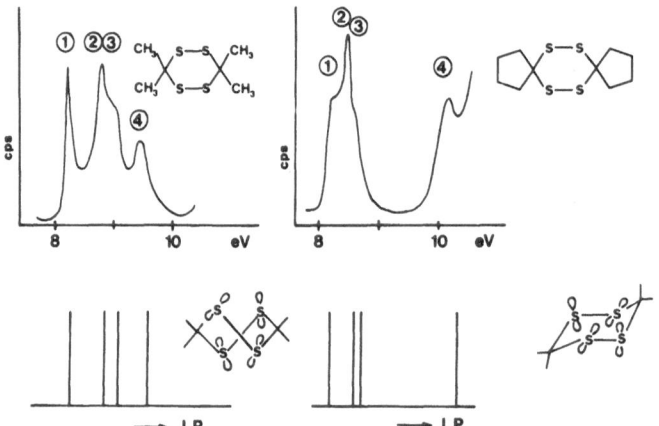

Fig. 24. General outline of the low energy PE bands of the alkyl-tetrathianes 21 and 22 [58]. Results of model calculations are indicated below [58]

zero. This indicates that the "half-chair" rather than the "envelope" conformation of *20* is preferred in the gas phase. The low energy region of the PE spectrum of *21* differs markedly from that of *22* and *23*, see Fig. 24. The analysis indicates that the preferential forms are "twist" for *21* and "chair" for *22* and *23*.

III Unsaturated Sulfides

1 Aryl-Alkyl-Sulfides

The "through space" interaction between a sulfide n_S orbital and a distant ethylenic π orbital has been briefly mentioned in the case of *14* (Fig. 16). In this paragraph we consider the much stronger conjugative interaction in the case of thioalkyl substituted π systems. PE spectroscopic investigation of mono-, di-, and tetra-thiomethyl substituted ethylenes suggests an effective β_{CS} parameter for the π interaction between the linked S3p and C2p atomic orbitals of -1.6 eV[59]. This value is consistent with the PE data for thioanisole[60−63] and α- and β-thiomethyl-naphthalene[61].

The interaction between the S3p atomic orbital and the highest occupied benzene π orbitals in the case of thioanisole is indicated in Fig. 25. The resulting three levels correspond to the first three main PE bands[60−63]. However, an additional complication is the non-rigidity of the molecule. Dewar et al.[62] suggested the existence of a conformational equilibrium with two predominant conformers, one essentially planar with maximum conjugation and the other with the S3p orbital orthogonal to the π system. The shoulder at $\sim$8.6 eV in the PE spectrum of thioanisole (Fig. 25) was attributed to the presence of ca. 10% of the latter conformer

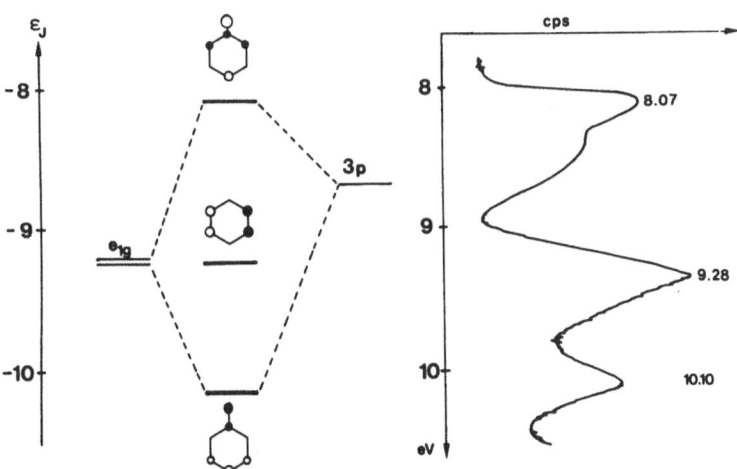

Fig. 25. Diagram indicating the interaction between a S3p orbital and the highest occupied benzene π orbitals to yield the three highest occupied orbitals of planar thioanisole. The PE spectrum is shown to the right[60]

at room temperature. The percentage of the non-planar conformer could be roughly estimated from the PE spectra; for thiophenol no non-planar conformer could be detected, but the percentage was found to increase with increasing bulk of the alkyl group in a series of phenyl-alkyl-sulfides, to about 95% for the *t*-butyl derivative. The results for thioanisole have later been refined by the application of variable temperature PE spectroscopy[63] (see Sect. VI).

2 π Systems

In this paragraph we consider compounds which can be thought of as being derived from conjugated cyclic hydrocarbons by replacement of –CH=CH– units by sulfur atoms. A few sulfur heterocycles which do not belong to this class are considered in Sect. V. Replacement of a double bond by a sulfur atom is a minor perturbation of the π electron system. This is clearly demonstrated in Fig. 26 which indicates the ionization energies of phenanthrene and five sulfur analogues. The remarkable similarity of the PE spectra can partly be explained by noting that the first ionization potentials of ethylene and hydrogen sulfide are essentially identical (10.51[2] and 10.48 eV[2, 14], respectively).

In this class of heterocycles, the "non-bonding" S3p orbital of the sulfide function is fully conjugated with the π system of the hydrocarbon moiety, giving rise to a delocalized π system to which the sulfur atom formally contributes two electrons. The concept of an n_S orbital which was so useful in the discussion of the saturated sulfides is no longer applicable; instead we resort to the well-known techniques of simple π electron theory [64–66].

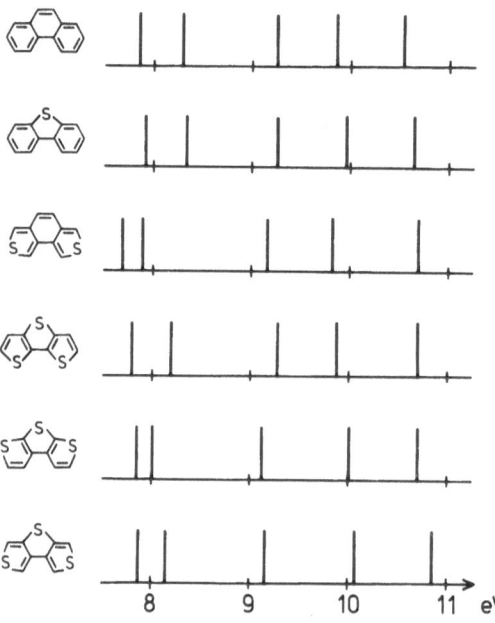

Fig. 26. Ionization energies of phenanthrene and five sulfur analogues[125, 126, 132]

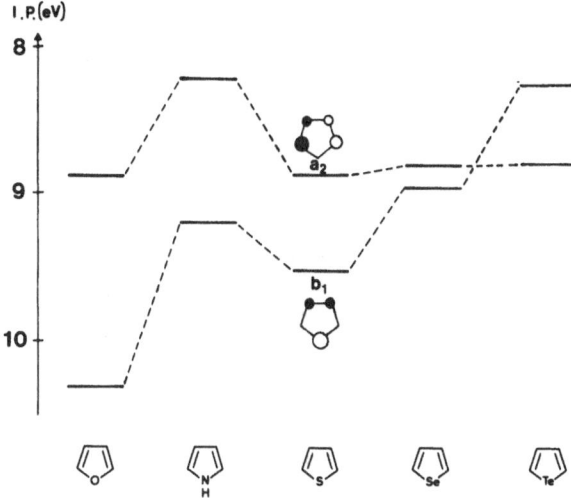

Fig. 27. Correlation of the two lowest ionization energies of furan, pyrrole, thiophene, selenophene, and telluro-phene[67, 73–75]

Thiophene is a key compound in heterocyclic chemistry and its PE spectrum has been studied extensively[67–72]. The two lowest ionization energies of thiophene at 8.9 and 9.5 eV can be assigned to the two highest π levels, $a_2(\pi)$ and $b_1(\pi)$, respectively, as indicated in Fig. 27. Replacement of the sulfur atom by another element of the 6th group (i.e. 0, Se or Te) tends to affect insignificantly the $a_2(\pi)$ level while the $b_1(\pi)$ level is shifted according to the electronegativity of the hetero-atom[67, 73–75]. This is consistent with the shape of the MOs (Fig. 27); according to simple perturbation theory, the $a_2(\pi)$ MO which has a node through the position of the heteroatom should be much less affected than the $b_1(\pi)$ MO which has large amplitude on the heteroatom. However, it is apparent that simple perturbation theory fails to predict the shifts observed when sulfur is replaced by NH, yielding pyrrole; the shift of the $a_2(\pi)$ level is much larger than the shift of the $b_1(\pi)$ level, and both shifts are towards lower binding energies. An analysis indicates that the shifts are governed by a balance between the influence of the increase of the "resonance integral" for the bonds involving the heteroatom and a strongly destab-ilizing inductive effect due to the polarity of the N–H bond[76].

The π-type ionization energies of thiophenes and selenophenes can be estimated by means of Hückel (HMO)[64–66] calculations with the following set of parameters: $\alpha_S = -9.4$, $\alpha_{Se} = -8.5$, $\alpha_C = -7.0$, $\beta_{CS} = -1.8$, $\beta_{CSe} = -1.5$, and $\beta_{CC} = -3.0$ eV[76]. In the case of compounds containing pyrrole-type nitrogen, special care must be taken to account for the effects mentioned above; in particular, the destabilizing inductive effect on the carbon atom frame must be represented in the HMO model through the choice of less negative α_C parameters. Application of a parameter set derived by Eland[67] for the pyrrole fragments in the series 24–28, together with

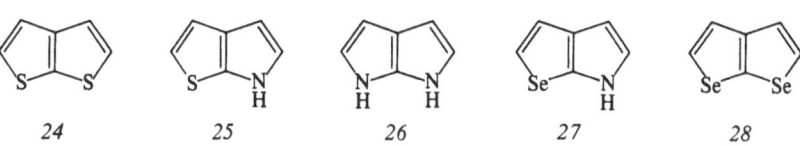

24   25   26   27   28

159

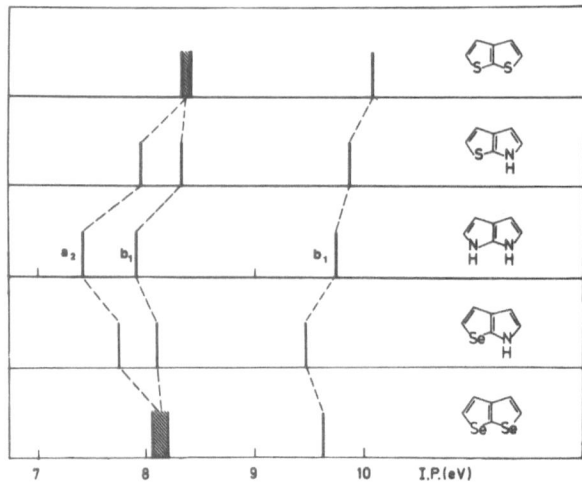

Fig. 28. Correlation of ionization energies in the series *24−28*

the parameters given above for the remaining portions, leads to good agreement with the trends of the PE data indicated in Fig. 28 [76].

29   30   31

32   33

Tetrathiafulvalene (TTF, *29*) and its derivatives have attracted particular interest due to the observation of unusual electric properties of their adducts with tetracyanoquinodimethane (TCNQ)[77, 78]. The PE spectrum of *29* is unusual for an

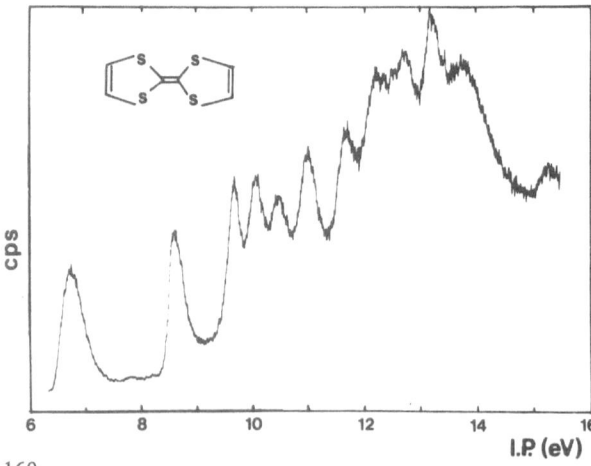

Fig. 29. He(I) PE spectrum of tetrathiafulvalene *29*[79]

organic compound of this size in that at least twelve individual peaks are clearly observed below 16 eV (Fig. 29)[79-82]. The π MOs of 29 are most simply derived in terms of a LCBO-type (Linear Combination of Bond Orbitals) model. The advantage of such an approach is that basis orbital energies can often be estimated from PE spectra of suitable reference compounds, leading to correlation of PE bands in complicated spectra with those of much simpler spectra. The basis orbitals in the case of 29 are conveniently taken as π and π^* orbitals of the three ethylenic double bonds

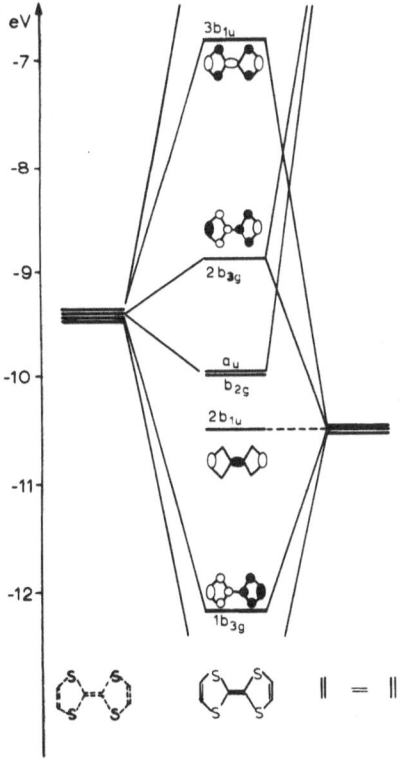

29

and the $3p\pi$ orbitals of the four sulfur centers. The resulting energy levels are indicated in Fig. 30, and correlation of bands in the series 29–33 is shown in Fig. 31[81]. The ordering of the π levels indicated in Fig. 30 is supported by correlation of PE

34   35

peaks in the series 29, cis/trans diselenadithiafulvalene (34), and tetraselenafulvalene (35)[82]. According to this work, the five lowest ionization energies of 29 can be assigned to π levels, and the sixth to a σ level.

eV

-7

3b₁ᵤ

-8

-9

2b₃g

-10

aᵤ
b₂g
2b₁ᵤ

-11

-12

1b₃g

Fig. 30. Diagram indicating the interaction between four S3p orbitals and three ethylene units to yield the highest occupied π orbitals of tetrathiafulvalene[81]

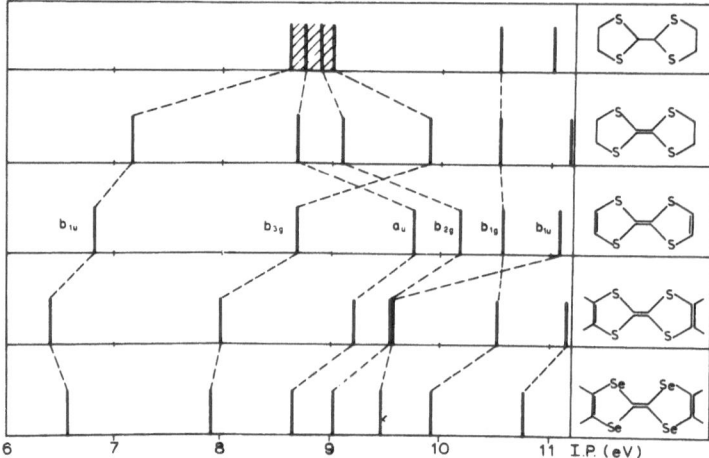

Fig. 31. Correlation of ionization energies in the series *29–33*[81]

As an example of a π system involving a disulfide unit we briefly consider naphthalene-1,8-disulfide (*36*). Also this species belongs to a series of π donors with remarkable electric properties[53, 54]. The π MO energy diagram can be constructed

36

on the basis of the highest occupied orbitals of the naphthalene moiety and the n_S^- and n_S^+ of the disulfide unit, as indicated in Fig. 32[53]. The splitting of n_S^- and n_S^+ for

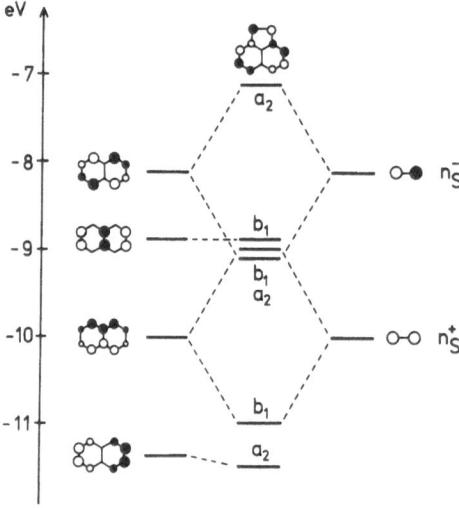

Fig. 32. Diagram indicating the derivation of the highest occupied π orbitals of naphthalene-1,8-disulfide from π orbitals of naphthalene and n_S^- and n_S^+ orbitals of the disulfide unit[53]

a disulfide with dihedral angle $\theta = 0°$ can be estimated to be 2.0 eV, consistent with a β_{SS} value of ~ -1.0 eV (see Sect. II). The interaction diagram in Fig. 32 leads to excellent agreement with the low energy region of the PE spectrum of 36[53]; the main features of the diagram have recently been supported by the results of a non-empirical Hartree-Fock calculation[55].

IV Thiocarbonyls

1 Simple Thiocarbonyls

Compared to the classes of compounds considered in the two preceding sections, relatively few simple thiocarbonyl compounds have been investigated by PE spectroscopy. This is mainly due to the relative instability of many of these species.

The simplest thiocarbonyl compound is thioformaldehyde; the low energy portion of its PE spectrum is shown in Fig. 33[48, 83]. The three lowest ionization energies at 9.4, 11.8, and 13.9 eV can be assigned to the MOs $2b_2(n_{CS})$, $1b_1(\pi_{CS})$, and $3a_1(\sigma_{CS})$, respectively[13, 83] as indicated in Fig. 33. The assignment has been supported by theoretical calculation of the vibrational structure of the PE bands[13]: The strongly Franck-Condon allowed character of the first band is consistent with the ejection of an electron from an essentially non-bonding MO, n_{CS}. This orbital is predominantly S3p, similar to the n_S orbital of a sulfide, but oriented in-plane. The out-of-plane S3p orbital interacts with the corresponding C2p orbital to form the π system of the thiocarbonyl group. Note that the π_{CS} orbital has similar amplitude on carbon and sulfur, in contrast to the case of the carbonyl π_{CO} orbital; this is a reflection of the similar electronegativities of carbon and sulfur[84].

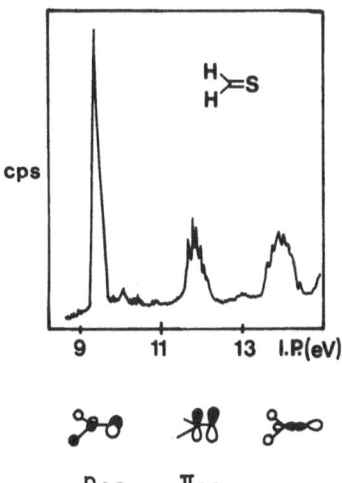

Fig. 33. The first three He(I) PE bands of thioformaldehyde with indication of the orbital assignment[13, 48, 83]

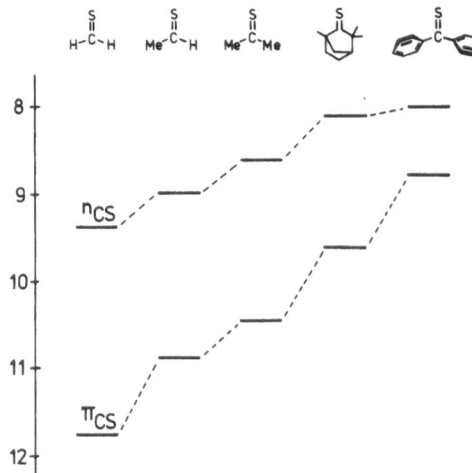

Fig. 34. Correlation of n_{CS} and π_{CS} levels in alkyl and phenyl derivatives of thioformaldehyde[33, 85, 86]

The correlation of n_{CS} and π_{CS} levels for alkyl and phenyl derivatives of thioform-aldehyde is shown in Fig. 34[33, 85, 86] (note that the benzene rings in thiobenzo-phenone are strongly twisted). The π_{CS} level is generally more sensitive to substitu-tion than the n_{CS} level. The relatively large shifts of the π_{CS} level can be understood in view of the considerable amplitude on carbon (Fig. 33) which renders the π_{CS} orbital sensitive to inductive and hyperconjugative effects.

The electronic structure of the dithione 37 is closely related to that of thio-acetone. The PE spectrum of 37[87] shown in Fig. 35 clearly displays two n_{CS} peaks separated by about 0.5 eV. The splitting of the n_{CS} levels is due to "through bond" interaction via high lying "Walsh" orbitals of the four-membered ring. A similar

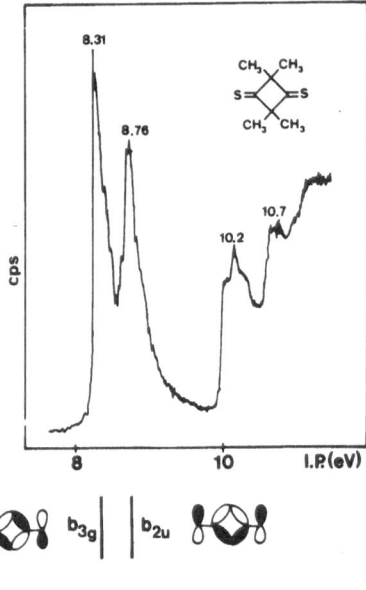

Fig. 35. Low energy region of the He(I) PE spectrum of 2,2,4,4-tetramethylcyclobutane-1,3-dithione 37 with indication of the orbital assignment of the n_{CS} bands[87]

Me   Me

S= $\diamond$ =S

Me   Me

37

effect but a reversed orbital sequence has been encountered in the case of the corresponding dicarbonyl compound[88, 89]. The third and the fourth band between 10 and 11 eV can probably be assigned to plus and minus combinations of the two π_{CS} orbitals.

2 π Systems

The π orbital of the thiocarbonyl group interacts strongly with neighbouring π type orbitals, e.g., in thioacetic acid[90], thioformamide[91], and thiocarbonates[85, 92–95], giving rise to delocalized π systems. The lowest π type ionization energy in these compounds is close to the corresponding n_{CS} level.

As an example of a conjugated π electron system involving the thiocarbonyl group we consider the cyclic trithiocarbonate 1,3-dithiole-2-thione (*38*). *38* and its selenium analogues *39–43* comprise an ideal series for the study of the valence

38   *39*   *40*   *41*   *42*   *43*

electronic structure since the perturbation introduced in replacing S by Se is relatively small. The correlation of the first four PE bands in the series *38–43* is shown in Fig. 36[95].

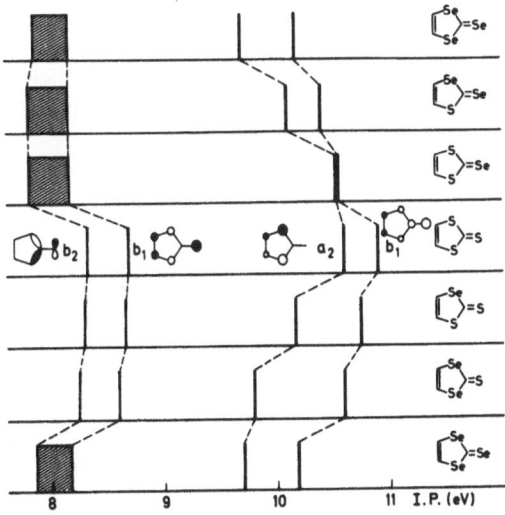

Fig. 36. Correlation of PE bands in the series *38–43*[95]

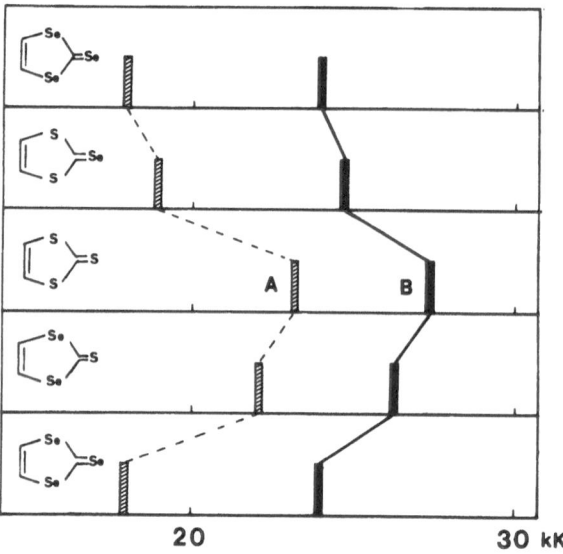

Fig. 37. Correlation of the two lowest electronic transitions in *38, 40, 41* and *43* measured in stretched polyethylene. Transition A is symmetry forbidden and transition B is polarized along the symmetry axes of the species[95]

The first two bands are shifted very similarly; they can be assigned to the MOs $5b_2(n_{CS}$, and $3b_1(\pi)$ as indicated in Fig. 36. The assignment of the ordering of these levels is rather tentative, but the parallel shifts are consistent with the very high density on the thione sulfur position in both MOs. The parallel shifts are reflected by the shifts of the two lowest optical transitions (Fig. 37), which can be explained by the assignment of these transitions to predominantly $5b_2(n_{CS}) \rightarrow 4b_1(\pi^*)$ and $3b_1(\pi) \rightarrow 4b_1(\pi^*)$, respectively[95].

The shifts of the third and the fourth PE band in the series *38–43* are distinctly different (Fig. 36), providing a definite clue to the assignment. The third level is insignificantly affected by replacement of the thione sulfur, consistent with the assignment to the $1a_2(\pi)$ MO which has zero amplitude on the thione fragment. Conversely, the shifts of the fourth level indicate rather high amplitude on the thione sulfur, consistent with the assignment to the $2b_1(\pi)$ MO.

The shifts of the bands in the series *38, 44*[94], and *45*[96] are much larger than in the series *38–43*, see Fig. 38. This is not surprising in view of the result displayed in

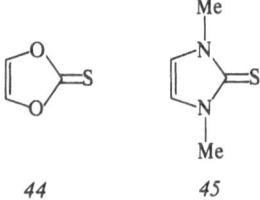

44   *45*

Figs. 8 and 27. Again, attempts to predict the shifts when replacing sulfur by pyrrole type nitrogen cannot be based on electronegativity considerations and simple perturbation theory[76]. Consider f.i. the third and the fourth level of *38*, $1a_2$ and $2b_1$, respec-

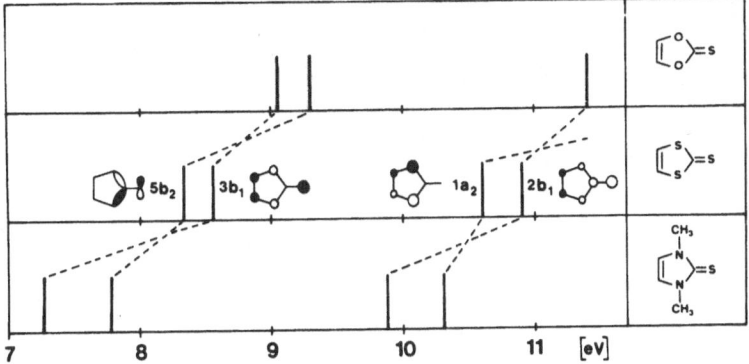

Fig. 38. Correlation of PE bands in the series *38, 44,* and *45* [95, 96]

tively (Fig. 38). Simple perturbation theory would predict larger sensitivity of the $1a_2$ level to replacement of the thiole sulfurs than the $2b_1$ level, due to the much larger amplitude of the former MO in these positions. This prediction leads to consistent results when replacing sulfur by selenium (Fig. 36) or oxygen, but in the case of pyrrole type nitrogen particular attention must be payed to the strongly destabilizing inductive effect on the remaining positions of the π system. As the $2b_1$ MO has much larger amplitude in these positions, this effect tends to destabilize the $2b_1$ level more than the $1a_2$ level, leading to a crossing of the bands in the PE spectra (Fig. 38). Consideration of the bond contributions indicates a stabilization of the $1a_2$ level relative to the $2b_1$ level, further contributing to the trend mentioned above, while the influence of the electronegativity of the heteroatom is probably of minor importance; the effective electronegativity of pyrrole type nitrogen can be considered to be quite similar to that of sulfur[64, 76].

V Polyvalent Sulfur Compounds

In this section we consider the PE spectra of some sulfur compounds with higher oxidation states than the compounds treated so far. Sulfoxides and sulfones are among the most simple compounds of this type. As an example, the PE spectra of dimethylsulfoxide (*46*) and dimethylsulfone (*47*)[97, 98] are shown in Fig. 39, together with an indication of the orbital assignments. The PE spectrum of *46* shows two bands at low energy, well separated from more compact structures at higher energy. The results of MO calculations indicate that the first band can be assigned to the sulfur "lone pair" orbital, and the second band to a π type orbital mainly localized on the SO bond. Analogously, the first four strongly overlapping bands in the PE spectrum of *47* can be assigned to MOs localized mainly in the sulfone group. A confirmation of this assignment is given by analysis of the PE spectra of sulfodiimides[98].

167

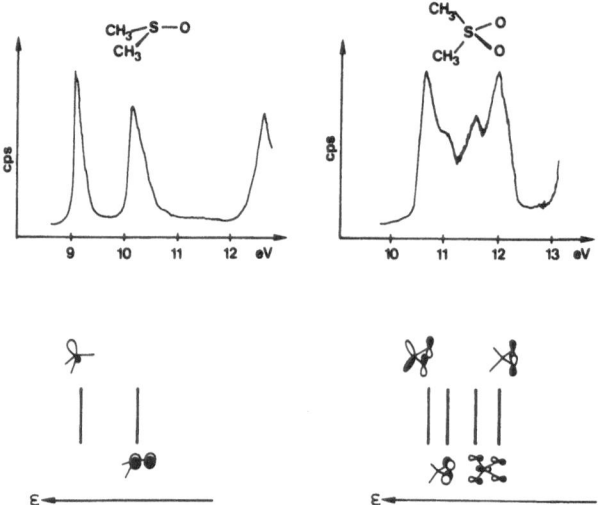

Fig. 39. He(I) PE spectra of dimethylsulfoxide (left) and dimethylsulfone (right) with indication of the orbital assignment[97, 98]

While it is fairly obvious that the compounds 46 and 47 involve polyvalent sulfur, the situation is less straightforward in case of the thiapentalene systems 48 and 49. It is not possible to draw classical bond structures for these systems with-

48

49

out invoking either polyvalent sulfur or ionic structures. In case of 48 it is possible to write "no-bond-resonance" structures, which is one attempt to describe the peculiar type of bonding in this species[99].

The correlation of PE bands in the series of trithiapentalene derivatives and analogues 50–54 is shown in Fig. 40[100, 101]. Note the assignment of a relatively

50

51

52

53

54

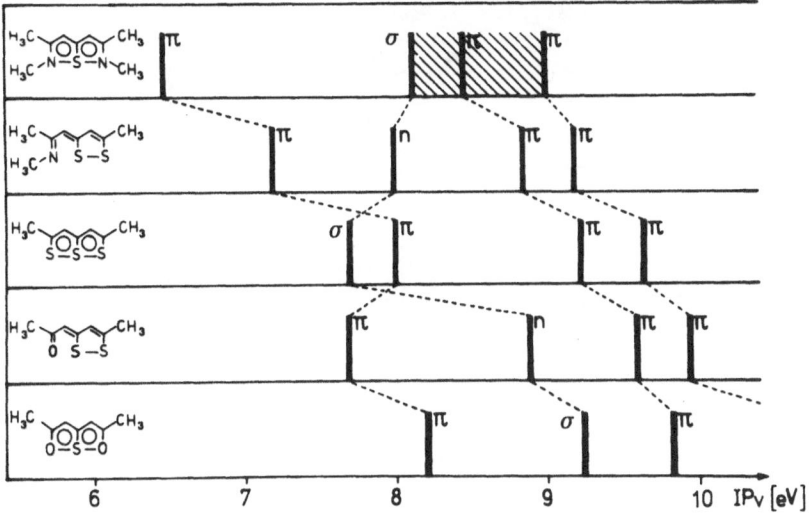

Fig. 40. Correlation of PE bands in the series $50-54$ [100, 101]

low energy σ band in case of 50, 52 and 54, which is consistent with the results of model considerations[99, 102]. A considerable lowering of the first ionization energy is observed when sulfur is replaced by pyrrole type nitrogen, an effect which has been discussed above.

The ionization energies of tetraphenylthieno[3,4-c]thiophene (55)[21, 103–105] (Fig. 7) and its aza-analogue 56[104, 105] are correlated in Fig. 41. The first ionization energy of 55 is extremely low (6.2 eV), consistent with the fact that it is possible to generate the radical cation of 55 in solution under relatively mild con-

55   56

ditions[104]. The sequence of the inner π levels "b_{1g}" and "b_{2u}" indicated in Fig. 41 is opposite to the one which is obtained on the basis of π electron calculations on the unsubstituted thienothiophene 49[103, 104]. The reversal of the sequence is due to the circumstance that interaction with the (twisted) phenyl groups destabilizes the b_{1g} orbital of the thienothiophene fragment more than the b_{2u} orbital. The reversed ordering is supported by comparison with the PE data for 56; aza substitu-

169

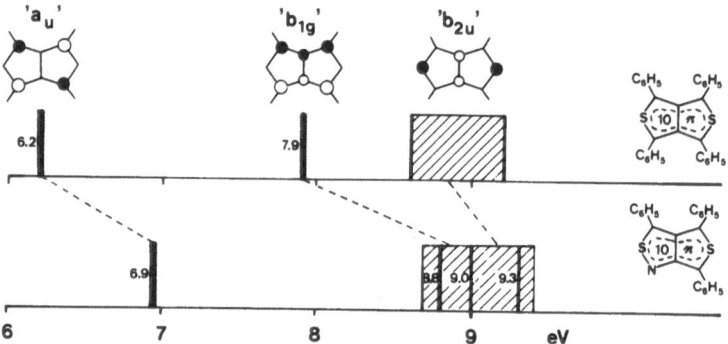

Fig. 41. Correlation of PE bands in 55 and 56 [103—105]

tion causes a large shift towards higher ionization energy for the first two bands but not for the one assigned to "b_{2u}", which is what one would expect on the basis of first order perturbation theory.

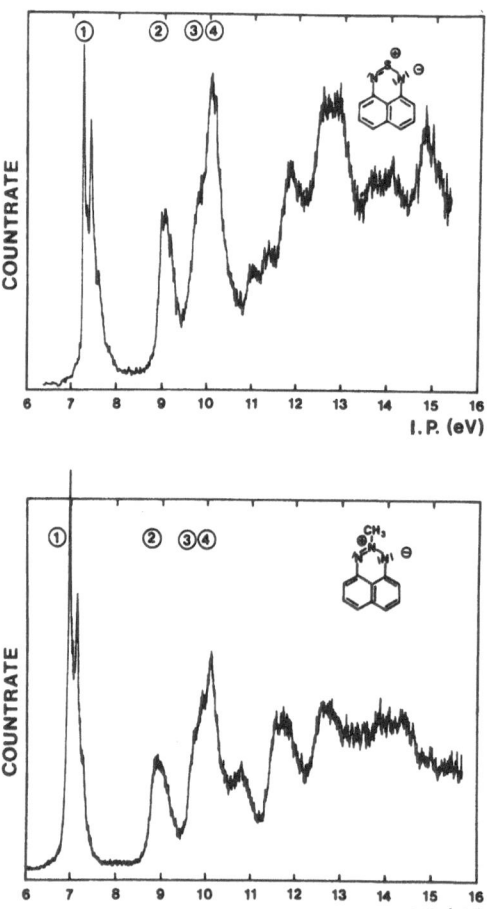

Fig. 42. He(I) PE spectra of 57 and and 58 [107]

PE spectroscopy does not provide conclusive evidence concerning the contribution of S3d orbitals to the bonding in the non-classical structures *48* and *49* and their derivatives, e.g., *50–56*. However, the ionization energies of these species can be reasonably well predicted by MO calculations without the consideration of S3d orbitals, and inclusion of these orbitals does apparently not lead to improved agreement[106]. The close similarity of the PE spectra (Fig. 42) and of the electronic absorption spectra of the non-classical compounds *57* and *58*[107] indicates that S3d orbitals participate insignificantly in the bonding in *57*.

57   *58*

VI Variable Temperature Photoelectron Spectroscopy

We shall conclude this survey by briefly considering some results obtained by means of the technique of variable temperature PE spectroscopy. In a PE spectrometer it is possible to detect short lived species generated by pyrolysis (or irradiation) and to explore reactions and equilibria in the gas phase over a wide temperature range[33, 48, 63, 83, 108–111]. This technique has been particularly useful in the field of sulfur organic chemistry, as illustrated by a few examples below.

The PE spectrum of thioformaldehyde is shown in Fig. 33. The compound was generated in the spectrometer by heating methylsulfenyl chloride to 860 K at reduced pressure[83]. Thioformaldehyde is formed quantitatively:

$$CH_3-S-Cl \xrightarrow{860\ K} CH_2 = S + HCl$$

The hydrogen chloride can be removed by stoichiometric ammonia injection forming an ammonium chloride deposit on the walls of the reaction chamber. Careful optimization of the reaction conditions yields pure monomeric thioformaldehyde.

Another example is shown in Fig. 43[109]. By heating di-*t*-butyldisulfide to 690 K the PE spectrum of a mixture of H_2S and isobutylene is recorded. At 1400 K the spectrum has completely changed and CS_2 and acetylene are the main components.

Application of variable temperature PE spectroscopy to ethane-1,2-dithiole revealed the presence of thiirane and hydrogen sulfide at elevated temperature, indicating the interesting reaction

$$HS-CH_2CH_2-SH \xrightarrow{\Delta} \triangledown S + H_2S$$

at low pressure ($\sim$100 mtorr)[111].

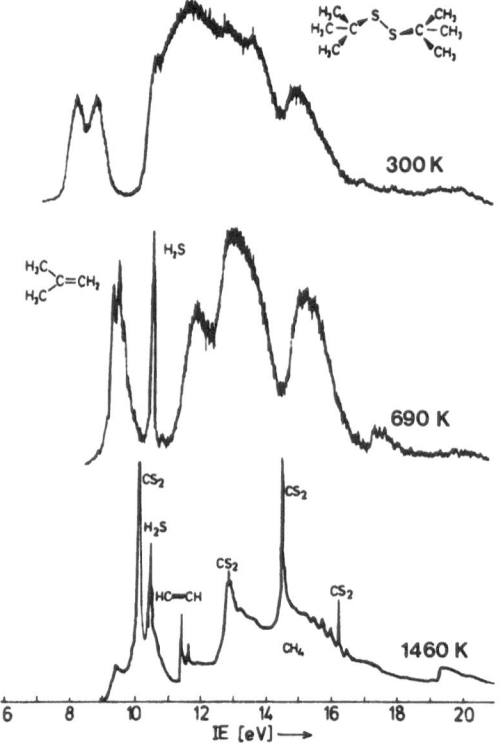

Fig. 43. He(I) PE spectra of di-*t*-butyldisulfide at 300 K and its pyrolysis products at 690 K and 1460 K [109]

The first bands of the PE spectrum of thioanisole at 20 °C and 500 °C are shown in Fig. 44[63]. The relative intensities of the bands are observed to be temperature dependent. This is explained by the assumption of an equilibrium between a planar and a non-planar rotamer, as indicated in Sec. III[62]. Bands ① and ④ can be assigned to the planar and band ② to the non-planar rotamer. Under the envelopes of band ③ and ⑤ ionization occur from both forms. From the measured intensity

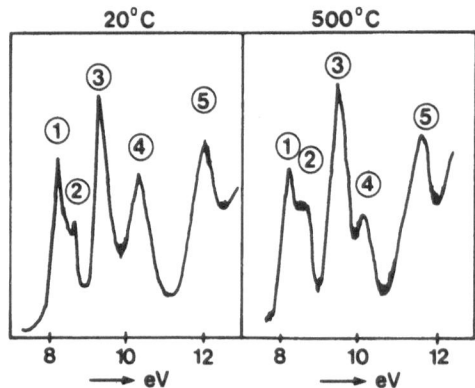

Fig. 44. He(I) PE spectra of thioanisole at 20 °C and 500 °C[63]

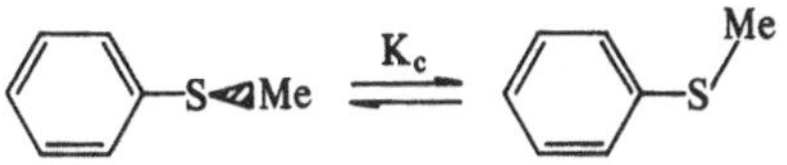

ratios I_4/I_3 the equilibrium constant K_c = [planar]/-[non-planar] could be obtained as a function of temperature $(K_c(20\,^\circ C) = 3.7, K_c(500\,^\circ C) = 1.2)^{63)}$. A plot of $\ln K_c$ versus $1/T$ indicated an energy difference between the conformers of -0.84 ± 0.07 kcal/mole.

VII Table of Ionization Energies

This section contains a compilation of PE data for 110 selected sulfur organic compounds. Only molecules composed of carbon, sulfur and hydrogen have been included; additional data concerning derivatives and hetero analogues are often found in the references given. "Handbook of Spectroscopy"[112] is also useful and contains, e.g., PE data for several derivatives of thiophene.
The following information is given in the table:

(a) Structural formula. Usual abreviations, like "Me" for methyl, "Et" for ethyl, etc., are employed.
(b) Molecular symmetry point group. An asterisk indicates that the symmetry is idealized.
(c) Vertical ionization energies I_v and their (sometimes tentative) orbital assignment. The vertical ionization energies were obtained by He(Iα) PE spectroscopy in the gas phase, and were generally estimated from the maximum of the PE band. In case of phenyl-alkylsulfides, data for two conformers are given.
(d) Leading references, providing a key to the literature.

R. Gleiter and J. Spanget-Larsen

174

Table

(a)	(b)	(c)					(d)
H–S–H	(C_{2v})	10.48	13.4	15.5	22.0	23.3	2, 14, 15, 23, 24
		$2b_1(n_S)$	$5a_1$	$2b_2$	$4a_1$ + "shake up"		
H–S–Me	(C_s)	9.41	11.90	13.50	14.90	15.5	26, 27. 29
		$a''(n_S)$	a'				
Me–S–Me	(C_{2v})	8.68	11.2	12.5			27–30
		$b_1(n_S)$	a_1	b_2			
Et–S–Et	(C_{2v})	8.48	10.7	11.6			30
		$b_1(n_S)$	a_1	a_1			
nPr–S–nPr	(C_{2v})	8.34	10.5	(11.9)			30
		$b_1(n_S)$	a_1				
iPr–S–iPr	(C_{2v})	8.26	10.4	11.0			30
		$b_1(n_S)$	a_1				

Compound	Symmetry										Ref
nBu–S–nBu	(C$_{2v}$)	8.22 $b_1(n_S)$		10.4 a_1		(11.7)					30)
tBu–S–tBu	(C$_{2v}$)	8.07 $b_1(n_S)$		9.9 a_1		(10.4)					30)
C$_6$H$_5$–CH$_2$–S–Me		8.42 n_S		9.01 π_1		9.2 π_2		10.9		11.6	61)
H–S–S–H	(C$_2$)	10.01 $a(\overline{n_S})$		10.28 $b(\overset{+}{n_S})$		(12.7) $a(\sigma_{SS})$		(14.1)		(15.1) $b(\sigma_{SH})$	30)
Me–S–S–Me	(C$_2$)	8.97 $a(\overline{n_S})$		9.21 $b(\overset{+}{n_S})$		(11.3) $a(\sigma_{SS})$		(12.3)		(13.4)	30, 48–50)
Et–S–S–Et	(C$_2$)	8.70 $a(\overline{n_S})$		8.92 $b(\overset{+}{n_S})$							30, 49–50)
nPr–S–S–nPr	(C$_2$)	8.62 $\overline{n_S}$		8.87 $\overset{+}{n_S}$							30)

175

R. Gleiter and J. Spanget-Larsen

176

Table (continued)

(a)	(b)	(c)					(d)
iPr–S–S–iPr	(C$_2$)	8.54	8.76				30, 50)
		n$_S^-$	n$_S^+$				
nBu–S–S–nBu	(C$_2$)	8.51	8.72				50)
		n$_S^-$	n$_S^+$				
tBu–S–S–tBu	(C$_2$)	8.17	8.82	(10.6)			30, 49, 50, 109)
		b(n$_S^-$)	a(n$_S^+$)				
HS∼∼SH		9.30	9.61				111)
		n$_S$	n$_S$				
HS∼S∼SH		8.85	9.52	11.32	12.33	13.2	35)
		n$_S$	{ n$_S$,n$_S$				
Me–S∼∼S–Me		8.65	8.90				29)
		n$_S$	n$_S$				

Structure	Symmetry							Ref.
Me-S, S-Me	(C_{3v})	(8.5) $a_2(n_S)$	(9.0) $e(n_S)$					34, 35)
Me-S / S-Me (cross)	(D_{2d}^*)	8.29 $b_1(n_S)$	8.69 $a_2(n_S)$	8.69 $e(n_S)$	10.0			45)
Me-S⁓S-Me		8.64 n_S	8.90 n_S					29)
△ S (triangle)	(C_{2v})	9.00 $b_1(n_S)$	11.30 b_2	11.86 a_1	13.5 a_2	15.2	16.4	113, 114)
◇ S (square)	(C_{2v}^*)	8.65	10.60	11.89	12.60	13.27	14.13 / 15.38	115)
⬠ S (pentagon)		8.42 n_S						116)
⬠ S-S	(C_2)	8.25 $a(n_S^-)$	9.97 $b(n_S^+)$	11.2	11.6	12.4		35)

177

R. Gleiter and J. Spanget-Larsen

Table (continued)

(a)	(b)	(c)				(d)
(5-membered dithiolane ring)		8.77 n_S^-	9.12 n_S^+			45)
(structure) "half chair"	(C_2)	8.72 $a(n_S)$	9.10 $b(n_S)$	10.19 $b(n_S)$	11.06 $b(n_S)$	50, 57)
(6-membered thiane ring)	(C_{2v}^*)	8.45 $b_1(n_S)$	(10.5)			32)
(6-membered dithiane ring)	(C_2)	8.36 $a(n_S^-)$	9.31 $b(n_S^+)$	10.9		30)
(6-membered dithiane ring)	(C_s)	8.54 $a''(n_S^-)$	8.95 $a'(n_S^+)$	(10.9)		32, 45)
(6-membered dithiane ring)	(C_{2h})	8.58 $b_u(n_S^+)$	9.03 $a_g(n_S^-)$	(10.7)		32)

Structure	Symmetry					Ref.
(1,3-dithiane ring)	(C_{3v})	8.76 $e(n_S)$	9.27 $a_1(n_S)$	(11.2)		32, 33)
(tetramethyl, "twist")	(D_2)	8.23 $b_2(n_S)$	8.85 $b_1(n_S)$	9.04 $a(n_S)$	9.44 $b_3(n_S)$	58)
(spiro "chair")	(C_{2h})	8.17 $b_g(n_S)$	8.39 $a_u(n_S)$	8.4 $a_g(n_S)$	10.03 $b_u(n_S)$	58)
("chair")	(C_{2h})	7.98 $b_g(n_S)$	8.34 $a_u(n_S)$	8.5 $a_g(n_S)$	(10.0) $b_u(n_S)$	58)
(7-membered ring)	(C_2)	8.20 $a(\bar{n}_S)$	8.59 $b(\overset{+}{n}_S)$			29)
(bicyclic)		8.55 $\bar{n}_S$	8.81 $\overset{+}{n}_S$	9.72 w	10.4	31)

179

R. Gleiter and J. Spanget-Larsen

Table (continued)

(a)	(b)	(c)		(d)
		8.41 n_S^-	8.82 n_S^+ 10.4	31)
		8.29 n_S^-	8.80 n_S^+ 10.5	31)
		8.22 n_S^-	8.63 n_S^+ 10.2	31)
		8.19 n_S^-	8.68 n_S^+ 10.3	31)
	(C_s)	8.07 $a'(n_S)$	9.06 $a'(\pi)$	40)

Compound	Symmetry							Ref.
(two fused dithiolane rings)	(C_{2h})	8.6–9.0 {$b_g(n_S)$, $a_u(n_S)$, $a_g(n_S)$, $b_u(n_S)$}		10.53	11.00			76)
(spiro bis-dithiolane)	(D_{2d}^*)	8.26 $a_2(n_S)$	8.71 $e(n_S)$	9.01	9.36 $b_1(n_S)$	10.77		44, 45)
(spiro bis-dithiane)	(D_{2d}^*)	8.09 $b_1(n_S)$	8.64 $e(n_S)$	8.85 $a_2(n_S)$	10.40			45)
S–Me (vinyl)	(C_s)	8.45 $a''(n_S)$	11.0 $a''(\pi)$	11.5	12.5	13.8	14.9	59)
S–Me / Me–S	(C_{2h})	7.85 $a_u^-(n_S)$	9.20 $b_g(n_S^-)$	10.85 $a_u(\pi)$	11.2 σ	11.75	12.6	59)
Me–S / S–Me	(C_{2v})	7.80 $b_1(n_S^+)$	9.15 $a_2(n_S^-)$	10.65 $b_1(\pi)$	11.15 σ	11.85	12.6	59)
S–Me / S–Me	(C_{2v})	8.2 $b_1(n_S^+)$	8.8 $a_2(n_S^-)$	10.4 $b_1(\pi)$	11.1 σ	11.8		59)

181

Table (continued)

(a)	(b)	(c)							(d)
Me–S, S–Me (Me–S, S–Me)	(D$_{2h}$*)	7.75 b$_{1u}$(n$_S$)	8.58 a$_u$(n$_S$)	8.85 b$_{2g}$(n$_S$)	9.20 b$_{3g}$(n$_S$)	10.30 σ	10.60 σ	11.40 b$_{1u}$(π)	59)
(bicyclic dithiafulvene structure)	(D$_{2h}$)	7.17 b$_{1u}$(n$_S$)	8.68 a$_u$(n$_S$)	9.08 b$_{2g}$(n$_S$)	9.88 b$_{3g}$(n$_S$)	10.53 σ			80, 81)
HC≡C–S–Me	(C$_s$)	8.81 a"(n$_S$)	10.34 a'(π)	11.62 a"(π)	12.59	(14.3)			35, 90)
Me–S–C≡C–S–Me	(C$_2$)	8.25 { a(n$_S$) b(n$_S$)	8.55 b(n$_S$)	10.80 b(π)	11.12 b(π)	12.08	12.34		35)
Me–S, S–Me (diene)	Planar ~ 100%	7.48 n$_S$							29)
H–S–(phenyl)	(C$_{2v}$*)	8.47 b$_1$(n$_S$)	9.40 a$_2$(π)	10.62 b$_1$(π)					62)

Compound	Conformation	Values				Ref.
Me–S–Ph	Planar ~90%	8.02	9.25	10.2	11.12	60–63)
	Non-planar ~10%	8.55	9.25	9.25	11.10	
	(C_{2v}^{*})	$b_1(n_S)$	$a_2(\pi)$	$b_1(\pi)$	σ	
Et–S–Ph	Planar ~40%	8.0	10.12			62)
	Non-planar ~60%	8.53	9.29			
	(C_{2v}^{*})	$b_1(n_S)$	$b_1(\pi)$			
iPr–S–Ph	Non-planar ~85%	8.46	9.24			62)
	(C_{2v}^{*})	$b_1(n_S)$	$b_1(\pi)$			
tBu–S–Ph	Non-planar ~95%	8.40	9.14			62)
	(C_{2v}^{*})	$b_1(n_S)$	$b_1(\pi)$			
Me–S–C$_6$H$_4$–S–Me		7.93	8.80	9.28	10.10	60, 61)
		π_1	n_S^{+}	π_2	π_3	

183

R. Gleiter and J. Spanget-Larsen

184

Table (continued)

(a)	(b)	(c)						(d)
(styryl–S–Me)		7.75 π_1	9.13 π_2	9.13 π_3	> 10.7			29, 61)
(naphthyl–S–Me)		7.67 π_1	8.72 π_2	8.72 π_3	10.35 π_4			61)
(naphthyl–S–Me)		7.71 π_1	8.38 π_2	9.39 π_3	10.03 π_4			61)
(thioxanthene)		7.95	8.52	9.33	10.45			117)
(dibenzothiepine)	(C_{2v}^*)	8.5 $b_1(\pi)$	8.9 $a_2(\pi)$	9.0 $a_2(\pi)$	9.1 $b_1(\pi)$	11.0	11.6	118)

Structure	Symmetry							Ref.
(thiophene)	(C_{2v})	8.87 $a_2(\pi)$	9.52 $b_1(\pi)$	12.1 $a_1(\sigma)$	12.7 $b_1(\pi)$	13.3 $b_1(\sigma)$	(13.9) $a_1(\sigma)$ 14.3 $b_2(\sigma)$	67–72)
(bithiophene)	(C_{2h}^{*})	8.05 $b_g(\pi)$	9.14 $a_u(\pi)$	9.37 $b_g(\pi)$	9.77 $a_u(\pi)$			119)
(bithiophene)	(C_{2h}^{*})	8.19 $b_g(\pi)$	9.15 $a_u(\pi)$	9.15 $b_g(\pi)$	10.03 $a_u(\pi)$			119)
(macrocycle)	(C_{2h})	7.9 $a_u(\pi)$	7.9 $b_g(\pi)$	8.2 $a_g(\pi)$	9.08 $b_u(\pi)$			120)
(bicyclic thiophene)	(C_2)	8.15 $a(\pi)$	8.65 $b(n_S)$ 8.75 $b(\pi)$	9.05 $a(\pi)$	9.8 $b(\pi)$	11.2		121)
(bicyclic thiophene)	(C_2)	8.4 $a(\pi)$	8.4 $b(n_S)$	8.9 $b(\pi)$	8.9 $a(\pi)$	9.78 $b(\pi)$	11.1	121)

185

R. Gleiter and J. Spanget-Larsen

Table (continued)

(a)	(b)	(c)					(d)
	(C_2)	8.40 $b(\pi)$	8.95 $a(\pi)$	9.45 $b(\pi)$	9.45 $a(\pi)$	10.15 $b(\pi)$	122)
	(C_2)	8.06 $b(\pi)$	9.16 $b(\pi)$	9.16 $a(\pi)$	9.50 $a(\pi)$	10.40 $b(\pi)$	122)
	(C_{2v})	8.32 $b_1(\pi)$	8.41 $a_2(\pi)$	10.08 $b_1(\pi)$	11.27 $a_2(\pi)$		76, 123)
	(C_{2h})	8.10 $a_u(\pi)$	8.61 $a_u(\pi)$	10.04 $b_g(\pi)$	11.5 $b_g(\pi)$		76, 123)
	(C_s)	8.20 $a''(\pi)$	8.76 $a''(\pi)$	10.08 $a''(\pi)$	11.03		67, 123)
	(C_{2v})	7.75 $a_2(\pi)$	8.90 $b_2(\pi)$	9.90 $a_2(\pi)$	11.33		123, 124)

(C$_{2v}$)	7.93 b$_1$(π)	8.34 a$_2$(π)	9.26 a$_2$(π)	9.96 b$_1$(π)	10.65 b$_1$(π)	11.38	11.66	(125)
(C$_{2v}$)	7.67 b$_1$(π)	7.86 a$_2$(π)	9.17 a$_2$(π)	9.83 b$_1$(π)	(10.7) b$_1$(π)			(126)
(D$_{3h}$)	7.76 e″(π) / (8.3) a$_1''$(π)	9.61 e″(π)	10.23 a$_2''$(π)					(126)
	7.14	8.5	8.7	9.0				(126)
(C$_{2v}^{*}$)	7.96	8.39	8.99	9.34	9.69	10.46	11.07	(118, 127)
	7.58	8.10	9.0	9.2				(117)

Table (continued)

(a)	(b)	(c)							(d)
	(D$_{2h}$)	6.83 b$_{1u}(\pi)$	8.69 b$_{3g}(\pi)$	9.76 a$_u(\pi)$	10.18 b$_{2g}(\pi)$	10.56 b$_{1u}(\pi)$	11.08 b$_{1g}(\sigma)$	11.7	79–82)
	(D$_{2h}$)	6.81 b$_{1u}(\pi)$	8.42 b$_{3g}(\pi)$	8.52 a$_u(\pi)$	8.86 b$_{2g}(\pi)$	9.66 b$_{1u}(\pi)$	(10.2)	10.4	128)
	(C$_{2v}$)	7.15 a$_2(\pi)$	8.95 b$_1(\pi)$	9.30 b$_1(\pi)$	9.44 a$_2(\pi)$	11.18 b$_1(\pi)$			53)
	(D$_{2h}$)	6.75	8.42	9.1	9.1	10.02			129)
	(D$_{2h}^{*}$)	6.19 a$_u(\pi)$	7.85 b$_{1g}(\pi)$	8.35 – 9.35 b$_{2u}(\pi)$, 8x$\pi\Phi$		10.75			103–105)

Compound	Symmetry	Ionization energies (eV) with assignments	Ref.
(dithiin structure)	(C_{2v})	8.11 $a_1(\sigma)$; 8.27 $a_2(\pi)$; 9.58 $b_1(\pi)$; 10.01 $b_1(\pi)$; 10.64; 11.10	100, 101, 106
H–C(=S)–H	(C_{2v})	9.38 $b_2(n_{CS})$; 11.76 $b_1(\pi_{CS})$; 13.85 $a_1(\sigma_{CS})$; 15.20 b_2; 19.9 a_1	13, 48, 83
Me–C(=S)–H	(C_s)	8.98 $a'(n_{CS})$; 10.87 $a''(\pi_{CS})$; 12.74	33
Me–C(=S)–Me	(C_{2v})	8.60 $b_2(n_{CS})$; 10.46 $b_1(\pi_{CS})$; 12.40	33
(thiocamphor structure)		8.1 n_{CS}; 9.6 π_{CS}	85
(thiobenzophenone structure)	(C_2)	8.0 $b(n_{CS})$; 8.77 $b(\pi_{CS})$; 9.2; 9.4; 10.6	86
(dithione cyclobutane structure)	(D_{2h})	8.31 $b_{3g}(n_{CS})$; 8.76 $b_{2u}(n_{CS})$; 10.2 $b_{3u}(\pi_{CS})$; 10.7 $b_{2g}(\pi_{CS})$	87

189

Table (continued)

(a)	(b)	(c)						(d)
Me–C(=S)–SH		(8.7)			(12.3)			90)
Me–S–C(=S)–S–Me		8.5 n_{CS}	8.97 π	8.97 π	11.40			85)
(four-membered ring)	(C$_{2v}$)	8.83 $b_1(\pi)$	9.09 $b_2(n_{CS})$	9.89 $a_2(\pi)$	11.3	12.2	12.8	93)
(five-membered ring)	(C$_{2v}^*$)	8.40 $b_1(\pi)$	8.87 $b_2(n_{CS})$	9.42 $a_2(\pi)$	11.42			85, 92, 93)
(six-membered ring)	(C$_{2v}^*$)	8.40 $b_1(\pi)$	8.56 $b_2(n_{CS})$	9.29 $a_2(\pi)$	10.84	11.46	11.68	85)
(ring)	(C$_{2v}$)	8.33 $b_2(n_{CS})$	8.56 $b_1(\pi)$	10.60 $a_2(\pi)$	10.90 $b_1(\pi)$	12.00 $b_2(\sigma)$	12.37 $a_1(\sigma)$	93–95)
(benzo ring)	(C$_{2v}$)	8.12 $b_2(n_{CS})$	8.36 $b_1(\pi)$	9.09 $a_2(\pi)$	10.04 $b_1(\pi)$	10.84 $a_2(\pi)$	11.76 $b_2(\sigma)$	94, 95)

190

Compound	Symmetry							Ref.
(thiophene-dithiole ring structure)	(C_s)	7.74	8.49	9.26	9.71			130)
(ring structure with =S)	(C_s)	8.42 $a''(\pi)$	8.42 $a'(n_{CS})$	10.52 $a''(\pi)$	10.79 $a''(\pi)$	12.42 $a'(\sigma)$	12.79 $a'(\sigma)$	131)
(benzo ring structure with =S)	(C_s)	8.10 $a''(\pi)$	8.33 $a'(n_{CS})$	9.32 $a''(\pi)$	9.82 $a''(\pi)$	10.81 $a''(\pi)$	11.41 $a''(\pi)$	131)
$CH_2{=}C{=}S$	(C_{2v})	8.89 $b_1(\pi)$	11.32 $b_2(\pi)$	12.14 $b_1(\pi)$	14.55 $a_1(\sigma)$	15.5 $b_2(\sigma)$		90)
$S{=}C{=}S$	$(D_{\infty h})$	10.09 π_g	12.83 π_u	14.47 σ_u	16.19 σ_g			2)
$S{=}C{=}C{=}C{=}S$	$(D_{\infty h})$	9.09 $\pi_u(n_S)$	11.24 $\pi_g(CS)$	12.87 $\pi_u(CC)$	14.47 $\sigma_g^+(n_S)$	14.87 $\sigma_u^+(n_S)$		17)
$C{\equiv}S$	$(C_{\infty v})$	11.33 7σ	12.79 2π	15.84 6σ	18.00 + "shake up"			5–9)

R. Gleiter and J. Spanget-Larsen

Acknowledgements. Financial support has been obtained by the Deutsche Forschungsgemeinschaft, the Fonds der Chemischen Industrie and the Otto Röhm Stiftung. We would like to express our most sincere thanks to the many collaborators whose names appear in the References, for their contributions. We are grateful to Profs. H. Bock and W. von Niessen for the communication of results prior to publication.

VIII References

1. See f. i. Senning, A. (ed.): Sulfur in organic and inorganic chemistry, Vol. 1. New York: Marcel Dekker 1971; Price, C. C., Oae, S.: Sulfur bonding. New York: Ronald Press 1962
2. Turner, D. W., Baker, A. D., Baker, C., Brundle, C. R.: Molecular photoelectron spectroscopy. London: Wiley-Interscience 1970
3. Brundle, C. R., Baker, A. D. (eds.): Electron spectroscopy: Theory, techniques and applications, Vol. 1. London: Academic Press 1977
4. Koopmans, T.: Physica *1*, 104 (1934); Bock, H.: Angew. Chem. *89*, 631 (1977); Angew. Chem. Int. Ed. Engl. *16*, 724 (1977)
5. Domcke, W., Cederbaum, L. S., von Niessen, W., Kraemer, W. P.: Chem. Phys. Letters *43*, 258 (1976)
6. Schirmer, J., Domcke, W., Cederbaum, L. S., von Niessen, W.: J. Phys. *B11*, 1901 (1978)
7. Jonathan, N., Morris, A., Okuda, M., Smith, D. J., Ross, K. J.: Chem. Phys. Letters *13*, 334 (1972); Jonathan, N., Morris, A., Okuda, M., Ross, K. J., Smith, D. J.: Discuss. Faraday Soc. *54*, 48 (1972)
8. King, G. H., Kroto, H. W., Suffolk, R. J.: Chem. Phys. Letters *13*, 457 (1972)
9. Frost, D. C., Lee, S. T., McDowell, C. A.: Chem. Phys. Letters *17*, 153 (1972)
10. Price, W. C.: Ref. 3, Chap. 4
11. Almlöf, J., Johansen, H., Roos, B., Wahlgren, U.: J. Electron Spectr. *2*, 51 (1973)
12. Von Niessen, W., Diercksen, G. H. F., Cederbaum, L. S.: Chem. Phys. *10*, 345 (1975); von Niessen, W., Kraemer, W. P., Diercksen, G. H. F.: Chem. Phys. *41*, 113 (1979)
13. Von Niessen, W., Cederbaum, L. S., Domcke, W., Diercksen, G. H. F.: J. Chem. Phys. *66*, 4893 (1977)
14. Potts, A. W., Price, W. C.: Proc. Roy. Soc. London *A326*, 181 (1972)
15. Sakai, H., Yamabe, S., Yamabe, T., Fukui, K., Kato, H.: Chem. Phys. Letters *25*, 541 (1974)
16. See f. i. Brundle, C. R., Robin, M. B., in Nachod, F., Zuckerman, G. (eds.): Determination of organic structures by physical methods, Vol. 3, p. 1. New York: Academic Press 1971
17. Ginsberg, A. P., Brundle, C. R.: J. Chem. Phys. *68*, 5231 (1978)
18. Cowan, D. O., Gleiter, R., Glemser, O., Heilbronner, E., Schäublin, J.: Helv. Chim. Acta *54*, 1559 (1971)
19. Dixon, R. N., Duxbury, G., Fleming, G. R., Hugo, J. M. V.: Chem. Phys. Letters *14*, 60 (1972)
20. DeKock, R. L., Lloyd, D. R., Breeze, A., Collins, G. A. D., Cruickshank, D. W. J., Lempka, H. J.: Chem. Phys. Letters *14*, 525 (1972)
21. Gleiter, R., Bartetzko, R.: Unpublished; cf. Sect. V
22. Heilbronner, E., Maier, J. P.: Ref. 3, Chap. 5
23. Roos, B., Siegbahn, P.: Theor. Chim. Acta *21*, 368 (1971)
24. Domcke, W., Cederbaum, L. S., Schirmer, J., von Niessen, W., Maier, J. P.: J. Electron Spectr. *14*, 59 (1978)
25. Elkel, S., Bergmann, H., Ensslin, W.: J. Chem. Soc., Faraday Trans. II *70*, 555 (1974)
26. Åsbrink, L., Fridh, C., Jonsson, B. Ö., Lindholm, E.: Int. J. Mass Spectrom. Ion Phys. *8*, 215 (1972)
27. Cradock, S., Whiteford, R. A.: J. Chem. Soc. Faraday Trans. II *68*, 281 (1972)
28. Frost, D. C., Herring, F. G., Katrib, A., McDowell, C. A., McLean, R. A. N.: J. Phys. Chem. *76*, 1030 (1972)

29. Bock, H., Wagner, G.: Angew. Chem. *84*, 119 (1972); Angew. Chem. Int. Ed. Engl. *11*, 150 (1972)
30. Wagner, G., Bock, H.: Chem. Ber. *107*, 68 (1974)
31. Kobayashi, M., Gleiter, R.: Z. Naturforsch. *33b*, 1057 (1978)
32. Schweigart, D. A. Turner, D. W.: J. Am. Chem. Soc. *94*, 5599 (1972)
33. Kroto, H. W., Landsberg, B. M., Suffolk, R. J., Vodden, A.: Chem. Phys. Letters *29*, 265 (1974)
34. Schulz, W.: Master Thesis, University of Frankfurt (1973)
35. Stein, U. C.: Master Thesis, University of Frankfurt (1975)
36. Hoffmann, R., Imamura, A., Hehre, W. J.: J. Am. Chem. Soc. *90*, 1499 (1968); Hoffmann, R.: Acc. Chem. Res. *4*, 1 (1971)
37. Gleiter, R.: Angew. Chem. *86*, 770 (1974); Angew. Chem. Int. Ed. Engl. *13*, 696 (1974)
38. Zefirov, N. S., Blagoveshchensky, V. S., Kazimirchik, I. V., Surova, N. S.: Tetrahedron *27*, 3111 (1971)
39. Gleiter, R., Kobayashi, M., Zefirov, N. S., Paliulin, K.: Doklady Akadem. Nauk USSR *235*, 347 (1977)
40. Böhm, M. C., Gleiter, R.: Tetrahedron *35*, 675 (1979)
41. Simmons, H., Fukunaga, T.: J. Am. Chem. Soc. *89*, 5208 (1967)
42. Hoffmann, R., Imamura, A., Zeiss, G. D.: J. Am. Chem. Soc. *89*, 5215 (1967)
43. Dürr, H., Gleiter, R.: Angew. Chem. *90*, 591 (1978); Angew. Chem. Int. Ed. Engl. *17*, 559 (1978), and literature cited
44. Baker, A. D., Bisk, M. A., Venanzi, T. S., Kwon, Y. S., Sadka, S., Liotter, D. C.: Tetrahedron Letters *1976*, 3415
45. Kobayashi, M., Gleiter, R., Coffen, D. L., Bock, H., Schulz, W., Stein, U.: Tetrahedron *33*, 433 (1977)
46. Winnewisser, M., Haase, J.: Z. Naturforsch. *23a*, 56 (1968)
47. Sutter, D., Dreizler, H., Rudolph, H. D.: Z. Naturforsch. *20a*, 1676 (1965)
48. Kroto, H. W., Suffolk, R. J.: Chem. Phys. Letters *15*, 545 (1972)
49. Baker, A. D., Brisk, M., Gellender, M.: J. Electron Spectr. *3*, 227 (1974)
50. Guimon, M.-F., Guimon, C., Pfister-Guillouzo, G.: Tetrahedron Letters *1975*, 441
51. Stroud, R. M., Carlisle, C. H.: Acta Cryst. *28*, 304 (1972)
52. Foss, O., Johnsen, K., Reistad, T.: Acta Chem. Scand. *18*, 2345 (1964)
53. Sandman, D. J., Ceasar, G. P., Nielsen, P., Epstein, A. J., Holmes, T. J.: J. Am. Chem. Soc. *100*, 202 (1978)
54. Riga, J., Verbist, J. J., Wudl, F., Kruger, A.: J. Chem. Phys. *69*, 3221 (1978)
55. Riga, J., Verbist, J. J., Lamotte, C., Andre, J.-M.: Bull. Soc. Chim. Belg. *87*, 163 (1978)
56. Bergson, G.: Ark. Kem. *12*, 233 (1958)
57. Guimon, M. F., Guimon, C., Metras, F., Pfister-Guillouzo, G.: Can. J. Chem. *54*, 146 (1976)
58. Guimon, M. F., Guimon, C., Metras, F., Pfister-Guillouzo, G.: J. Am. Chem. Soc. *98*, 2078 (1976)
59. Bock, H., Wagner, G., Wittel, K., Sauer, J., Seebach, D.: Chem. Ber. *107*, 1869 (1974)
60. Bock, H., Wagner, G., Kroner, J.: Tetrahedron Letters *1971*, 3713
61. Bock, H., Wagner, G., Kroner, J.: Chem. Ber. *105*, 3850 (1972)
62. Dewar, P. S., Ernstbrunner, E., Gilmore, J. R., Godfrey, M., Mellor, J. M.: Tetrahedron *30*, 2455 (1974)
63. Schweig, A., Thon, N.: Chem. Phys. Letters *38*, 482 (1976)
64. Streitwieser, A., Jr.: Molecular orbital theory for organic chemists. New York: John Wiley 1961
65. Heilbronner, E., Bock, H.: Das HMO Modell und seine Anwendung. Weinheim/Bergstr.: Verlag Chemie 1968
66. Dewar, M. J. S., Dougherty, R. C.: The PMO theory of organic chemistry. New York: Plenum Press 1975
67. Eland, J. H. D.: Int. J. Mass Spectrom. Ion Phys. *2*, 471 (1969)
68. Baker, A. D., Turner, D. W., Phil. Trans. Roy. Soc. *A268*, 131 (1970)
69. Dewar, M. J. S., Harget, A. J., Trinajstić, N., Worley, S. D.: Tetrahedron *26*, 4505 (1970)

70. Derrick, P. J. Asbrink, L., Edqvist, O., Jonsson, B.-Ö., Lindholm, E.: Int. J. Mass Spectrom. Ion Phys. *6*, 161, 177 (1971)
71. Gelius, U., Allan, C. J. Johansson, G., Siegbahn, H., Allison, D. A., Siegbahn, K.: Physica Scripta *3*, 237 (1971)
72. von Niessen, W., Kraemer, W. P., Cederbaum, L. S.: J. Electron Spectr. *8*, 179 (1976)
73. Schäfer, W., Schweig, A., Gronowitz, S., Taticchi, A., Fringuelli, F.: J. Chem. Soc., Chem. Commun. *1973*, 541
74. Distefano, G., Pignataro, S., Innorta, G., Fringuelli, F., Marino, G., Tatticchi, A.: Chem. Phys. Letters *22*, 132 (1973)
75. Fringuelli, F., Marino, G., Tatticchi, A., Distefano, G., Colonna, F. P., Pignataro, S.: J. Chem. Soc. Perkin II *1976*, 276
76. Gleiter, R., Kobayashi, M., Spanget-Larsen, J., Gronowitz, S., Konar, A., Farnier, M.: J. Org. Chem. *42*, 2230 (1977)
77. Ferraris, J. P., Cowan, D. O., Walatka, V., Perlstein, J. H.: J. Am. Chem. Soc. *95*, 948 (1973)
78. Garito, A. F., Heeger, A. J.: Acc. Chem. Res. *7*, 232 (1974)
79. Gleiter, R., Schmidt, E., Cowan, D. O., Ferraris, J. P.: J. Electron Spectr. *2*, 207 (1973)
80. Berlinsky, A. J., Carolan, J. F., Weiler, L.: Can. J. Chem. *52*, 3373 (1974)
81. Gleiter, R., Kobayashi, M., Spanget-Larsen, J., Ferraris, J. P., Bloch, Aa. N., Bechgaard, K., Cowan, D. O.: Ber. Bunsenges. Phys. Chem. *79*, 1218 (1979)
82. Schweig, A., Thon, N., Engler, E. M.: J. Electron Spectr. *12*, 335 (1977)
83. Solouki, B., Rosmus, P., Bock, H.: J. Am. Chem. Soc. *98*, 6054 (1976)
84. Pauling, L.: The nature of the chemical bond. Ithaca: Cornell University Press 1960
85. Guimon, C., Gonbeau, D., Pfister-Guillouzo, G., Åsbrink, L., Sandström, J.: J. Electron Spectr. *4*, 49 (1974)
86. Bernardi, F., Colonna, F. P., Distefano, G., Maccagnani, G., Spunta, G.: Z. Naturforsch. *33a*, 468 (1978)
87. Gleiter, R., Kobayashi, M.: Unpublished results
88. Cowan, D. O., Gleiter, R., Hashmall, J. A., Heilbronner, E., Hornung, V.: Angew. Chem. *83*, 405 (1971); Angew. Chem. Int. Ed. Engl. *10*, 401 (1971)
89. Brint, P., Wittel, K., Hochmann, P., Felps, W. S., McGlynn, S. P.: J. Am. Chem. Soc. *98*, 7980 (1976)
90. Bock, H., Solouki, B., Bert, G., Rosmus, P.: J. Am. Chem. Soc. *99*, 1663 (1977)
91. Kimura, K., Katsumata, S., Ishiguro, T., Hirakawa, A. Y., Tsuboi, M.: Bull. Chem. Soc. Japan *49*, 937 (1976)
92. Schweigart, D. A., Turner, D. W.: J. Am. Chem. Soc. *94*, 5592 (1972)
93. Wittel, K., Astrup, E. E., Bock, H., Graeffe, G., Juslén, H.: Z. Naturforsch. *30b*, 862 (1975)
94. Guimon, C., Pfister-Guillouzo, G., Arbelot, M.: J. Mol. Struct. *30*, 339 (1976)
95. Spanget-Larsen, J., Gleiter, R., Kobayashi, M., Engler, E. M., Shu, P., Cowan, D. O.: J. Am. Chem. Soc. *99*, 2855 (1977)
96. Guimon, C., Pfister-Guillouzo, G., Arbelot, M., Chanon, M.: Tetrahedron *30*, 3831 (1974)
97. Bock, H., Solouki, B.: Chem. Ber. *107*, 2299 (1974)
98. Solouki, B., Bock, H., Appel, R.: Chem. Ber. *108*, 897 (1975)
99. Gleiter, R., Gygax, R.: Topics in Curr. Chem. *63*, 49 (1976)
100. Gleiter, R., Hornung, V., Lindberg, B., Högberg, S., Lozac'h, N.: Chem. Phys. Letters *11*, 401 (1971)
101. Gleiter, R., Gygax, R., Reid, D. H.: Helv. Chim. Acta *58*, 1591 (1975)
102. Gleiter, R., Hoffmann, R.: Tetrahedron *24*, 5899 (1968)
103. Müller, C., Schweig, A., Cava, M. P., Lakshmikantham, M. V.: J. Am. Chem. Soc. *98*, 7187 (1976)
104. Gleiter, R., Bartetzko, R., Brähler, G., Bock, H.: J. Org. Chem. *43*, 3893 (1978)
105. Gotthardt, H., Reiter, F., Gleiter, R., Bartetzko, R.: Chem. Ber., in press
106. Palmer, M. H., Findlay, R. H.: J. Chem. Soc. Perkin II *1974*, 1885; J. Mol. Struct. *37*, 229 (1977)
107. Bartetzko, R., Gleiter, R.: Angew. Chem. *90*, 481 (1978); Angew. Chem. Int. Ed. Engl. *17*, 468 (1978)

108. Koenig, T., Wielesek, R., Snell, W., Balle, T.: J. Am. Chem. Soc. *97*, 3225 (1975); Koenig, T., Southworth, S.: J. Am. Chem. Soc. *99*, 2807 (1977); Houle, A., Beauchamp, J. L.: J. Am. Chem. Soc. *100*, 3290 (1978)
109. Bock, H., Mohmand, S.: Angew. Chem. *89*, 105 (1977); Angew. Chem. Int. Ed. Engl. *16*, 104 (1977)
110. Bock, H., Solouki, B., Bert, G., Hirabayashi, T., Mohmand, S., Rosmus, P.: Nachr. Chem. Tech. Lab. *26*, 634 (1978), and literature cited
111. Schäfer, W., Schweig, A.: Z. Naturforsch. *30a*, 1785 (1975)
112. Robinson, J. W. (ed.): Handbook of spectroscopy, Vol. I. Cleveland: CRC Press 1974
113. Frost, D. C., Herring, F. G., Katrib, A., McDowell, C. A.: Chem. Phys. Letters *20*, 401 (1973)
114. Schweig, A., Thiel, W.: Chem. Phys. Letters *21*, 541 (1973)
115. Mollere, P. D., Houk, K. N.: J. Am. Chem. Soc. *99*, 3226 (1977)
116. Pignataro, S., Distefano, G.: Chem. Phys. Letters *26*, 356 (1974)
117. Gleiter, R.: Unpublished results
118. Sondergeld, W.: Master Thesis, Technische Hochschule Darmstadt (1976)
119. Meunier, P., Coustale, M., Guimon, C., Pfister-Guillouzo, G.: J. Mol. Struct. *36*, 233 (1977)
120. Bernardi, F., Bottoni, A., Colonna, F. P., Distefano, G., Folli, U., Vivarelli, P.: Z. Natur-forsch. *33a*, 959 (1978)
121. Meunier, P., Pfister-Guillouzo, G.: Can. J. Chem. *55*, 2867 (1977)
122. Meunier, P., Pfister-Guillouzo, G.: Can. J. Chem. *55*, 3901 (1977)
123. Clark, P. A., Gleiter, R., Heilbronner, E.: Tetrahedron *29*, 3085 (1973)
124. Palmer, M. H., Kennedy, S. M. F.: J. Chem. Soc. Perkin II *1976*, 81
125. Ruščić, B., Kovač, B., Klasinc, L., Güsten, H.: Z. Naturforsch. *33a*, 1006 (1978)
126. Gleiter, R., Hart, H., Wudl, F.: Unpublished results
127. Güsten, H., Klasinc, L., Táth, T., Knop, J. V.: J. Electron Spectr. *8*, 417 (1976)
128. Spanget-Larsen, J., Gleiter, R., Hünig, S.: Chem. Phys. Letters *37*, 29 (1976)
129. Gleiter, R., Wudl, F.: Unpublished results
130. Gleiter, R., Engler, E. M.: Unpublished results
131. Gonbeau, D., Guimon, C., Deschamps, J., Pfister-Guillouzo, G.: J. Electron Spectr. *6*, 99 (1975)
132. Mellink, W. A., Janssen, M. J.: J. Chem. Res. (S) *1978*, 422; J. Chem. Res. (M) *1978*, 4966

Received January 19, 1979

Photoelectron Spectra and Bonding in Small Ring Hydrocarbons

Rolf Gleiter

Institut für Organische Chemie der Universität, D-6900 Heidelberg, Germany

Dedicated to Professor Dr. W. Lüttke on the occasion of his 60th birthday

Table of Contents

I Introduction

The last decade has seen the application of He(I) photoelectron (PE) spectroscopy to a large number of organic compounds. Numerous reviews have appeared, describing the technique[1−4] and the different approaches to interprete[1−11] the PE spectra.

In this chapter only those aspects shall be commented on which are necessary to understand the interpretation given in the following paragraphs:

a) Technique of He(I) PE spectroscopy. A conventional light source to study the photoionization of molecules M is an open helium discharge lamp as discussed by Turner[1, 12]. This light source emits photons with an energy of 21.21 eV (He(I)$_\alpha$-

$$M + h\nu \longrightarrow M^{\ddagger} + e \qquad (1)$$

line), and thus all ionization potentials can be measured up to this energy-value. The emitted electrons are sorted according to their kinetic energy and are counted (counts/s or C/S). From the known energy of the light source [$E(h\nu)$] and the measured kinetic energy of the emitted electrons [$E_{kin}(e)$] the vertical ionization potential ($I_{V, J}$) is calculated according to (2)

$$I_{V, J} = E(h\nu) - E_{kin}(e). \qquad (2)$$

b) Interpretation. The peaks observed in a PE spectrum correspond to energy differences between the ground state Γ of M and the different doublet states, $^2\psi_J$, of M‡ generated by ejecting electrons (see Fig. 1, left). Neglecting configuration interaction, the ground state of M can be described by a single determinant wave function composed of the canonical[13] SCF molecular orbitals, ψ_i, of M which minimize the energy of M.

$$\Gamma = |\psi_i| \qquad (3)$$

The ionization of M leads to different doublet states, $^2\psi$. Approximately, each of these states can be represented by a single configuration, i.e., one single determinant wave function, $^2\psi$, is sufficient to describe these states. The molecular

$$^2\psi = |\Phi_j| \qquad (4)$$

orbitals, Φ_j, which are different from those of ψ_i, were obtained by minimizing the energy of the corresponding states of the cation.

This treatment will give the energies of the ground and different ionic states and thus an interpretation according to (5) is possible:

$$I_{V, J} = E(^2\psi_J) - E(\Gamma). \qquad (5)$$

The calculation of $E(^2\psi_J)$ for the different ionic states is very tedious and thus simpler approximations are usually looked for. Moreover, the calculation of the dif-

199

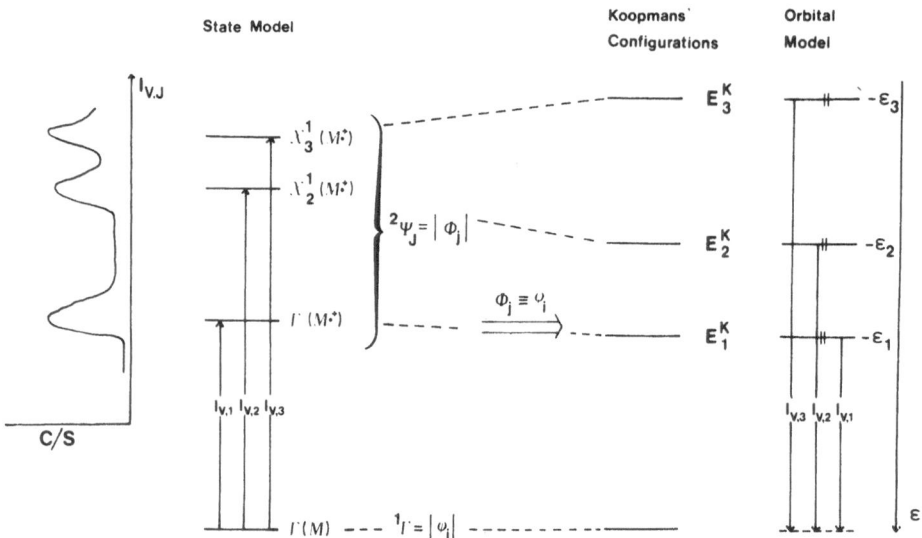

Fig. 1. Observable states of a radical cation $M^{\ddagger}$ (*left*) opposed by the nonobservable Koopmans' states and the negative values of the orbital energies (*right*)

ference in (5) will not yield any simple relation to model properties (wave functions, orbital energies) familiar from molecular orbital theory.

An approximation which allows PE spectroscopic data to correlate with molecular orbital energies derived for the ground state is Koopmans' theorem[14]. This strong simplification can be derived if one assumes that the Φ_j's of (4) can be replaced by the ψ_i's of (3) (see Fig. 1, right). This means that the molecular orbitals of M are not affected by removing an electron. The corresponding electronic configuration is called "Koopmans' configuration". In such a case the energy difference between a Koopmans' configuration E_J^K and the ground state represents according to (6) the

$$I_{V,J} \approx E_J^K - E\,(\Gamma) = -\,\epsilon_J \tag{6}$$

negative value of the orbital energy. This is called Koopmans' theorem[14] (see Fig. 1, right). This approximation is used quite often to interprete PE spectra for the reasons already mentioned: only one calculation on M is needed and it allows the correlation with model values from MO theory. The price one pays is that electron correlation and electron reorganization are neglected. In most cases the energy terms for both effects seem to cancel each other, and thus Koopmans' theorem[14] works rather well for most medium sized and large organic molecules. A breakdown of the validity of Koopmans' theorem has been found whenever the reorganization effects or correlation effects do not cancel. Examples are N_2[15], C_2N_2[16], CS[17] and metallocenes[8]. In the accompanying review on the PE spectra of organic sulfur compounds this is discussed in the case of CS explicitly.

In Fig. 2 we compare the photoelectron spectrum of cyclopropane[1] with results obtained from calculations using Koopmans' theorem ($-\epsilon_J{}^{SCF}$) and by considering

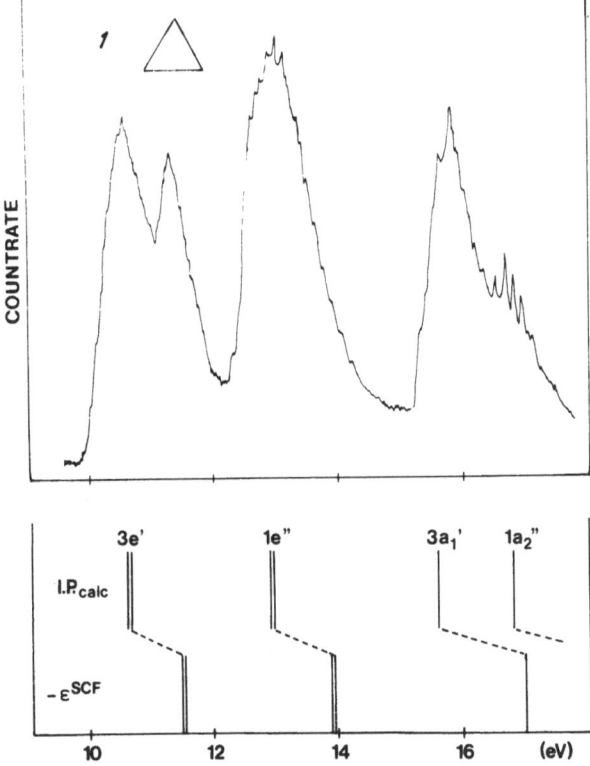

Fig. 2. PE spectrum of cyclopropane (*top*). Comparison
with orbital energies $(-\epsilon_j^{SCF})$
assuming Koopmans' theorem
and by considering electron
reorganization and correlation effects (*I.P.*)

electron reorganization and correlation effects by using a many-body Greens functions method[18]: This comparison demonstrates that Koopmans' theorem is a fairly
good approximation. Aside from the comparison between calculated and observed
ionization energies there are other criteria to aid the interpretation of PE spectra:
band shape and intensity. Usually it is possible to single out bands in the region below
12 eV. The presence of vibrational fine structure in some cases is very helpful.

Small strained hydrocarbon ring systems very often show Gaussian shaped bands
which are relatively broad (see Fig. 2) and fine structure is missing. This can be understood by invoking the Franck-Condon principle[19]. Ionization out of the valence orbitals of a strained hydrocarbon means in the simple picture outlined above ejection
of electrons from bonding σ-orbitals and thus changes in the geometry between ground
and ionic state are expected. As indicated in Fig. 3 transitions to close lying vibronic
states (C-C and C-H vibrations) of the cation occur and thus a Gaussian shape with
strongly overlapping vibrational fine structure is found.

In view of the shortcomings of Koopmans' theorem and the complications encountered in the PE spectra of large organic molecules, PE spectroscopy seems to
be useful only for a few small molecules. However, these limitations can be partly sur-
mounted by studying a series of closely related compounds. Correlation of PE bands
in combination with the perturbation theory leads to a consistent picture. A necessary condition for this kind of treatment is, that individual bands can be identified.

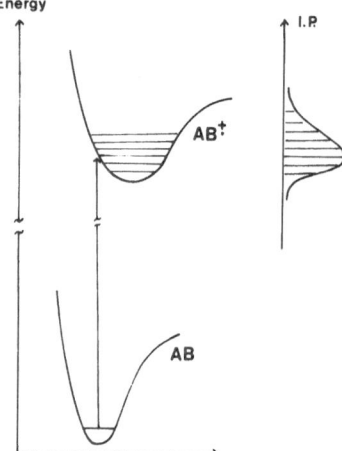

Fig. 3. Qualitative explanation of the origin of a band with Gaussian shape in the PE spectrum of a diatomic molecule

Examples for such a treatment are therefore those molecules containing π bonds, lone pairs or high lying σ orbitals, i.e., orbitals which give rise to bands well separated from the strongly overlapping σ bands above 12 eV.

Examples for compounds providing high lying σ orbitals are strained hydrocarbons. This makes the analysis of their PE spectra possible, using the criteria mentioned above. The PE studies on strained hydrocarbons are of special interest in connection with other physical measurements carried out on these species in the last twenty years. These studies carried out mainly on cyclopropane derivatives indicate deviations from what is expected to be a "normal" hydrocarbon. This has led to the development of several bonding models. In testing these concepts, PE spectroscopy based on Koopmans' theorem has proven to be quite successful.

In the following we will review the PE work carried out on hydrocarbons containing three- and four-membered rings.

II Cyclopropane Rings

1 Bonding Models of Cyclopropane

Two models have been advanced to describe the bonding in the cyclopropane ring: the bent bond model advocated by Coulson, Moffitt[20], and Förster[21] and the model proposed by Walsh[22]. Both descriptions are natural extensions to three centers of theoretical descriptions of the ethylenic two center double bond.

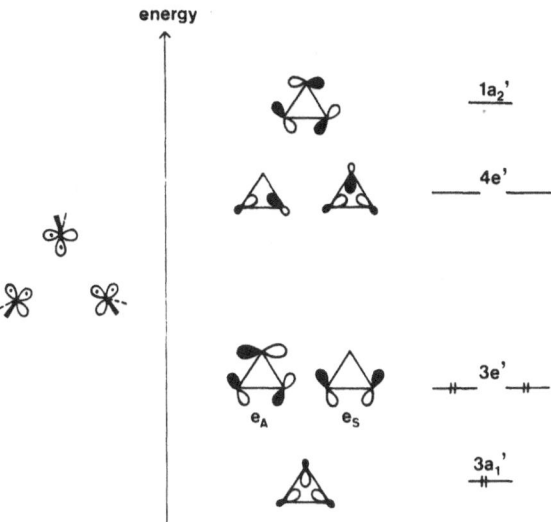

Fig. 4. Derivation of the Walsh representation of the molecular orbitals of cyclopropane from three methylenes

The bent bond picture, later restated in terms of localized molecular orbitals, extends the model presented by Pauling and Slater[24] for ethylene to a system with three centers. Walsh's model parallels that of Mulliken[25] for the σ, π model of the double bond widely used in Hückel theory. Since the canonical orbitals are good models for the interpretation of photoelectron spectra[13] (see Introduction) we will discuss the Walsh model briefly.

Three trigonal methylene groups are brought together as shown in Fig. 4. Each group provides an sp^2 σ-orbital and a 2p orbital which will interact with each other. From the three σ orbitals a strongly bonding ($3a_1'$) and two antibonding ($4e'$) linear combinations result. The three 2p orbitals interact less strongly and yield two bonding ($3e'$) and one antibonding ($1a_2'$) linear combination.

The resulting level scheme and wave functions explain nicely the unusual conjugating properties of cyclopropane. The overlap in the $3e'$ set is partly σ partly π and thus intermediate properties between π and σ systems are expected. The cyclopropane $3e'$ orbitals are expected to be at higher energy than the C—C σ orbitals of unstrained hydrocarbons but lower than the ethylene π orbital.

Theoretical calculations confirm the essential features of this picture[26, 27]. In addition these calculations suggest that a linear combination of pseudo π character ($1e''$ and $1a_2''$ shown below), and a low lying $\sigma*$ orbital, $4a_1'$, has to be taken into account.

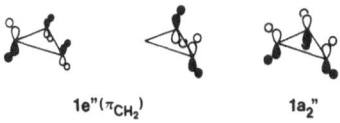

$1e''(\pi_{CH_2})$ $\qquad$ $1a_2''$

R. Gleiter

2 PE Spectrum of Cyclopropane

The PE spectrum of cyclopropane (1)[1, 26, 28] is shown in Fig. 2. It exhibits a double peak centered around 11 eV, a broad band at 13.2 eV, and two strongly overlapping bands at 15.7 eV and 16.5 eV. Below the spectrum the values for the calculated spectrum are shown as discussed in the Introduction. This comparison suggests the assignment of the first two bands to a $^2E'$ state, band three to a $^2E''$ state, and band four and five to the states $^2A_1'$ and $^2A_2''$. The large split of the $^2E'$ is of special interest. The observed energy difference (~ 0.8 eV) is explained as due to a Jahn-Teller split[1–18, 26, 28, 29]. This is confirmed by the optical Rydberg spectrum of 1 which shows a similar splitting $(6400\ cm^{-1})$[26].

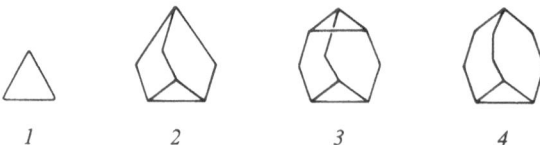

$\qquad 1 \qquad\qquad\qquad 2 \qquad\qquad\qquad 3 \qquad\qquad\qquad 4$

In Fig. 5 we have shown schematically the potential energy of 1 and its cation in their ground states. Calculations show[29, 30] that the ground state of radical cation, $1^{\ddagger}$, is stabilized by 9.5 and 9.2 kcal/mole, respectively, relative to a hypothetical $^2E'$ state with D_{3h} symmetry. The assignment of the first band in the PE spectrum of 1 is confirmed by the PE spectra of nortricyclane (2)[31] triasterane (3)[31] and hexahydrobullvalene (4)[32]. The PE spectra of 2 to 4 exhibit a Jahn-Teller split in the same order of magnitude as encountered in the PE spectrum of 1, namely 0.6 to 0.8 eV. A comparison between the Jahn-Teller split in 4 (0.6 eV) and bull-valene (5) (0.4 eV)[32] shows that the size of the split decreases as the delocalization of e' increases. The assignment of the first six bands in the PE spectrum of 1 has also been discussed by comparison with the spectra of ethyleneoxide, ethylenesulfide and ethyleneimine using arguments from perturbation theory[33].

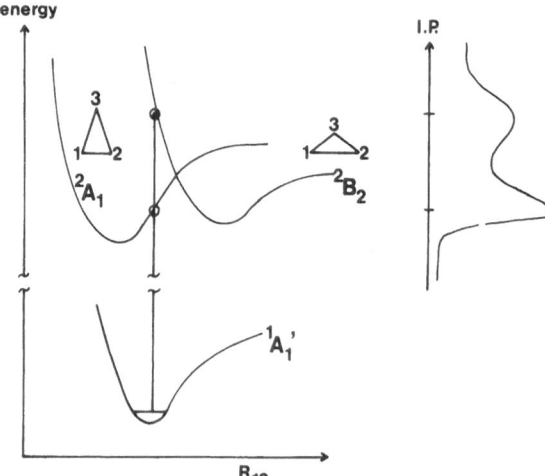

Fig. 5. Potential energy diagram for cyclopropane (1) in its ground state $(^1A_1')$ and the two lowest states of its cation $(^2A_1$ and $^2B_2)$. The vertical transition yields two bands (Jahn-Teller split) in the PE spectrum

In the Table in the Appendix the first ionization potentials of alkyl substituted cyclopropanes are listed. These data demonstrate that substitution of the cyclopropane ring by sp^3 hybridized carbon atoms shifts the center of gravity of the first two bands towards lower ionization potentials.

3 Vinylcyclopropanes

According to the Walsh model discussed before, one linear combination of the $3e'$ orbital of cyclopropane is capable of interaction with a π system in the so called bisected conformation a, the most stable conformation (s-trans) of vinylcyclopropane[34]. The same model predicts for the perpendicular conformation b no interaction between the three-membered ring and the double bond.

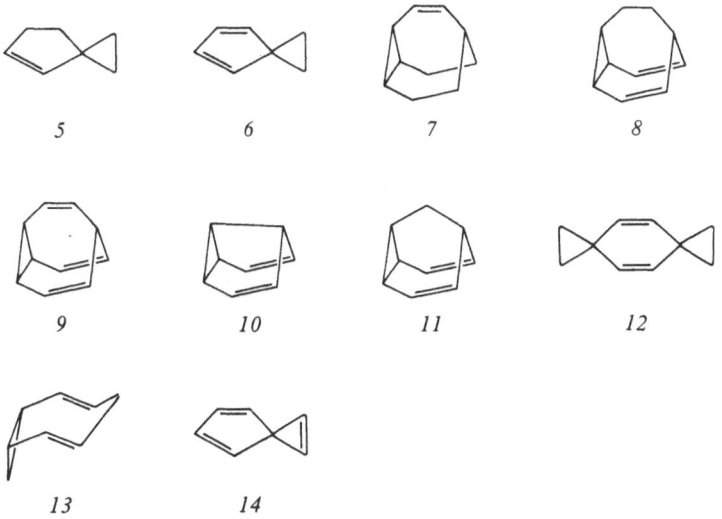

a b

Conformation a is fixed in spiro[2.4]hept-4-ene (5), homofulvene (6), the bullvalene derivatives 7 to 9, semibullvalene (10), barbaralene (11), and dispiro[2.2.2]-deca-4,9-diene (12). Homotropilidene (13) is an example for conformation b.

Besides electron absorption spectroscopy, PE spectroscopy is the method of choice to study the interaction between a double bond system and a three-membered ring.

      5   6   7   8

      9   10   11   12

      13   14

In the following we will discuss in more detail the case of 6. In Fig. 6 the PE spectrum of fulvene[35] and 6[36] are compared. Both spectra exhibit two bands below 10 eV and several closely overlapping bands between 11.5 and 14 eV with a similar pattern. The main difference between the two spectra is a gap in the case of

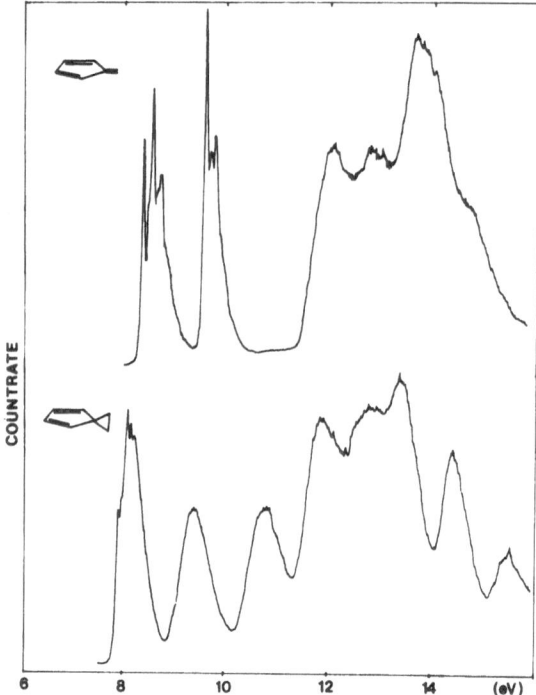

COUNTRATE

6   8   10   12   14   (eV)

Fig. 6. Comparison between the He(I) PE spectrum of fulvene with that of homofulvene (*6*)

fulvene between 10 eV and 11.5 eV and a band at 10.9 eV in the case of *6*. The similarity and the difference can be explained on the basis of a zero differential overlap (ZDO) model and more sophisticated treatments.

We start our discussion by considering the π MO's of fulvene. In Fig. 7 we show on the left side the π MO diagram of fulvene which has been derived using a ZDO model from the basis orbitals [$1a_2(\pi)$ ($\epsilon = -8.8$ eV) and $1b_2(\pi)$ ($\epsilon = -11.6$ eV)] of cis-butadiene and of an exocyclic double bond ($\epsilon = -10.3$ eV), using a resonance integral $\beta = -2.4$ eV for the newly formed bonds. The orbital energies predicted by this model are in excellent agreement with the experimental PE data. In case of homofulvene the model has to be altered only slightly. Instead of the exocyclic double bond we have to introduce the two Walsh orbitals e_A and e_S of the cyclopropane ring with an estimated basis orbital energy of -10.7 eV. Also the resonance integral has to be changed. By choosing $\beta = -1.9$ eV the agreement between experiment and calculation is achieved. The result of this model is shown on the right side of Fig. 7. The main difference between the two spectra, the band at 10.9 eV, is due to the ejection of electrons out of the symmetrical Walsh orbital which does not interact with the π system.

The relatively low first ionization potential of *6* compared to fulvene can be rationalized by assuming a charge transfer due to the interaction of the symmetric $1e''$ orbital of the cyclopropane ring with the corresponding σ^* orbital of the five-membered ring as discussed in case of spiro[4.2]heptatriene (*14*)[37].

The derived value for the resonance integral has been tested in case of *7* to *12*[38, 39]. One obtains an excellent agreement between ZDO calculation and experi-

ε(eV)

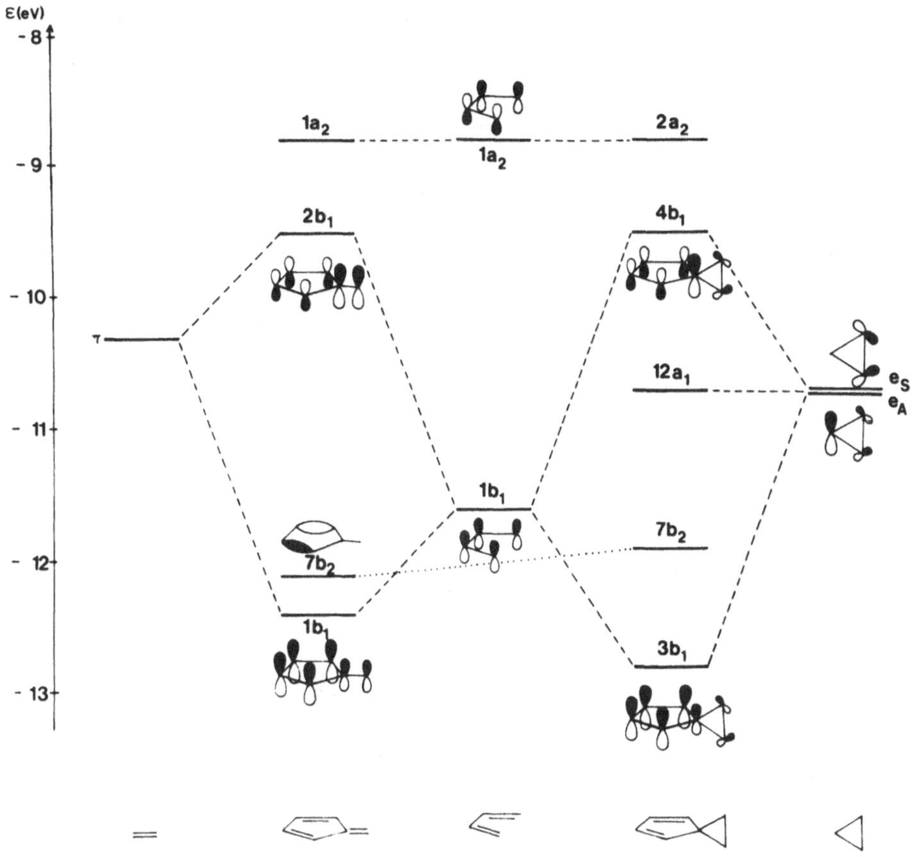

Fig. 7. Interaction diagram between the highest occupied orbitals of ethylene and butadiene to form those of fulvene (*left*). Analogous interaction diagram for homofulvene

ment by assuming a resonance integral of -1.9 eV. In Fig. 8 we have correlated the first ionization potentials of cycloheptadiene(1,4) with the first PE bands of *8, 10, 11* and *13*. The interpretation given assumes the validity of Koopmans' theorem. The MO's were derived on a ZDO model and semiempirical calculations[38]. This correlation diagram demonstrates a close similarity between cycloheptadiene(1,4) and homotropilidene and a strong interaction between the Walsh $e_s(a')$ orbital in *8, 10* and *11* with the a' combination of the diene system.

In Fig. 9 we have correlated the first bands of the PE spectra of vinylcyclopropane $(15)^{[36, 40]}$ with those of $5^{[36, 40]}$, $7^{[38]}$ and dihydrosemibullvalene $(16)^{[38]}$. This diagram indicates in all four compounds a similar interaction and thus confirms other

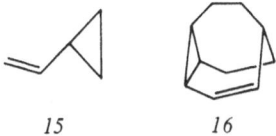

15   *16*

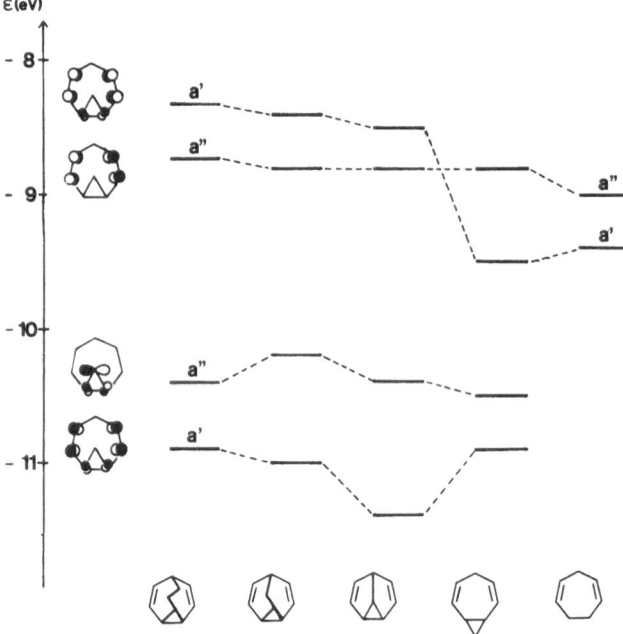

Fig. 8. Correlation between the highest occupied molecular orbitals of dihydrobullvalene, barbaralane, semibullvalene, homotropylidene and cycloheptatriene. The energy values are taken from experiment

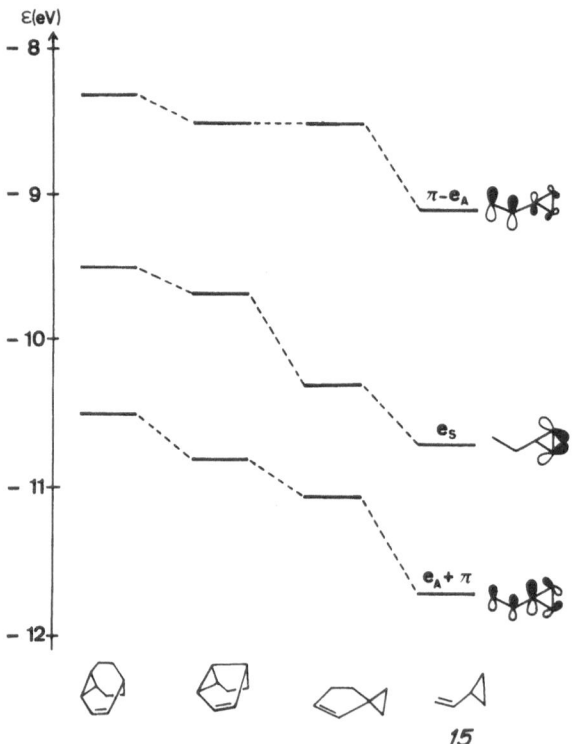

Fig. 9. Correlation between the highest molecular orbitals of vinylcyclopropane, spiroheptene (2), dihydrosemibull-valene and tetrahydrobull-valene. The energy values are taken from experiment

data[34] which show that in *15* the s-trans conformation predominates. The observed shift of all levels in going from *15* to *7* is due to inductive effects caused by the methylene groups.

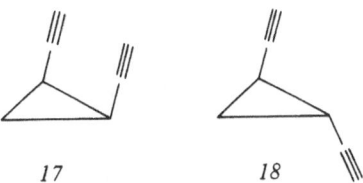

17   18

Strongly related to the divinylcyclopropanederivatives *8, 10* and *11* are cis and trans diethinylcyclopropane (*17*) and (*18*). The PE spectra of both compounds[41] provide evidence for a strong interaction between the acetylenic π bond and the Walsh orbitals. The assignment given (see Appendix) is based on semiempirical calculations. Those bands which are due to ejection of electrons from an orbital which can be described as a linear combination of Walsh and π orbitals are found to be broadened. The pure π bands in contrast are relatively sharp. This behavior is anticipated and confirms the assignment.

Other examples belonging into this chapter are the PE spectra of phenyl cyclopropane (*19*) and different derivatives[40, 42, 43]. An analysis of those measurements shows that the first peak of *19* (8.66 eV) is due to an ionization out of a linear combination of the benzene b_1 (π) orbital and the e_A Walsh orbital. The second peak of the PE spectrum of *19* (9.21 eV) is due to ionization out of the benzene a_2 (π) orbital. Introduction of alkyl substituents in o-position to the three-membered ring causes no significant change. This suggests the bisected conformation for the substituted and unsubstituted products which is only partly in agreement with solution studies[44].

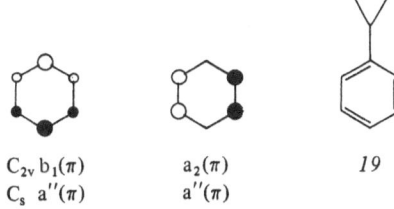

C_{2v} $b_1(\pi)$ $a_2(\pi)$ 19
C_s $a''(\pi)$ $a''(\pi)$

Attempts have also been made to demonstrate the interaction between the π* orbital of a carbonyl group and the antisymmetric Walsh orbital (e_A)[45]. The model compounds studied were 8,9-dehydro-2-adamantanone (*20*) and 2,4-dehydro-5-homoadamantanone (*21*). A comparison between the PE spectra of the corresponding hydrocarbons *22* and *23* reveals that the inductive effect of the carbonyl overrules any conjugative effect.

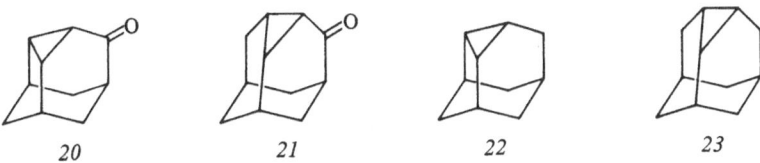

20   21   22   23

3.1 Conclusions

The PE spectra of vinylcyclopropyl derivatives possessing a bisected conformation can be described within a ZDO model by using an interaction parameter of -1.9 eV. This interaction parameter is close to the one found for conjugated π-systems $(-2.3$ eV$)$ and it suggests a rather strong interaction between the three-membered ring and the double bond. Such a strong interaction is also in line with more sophisticated calculations on the ground state. This indicates that Koopmans' theorem is valid for these compounds.

4 Bicyclopropylderivatives

So far the PE spectra of nearly twenty bicyclopropylderivatives have been studied[46–54]. Common of 25 to 38 is that the twist angle θ (for the definition of θ see 24) between the two cyclopropyl groups is 0° or deviates only slightly from

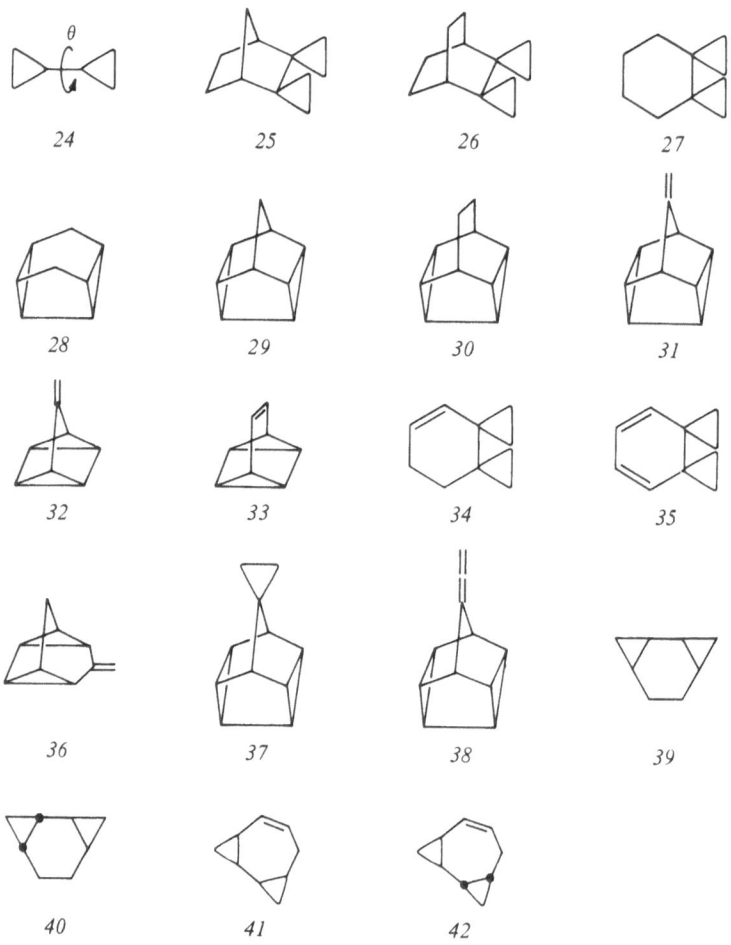

this value. For *24* two conformers are found to be present in the gasphase: a gauche ($\theta = 45°$) and a s-trans ($\theta = 180°$) conformer in the ratio 60:40[55, 56]. For *39* and *40* electron diffraction measurements reveal for the cis compound (*39*) a torsional angle of 18°, whereas the trans compound (*40*) exists as a mixture of two conformers, the one with a torsional angle of 56° being predominant ($87 \pm 5\%$)[57].

4.1 ZDO Model for Bicyclopropyl

To interprete the PE spectra of these bicyclopropyl derivatives the Walsh model is suitable. It has been shown recently[54] that for compounds with θ values of zero degree the consideration of the tangential p orbitals only (see below) in the three-membered rings is sufficient. For those compounds, however, in which θ also deviates considerable from zero the radially oriented 2p orbitals (see below) have to be con-

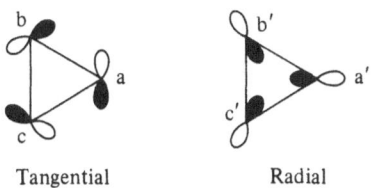

Tangential   Radial

sidered. Approximations to the frontier orbitals of cyclopropane, 3e' and 4e' (see Fig. 4), can be obtained by constructing the following linear combinations of the two basis sets:

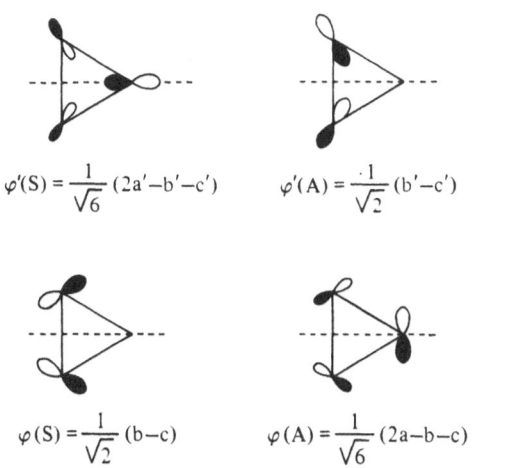

$$\varphi'(S) = \frac{1}{\sqrt{6}} (2a'-b'-c') \qquad \varphi'(A) = \frac{1}{\sqrt{2}} (b'-c')$$

$$\varphi(S) = \frac{1}{\sqrt{2}} (b-c) \qquad \varphi(A) = \frac{1}{\sqrt{6}} (2a-b-c)$$

These four MO's represent the highest occupied [$\varphi(S)$ and $\varphi(A)$] and the lowest unoccupied [$\varphi'(S)$ and $\varphi'(A)$] MO's of cyclopropane[27]. The letters S and A designate the symmetry with respect to a mirror plane indicated by the broken line. As indicated in Fig. 4 a more adequate description of the LUMO's of cyclopropane would be obtained by constructing them from carbon sp hybrids.

Our simple description, however, is sufficient for most qualitative applications. To use the Walsh model for quantitative description the mixing of HOMO and LUMO, both of the same symmetry, has to be considered[54]. This amounts essentially to an admixture of radially oriented components into the HOMOs. As will be shown below this type of mixing plays an important role in reproducing the orbital pattern of bicyclopropyls and higher homologues. The mixing can be expressed as follows:

$$w\,(S) = \sqrt{1 - \chi^2}\;\varphi\,(s) + \chi\,\varphi'\,(S)$$
$$w\,(A) = \sqrt{1 - \chi^2}\;\varphi\,(A) - \chi\,\varphi'\,(A).$$

The mixing parameter χ is a small positive quantity. The resulting $w\,(S)$ and $w\,(A)$ represent the HOMOs of cyclopropane .

4.2 Model Calculation on Bicyclopropyl

For bicyclopropyl the tangential and radial p orbitals are labeled in the following way:

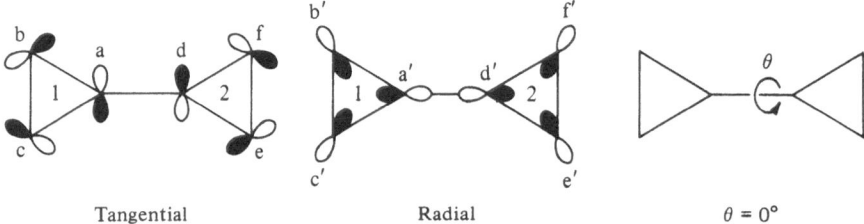

| Tangential | Radial | $\theta = 0°$ |

The relative phases indicated above correspond to the eclipsed conformation, corresponding to a twist angle $\theta = 0°$.

From this set of p-orbitals, the following four Walsh orbitals are constructed. The small components corresponding to the admixture of $\varphi'\,(S)$ and $\varphi'\,(A)$ are not drawn.

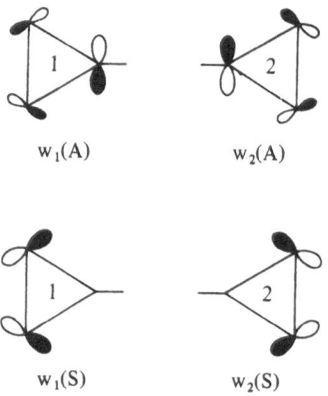

| $w_1(A)$ | $w_2(A)$ |

| $w_1(S)$ | $w_2(S)$ |

From these four basis orbitals we can construct the following four symmetry adapted linear combinations. The subscripts a and b indicate the irreducible representation in the C_2 point group:

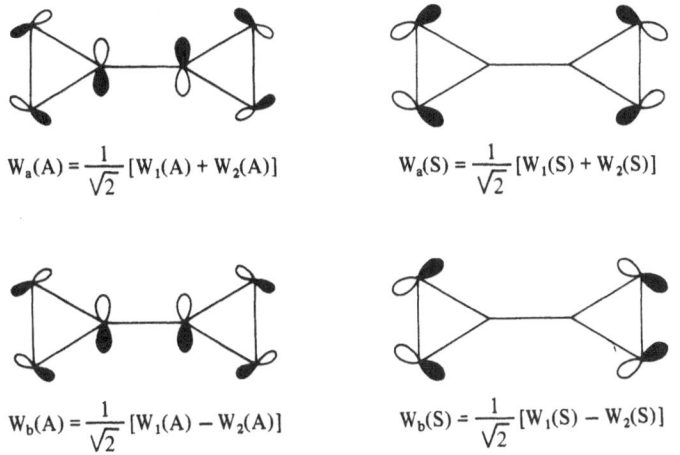

$$W_a(A) = \frac{1}{\sqrt{2}}[W_1(A) + W_2(A)]$$

$$W_a(S) = \frac{1}{\sqrt{2}}[W_1(S) + W_2(S)]$$

$$W_b(A) = \frac{1}{\sqrt{2}}[W_1(A) - W_2(A)]$$

$$W_b(S) = \frac{1}{\sqrt{2}}[W_1(S) - W_2(S)]$$

To derive the energies corresponding to these wave functions one has to estimate the matrix elements of the Hamiltonian energy operator. This has been reported in the literature[54].

In Fig. 10 we present the orbital energies, E_a, E_b, of the wave functions indicated above as a function of θ in terms of $\chi = (\alpha_W - E)/\beta_{WW}$. In this expression α_W is the basis orbital energy of the Walsh orbital and β_{WW} is the resonance integral between two three membered rings.

In Fig. 10a no admixture of radial contributions is assumed ($\chi = 0$), in Fig. 10b we allow 20% admixture. The main difference between the ZDO result of Fig. 10b and the result of an extended Hückel calculation (Fig. 10c) is the asymmetry of the plot in the latter case. Different orbital energy patterns are predicted for the s-cis ($\theta = 0°$) and the s-trans ($\theta = 180°$) conformation. This asymmetry can be traced back to the interaction of the central C-C σ-bond with the Walsh orbitals of the two cyclopropane units in bicyclopropyl. It turns out that only the linear combinations w_a (S) and w_b (S) will be influenced by such an interaction[54]. This can be represented in the ZDO model by raising the energy of the w_a (S) orbital and lowering the energy of the w_b (S) orbital by a roughly similar amount. This refinement predicts that the difference in energy between the second and third level from the top should be smaller for the s-cis than for the s-trans conformation.

A comparison between Fig. 10a and c shows that for angles close to $0°$ the different models predict similar results and thus the model using tangential Walsh orbitals only can be expected to be satisfactory.

An analysis of the PE spectra of 25 to 30 shows that the first four bands can be interpreted in terms of the ionization out of ϵ_1 to ϵ_4 (see Fig. 10) assuming dihedral angles around $0°$. This interpretation is supported by the study of the PE spectra of 31 to 38. In the latter cases the orbital sequence could be tested by interacting the

R. Gleiter

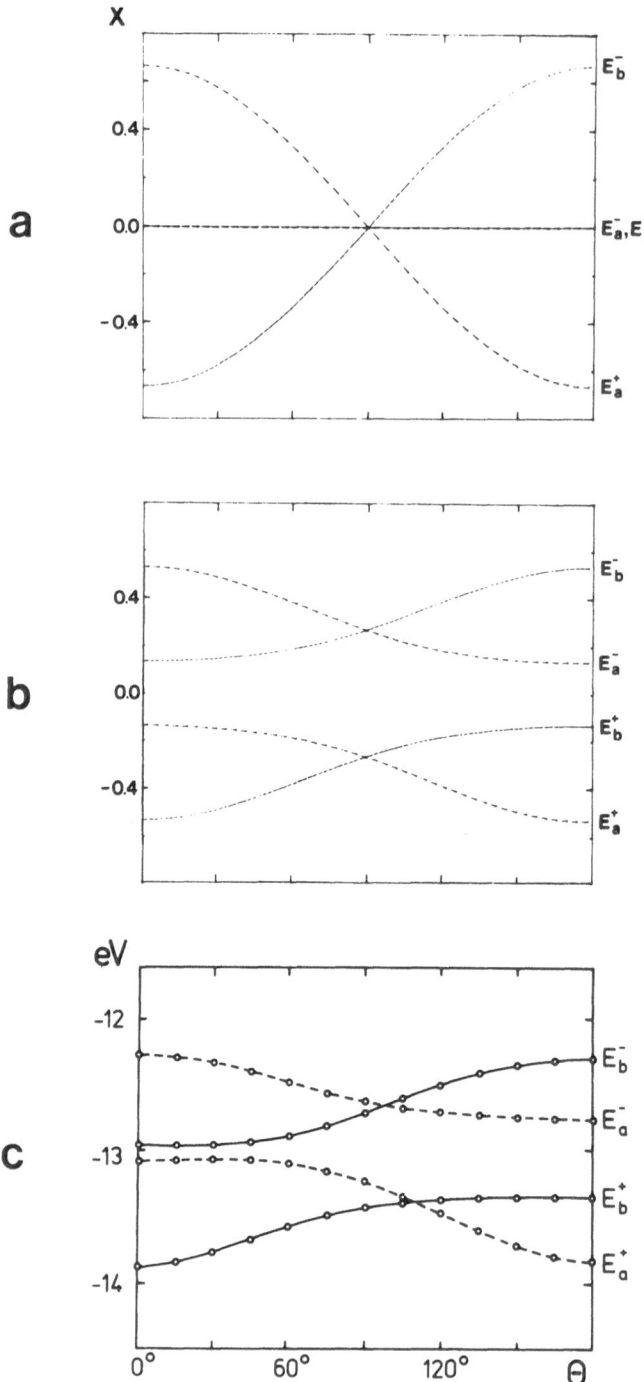

Fig. 10 a–c. Orbital energies $\chi = (\alpha_w - E)/\beta_{ww}$ obtained by the ZDO model (**a** and **b**) and the extended Hückel method (**c**) for bicyclopropane as a function of the twist angle θ. In **a** no admixture of radial contributions are assumed ($\chi = 0$), in **b** the admixture is 20%

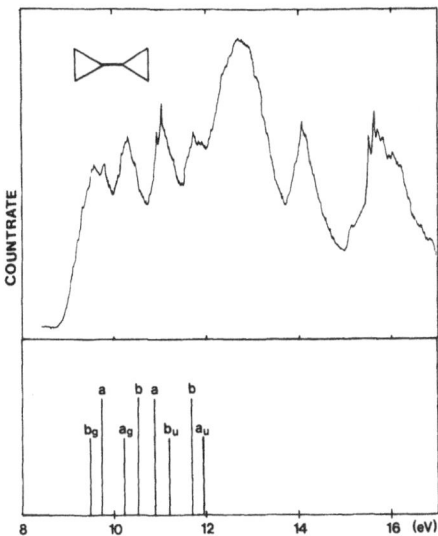

Fig. 11. Comparison between the PE spectrum of bicyclopropyl and calculated orbital energies according to the ZDO model with 20% admixture of radial contributions. A ratio of gauche/trans of 60:40 is assumed

appropriate Walsh-combination with π-orbitals. In case of *33* this has been demonstrated[46, 48, 52, 53].

For the bicyclopropylderivatives *41* and *42* as well as for *24* it has been shown[54] that the agreement between measured ionization potentials and calculated orbital energies can be improved in the frame of the ZDO model by admixing of radial p orbitals into the highest occupied tangential Walsh orbitals and by considering the effect of the C-C σ- and σ^* orbitals of the bond between the cyclopropyl rings.

In Fig. 11 the PE spectrum of bicyclopropane[50, 51] is compared with the results of a ZDO model calculation[54] with the improvements just mentioned. For bicyclopropane a ratio of gauche/s-trans $\approx$ 60:40[55, 56] was assumed. These results suggest that the first four bands of the spectrum are due to transitions from the four highest occupied MO's of bicyclopropyl in the gauche (long bars) and s-trans (short bars) conformation.

4.3 Conclusions

The research on bicyclopropylderivatives has led to two corollarys: (1) The purely "tangential" Walsh model works only if the dihedral angle, θ, between both cyclopropane rings does not deviate much from zero degree. (2) The interaction integral β_{WW} is found to be -1.73 eV, a value close to that found for vinylcyclopropane.

5 Tricyclopropylderivatives

As in the case of the bicyclopropylderivatives the PE spectra of compounds with three adjacent three-membered rings are discussed in terms of the twist angles θ_1 and

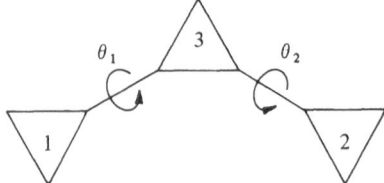

θ_2 defined below. For the examples examined so far by PE spectroscopy[54, 58, 59] the estimated values for θ_1 and θ_2 (from Dreiding models) are given below the formulas.

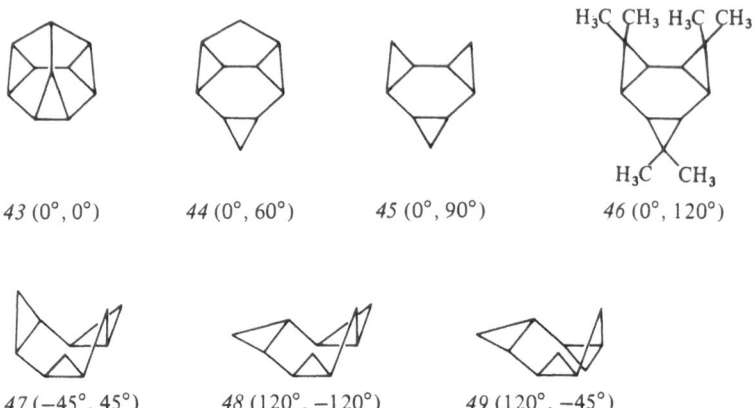

43 (0°, 0°)   44 (0°, 60°)   45 (0°, 90°)   46 (0°, 120°)

47 (−45°, 45°)   48 (120°, −120°)   49 (120°, −45°)

Application of the ZDO model to these species is similar to the procedure given in the previous chapters. Again the admixture of radial components is essential to understand the PE spectra of 44 to 49, those compounds whose torsional angles

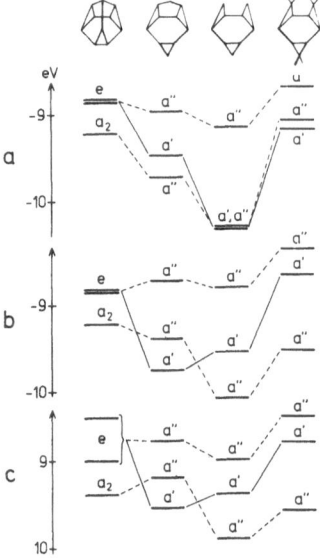

Fig. 12 a–c. Correlation of the three highest occupied levels in the series 43 to 46: In (a) no admixture of radial contributions are assumed, in (b) the admixture is 20%. In (c) the ionization energies are assigned according to results (b)

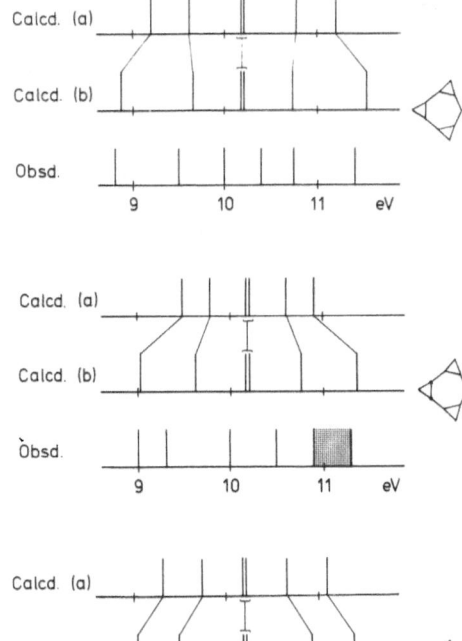

Fig. 13. Calculated and observed ionization energies of *47* (*a*) $x = 0$, (*b*) $x = 1\sqrt{5}$

deviate strongly from $0°$. To demonstrate this we compare in Fig. 12c the PE data of *43* to *46* with the orbital sequence obtained by a model which includes radial contributions (12b) and a model which considers the tangential Walsh orbitals only (12a)[59]. In Fig. 13 a similar comparison is made for *47* to *49*.

5.1 Conclusions

The main outcome of the PE studies on tricyclopropylderivatives is similar to that derived in the previous chapters: (1) For twist angles θ which deviate considerably from $0°$ the simple "tangential" Walsh model has to be extended and the "radial" contributions have to be considered. (2) For twist angles θ around $0°$ the resonance integral is a maximum, $\beta_{WW} = -1.73$ eV.

6 Rotanes

In the [n] rotanes two or more cyclopropane rings are tied together via one center only. So far the PE spectra of the first members of the series have been reported, namely the PE spectra of *50*, *51* and *52*[60]. The PE spectra of these three compounds are shown in Fig. 14. The preliminary interpretation suggests a considerably larger

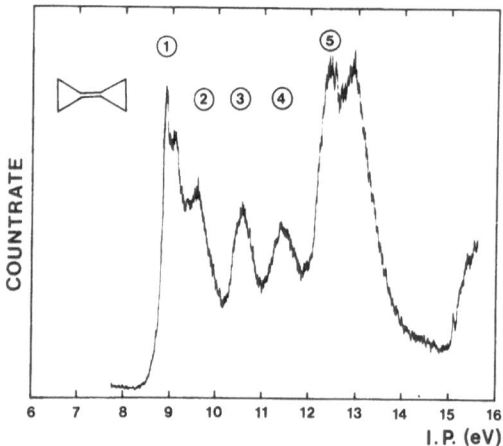

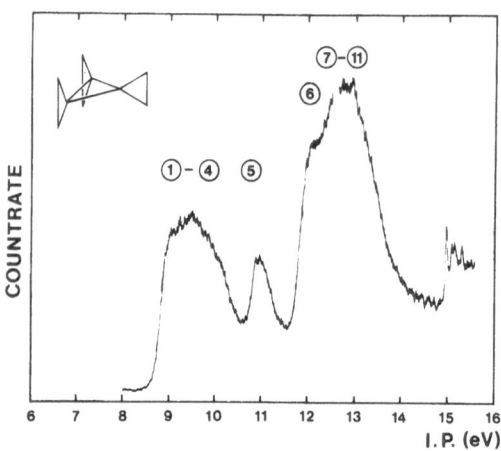

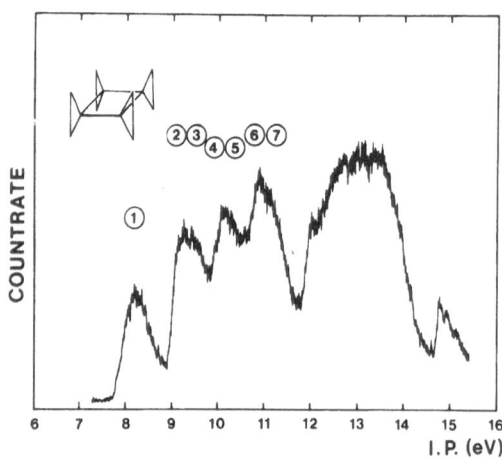

Fig. 14. PE spectra of cyclopropyl-idene (*50*), [3] rotane (*51*) and [4] rotane (*52*)

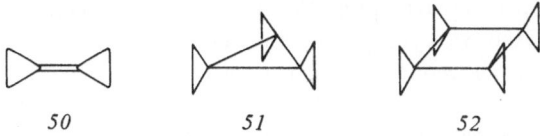

50   51   52

interaction parameter ($\beta = -2$ eV) for *50* and *51* compared to the value found for
the bicyclopropyl and tricyclopropyl derivatives discussed in the previous chapters.
In case of *52*, however, the interaction parameter found is in the same order of magnitude as reported for *24* to *49*.

7 Tricyclo[3.1.0.0^{2,4}]hexane Derivatives

In the tricyclo[3.1.0^{2,4}]hexanes *53* and *54* the two cyclopropane rings are connected
with each other via two centers. So far only *53*[61] and quadricyclane *55* as well as the

53   54   55

exomethylenequadricyclane (*32*)[47] have been synthesized and their PE spectra have
been investigated.

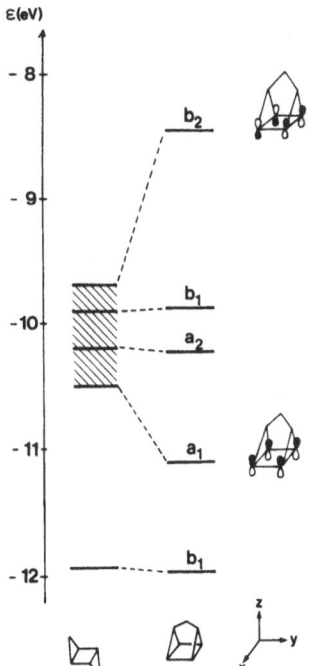

Fig. 15. Comparison between the highest occupied molecular orbitals of *53* and *54*. The energy values are taken
from experiment

219

The PE spectra of *53* and *55* are compared in Fig. 15. This comparison demonstrates nicely that the overlap between two cyclopropane rings depends very much on the dihedral angle between them. The sequence in *55* has been confirmed by the comparison between *55* and *32* on the basis of symmetry arguments. In connection with the work on *53* and *55* the investigations of the PE spectra of endo- and exo-cyclopropanonorbornenes (endo- and exo-tricyclo[3.2.1.0^{2,4}]octa-6-enes) *56* and *57* are of interest. It is found[62] that the interaction between π and Walsh e_s-orbital in *57* is relatively large (split 0.7 eV). In case of *56* the measured split of 0.45 eV indicates a smaller interaction[62].

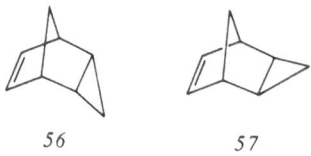

56   *57*

8 Miscellaneous Cyclopropane Compounds

In this chapter we mention results on PE spectra of compounds containing one or more cyclopropane rings which do not fit into the forgoing chapters.

Cyclopropylallene (*58*)[63] and a number of alkyl substituted products[64] have been investigated. In Fig. 16 we derive the highest occupied MO's of *58* from those of an allenic moiety and a cyclopropylmoiety. As anticipated there is essentially the same interaction between the cyclopropane ring and the adjacent double bond

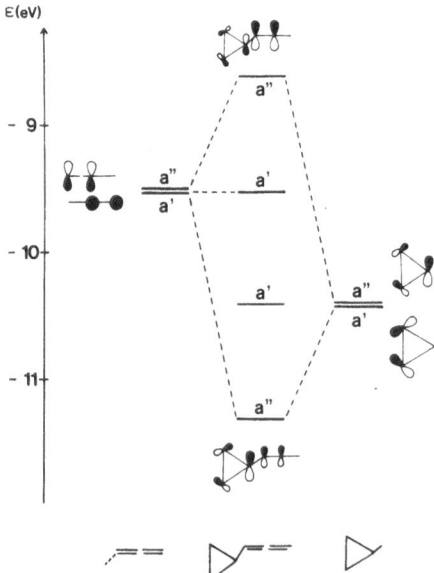

Fig. 16. Interactiondiagram between a cyclopropyl and allenyl fragment to yield the highest occupied molecular orbitals of cyclopropylallene

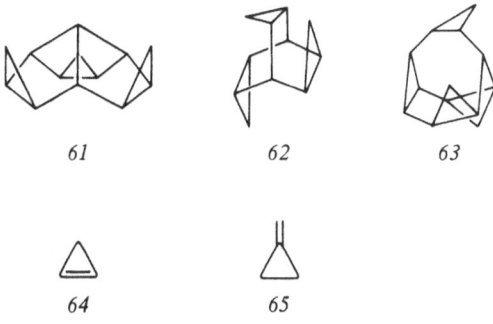

as in vinylcyclopropane, but essentially no interaction between the cyclopropane ring and the nonadjacent double bond. In alkylidenecyclopropane (*59*) and a number of its substitution products[65] only a small interaction between the π system and the three-membered ring is found. This is expected from the geometry of the compounds.

Further PE spectra which belong to this chapter are those of spiropentane (*60*)[50] and those of the homoderivatives of triquinacene, *61*, barrelene, *62*, and bullvalene, *63*[66]. The first ionization potentials of these and related species are listed in the Table 1 at the end of this report. While the interpretation of the PE spectra

of the latter compounds suffers from the strong overlap of bands, in the PE spectra of cyclopropene (*64*)[67, 68] exomethylenecyclopropane (*65*)[68] the first bands are easy to discriminate and to assign. A listing is given in the Appendix.

In this chapter we also should mention the PE spectra of such elusive species as [3]radialene (*66*)[69] and hexamethyl[3]radialene (*67*)[69]. In the latter compound

a Jahn-Teller effect of 0.74 eV in the $^2E'$ state is observed, a similar splitting as encountered in *1*. The PE spectroscopic investigation of benzocyclopropene (*68*)[71] leads to the conclusion[71] that the CH_2 group in *68* must be assigned a negative inductive effect and a strong hyperconjugative effect.

221

III Cyclobutane Rings

1 Bonding Models of Cyclobutane

As in the case of cyclopropane there are two descriptions used for the valence orbitals of cyclobutane: One following the Coulson-Moffitt description of cyclopropane which has been discussed by several authors[20, 23, 72] and most explicitly by Salem and Wright[73]. The other one which follows the Walsh description for cyclopropane has been used in detail by Hoffmann and Davidson[74]. In the latter model one starts out with four interacting methylene groups oriented as shown above. The radial σ orbitals interact strongly with each other to yield four linear combinations, a bonding a_{1g}, a nonbonding e_u pair and an antibonding b_{2g} combination as shown

on the left of Fig. 17. The tangential p orbitals interact weakly to form a bonding b_{1g}, a nonbonding e_u and an antibonding a_{2g} linear combination (see Fig. 17 right). For reasons of symmetry the two nonbonding e_u combinations interact with each other as shown in the middle of Fig. 17, resulting in a slightly bonding ($3e_u$) and a slightly antibonding ($4e_u$) linear combination.

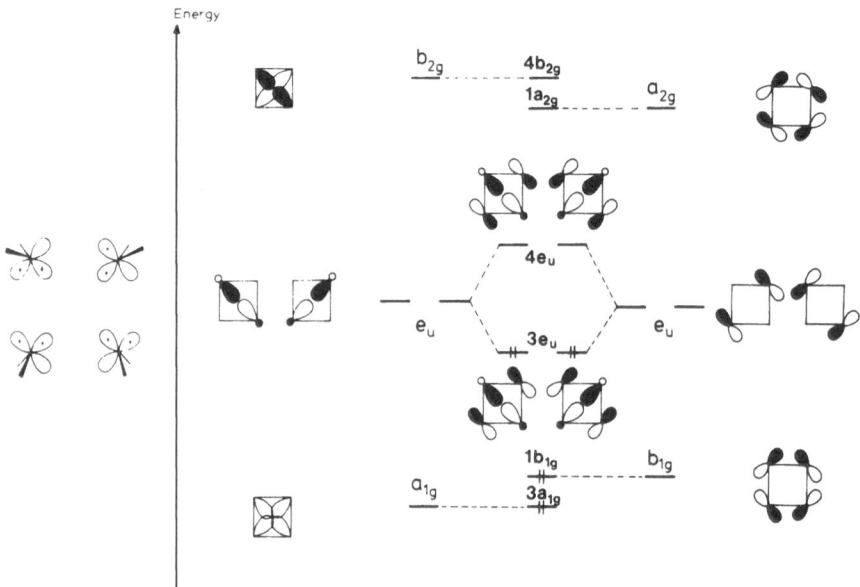

Fig. 17. Derivation of the Walsh representation of the molecular orbitals of cyclobutane from three methylenes

Semiempirical and ab initio calculations support the MO scheme just derived. The only addition which has to be made is a high lying b_{1u} level centered on the C-H bonds, analogous to the cyclopropane case (see below).

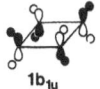

1b$_{1u}$

2 PE Spectrum of Cyclobutane

The PE spectrum of cyclobutane (69) exhibits (Fig. 18) two close lying intense bands at 10.7 and 11.3 eV, followed by a weak but clearly separated band at 12.5 eV and another broad band at 13.39 eV. The first two bands have been assigned as due to a Jahn-Teller split of the ^2E_u state[75]. The split encountered (0.6 eV) is similar to the one found in the PE spectrum of cyclopropane. The third band is assigned to the $^2B_{1u}$ state according to a comparison between a molecular orbital calculation

69

and the PE spectrum (Fig. 18)[75]. In line with the assignment discussed for 69 are the PE spectra of several alkyl substituted cyclobutane derivatives (see Appendix). It is interesting to note that in all cases listed, the split between the first two bands

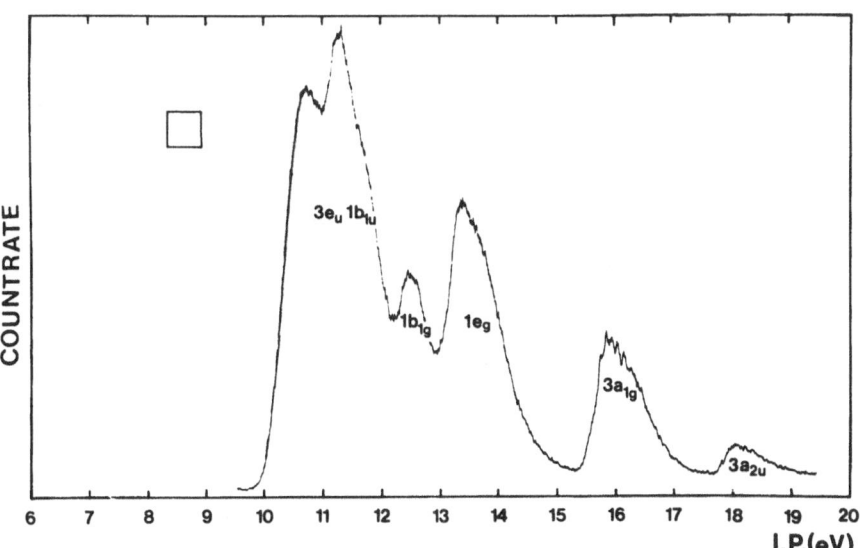

Fig. 18. PE spectrum of cyclobutane (*top*). Comparison with molecular orbital energies obtained from a SCF calculation (*bottom*)

amounts to 0.5 to 0.7 eV. A further confirmation of the assignment of the first three bands in the PE spectrum in *69* is found by comparing it with the PE spectra of heterocyclobutanes[76].

3 Vinylcyclobutanes

Semiempirical[74, 77] and ab initio[78] calculations indicate a sizable interaction between a double bond system and a four-membered ring similar to the interaction discussed for vinylcyclopropanes. In the cyclobutane case the energy difference be-

a   b $e_u(A)$ $e_u(S)$

tween conformations a and b is predicted to be smaller than in the corresponding vinylcyclopropane cases since in either conformation there is a significant interaction between π system and HOMO of the cyclobutane system possible. In the "bisected"

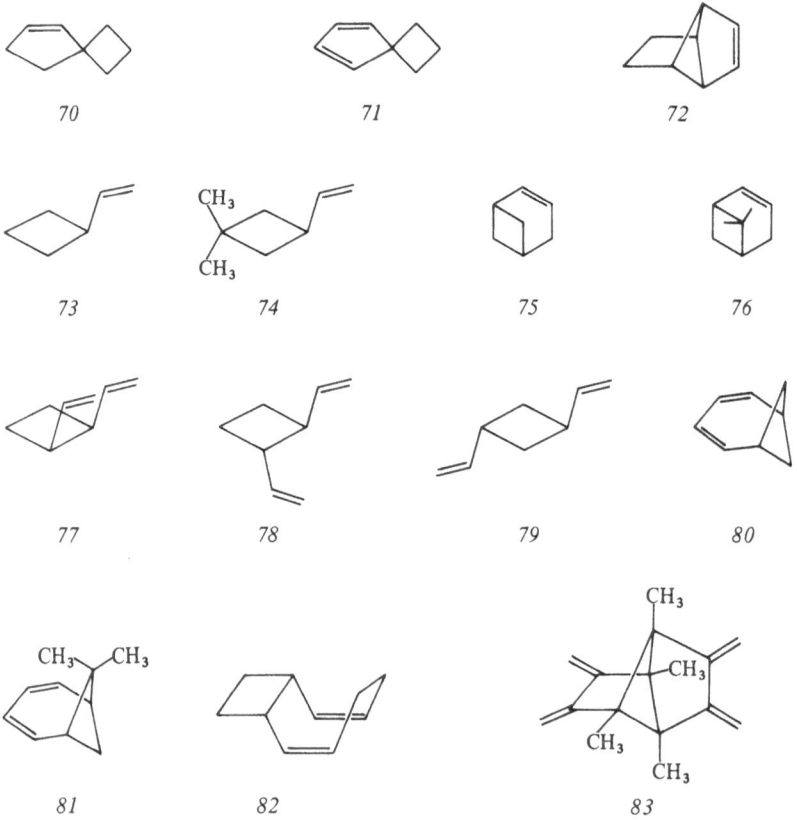

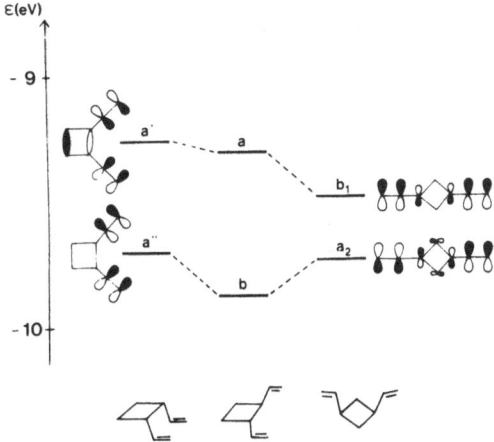

Fig. 19. Splitting pattern of the first two bands of the PE spectra of syn- and anti-1,2-divinylcyclobutane 77 and 78 and syn-1,3-divinylcyclobutane (79)

conformation a the $e_u(A)$ component interacts favorably, in the perpendicular conformation the $e_u(S)$ component of the HOMO interacts. The π type interaction in a is predicted to be still somewhat greater than in b.

In 70 to 83 we show some vinylcyclobutane systems whose PE spectra have been investigated[79−85] to explore the interaction between the π and Walsh system. The assignment of the PE spectra is more difficult than in the case of vinylcyclopropane derivatives for the following reasons: (1) In the cyclobutane case we have to consider the e_u and the b_{1g} orbitals, while in cyclopropane a consideration of e' is sufficient. (2) We observe strongly overlapping bands in the vinylcyclobutane derivatives but mostly separate bands in the PE spectra of cyclopropyl compounds. The PE spectra of 70[79, 81] and 72 to 76[80, 83, 84] exhibit one band around 9 eV clearly separated from strongly overlapping bands (see Appendix). This band is interpreted as due to the ionization out of the orbital which can be described as the antibonding linear combination between π and antisymmetric Walsh orbital of the four-membered ring. In Fig. 19 the splitting pattern of the first two bands for the divinylcyclobutanes 77 to 79 is shown[84]. In the first moment it is surprising that the split between the first two peaks in the PE spectrum of the trans-isomer 78 is larger than of the cis-isomer 77. Thinking only of spatial interactions of the two double bonds one would expect no splitting of the first two bands for 78 and 79, but a sizable energy difference for 77. The reason for the very similar pattern found in all three cases is the dominance of through bond interaction[87] (see Fig. 19). In connection with the splitting pattern shown in Fig. 19 it is interesting to note that in the PE spectrum of 82 the split between the first two peaks is minute (see Table in Appendix). This is probably due to a stronger through space interaction of the two π orbitals due to geometrical factors. Thus, the through bond interaction between them is about outbalanced.

In the case of 80[82], 81 and 83[85] the analysis of the PE spectra is simpler than in the foregoing examples. Here the b_1 and a_2 Walsh orbitals of the four-membered ring have only one partner to interact with. For 80 the interaction diagram is shown in Fig. 20[82].

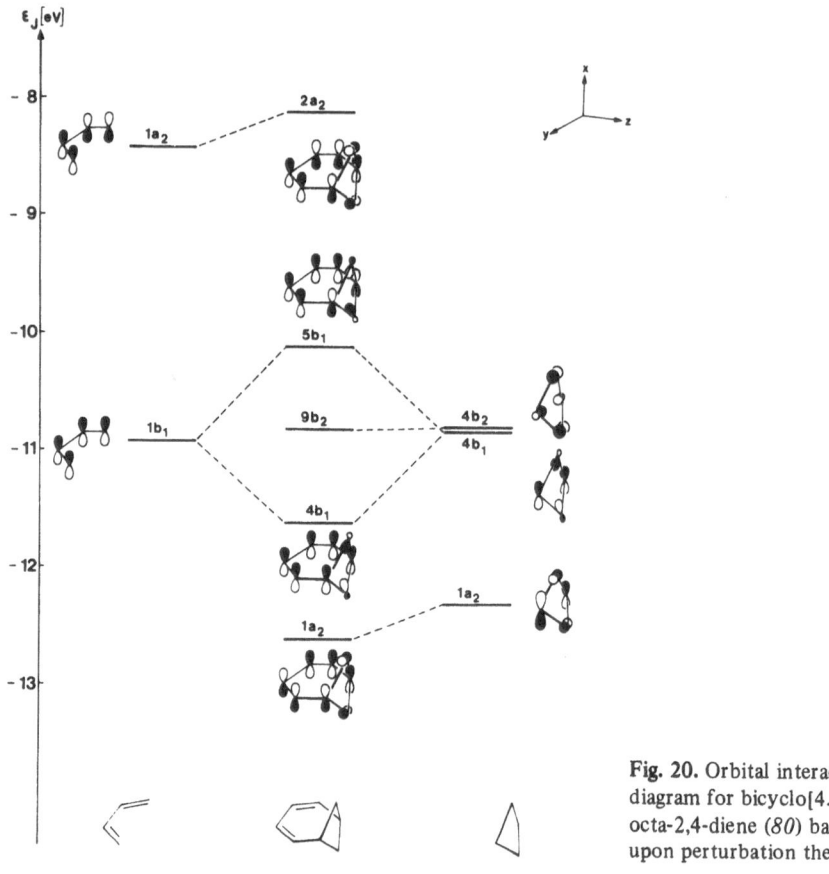

Fig. 20. Orbital interaction diagram for bicyclo[4.1.1]-octa-2,4-diene (*80*) based upon perturbation theory

Related to the PE spectra of vinyl and divinyl cyclobutane are those of ethinyl and cis- and trans-diethinyl-cyclobutane *84* to *86* and derivatives of them. The PE spectra of all three compounds have been recorded[83, 84]. A strong interaction is found between the π system and the four-membered ring. In the case of *85* and *86*

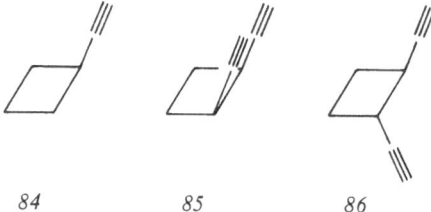

84   *85*   *86*

the through bond interaction between the π system and the four-membered ring dominates the through space interaction as anticipated from the analysis of the PE spectra of *17* and *18*.

To end this chapter it should be mentioned also that the PE spectrum of cyclo-butylbenzene (*87*) has been investigated[81]. Similarly to the case of cyclopropyl

87

benzene it has been found that the first ionization potential is due to the ionization out of the linear combination of $a_2(\pi)$ of benzene and Walsh orbital of cyclobutane. The second IP corresponds to the unperturbed $b_1(\pi)$ orbital of the phenyl part.

3.1 Conclusions – Comparison Between Vinylcyclobutanes and Vinylcyclopropanes

Most of the vinylcyclobutane derivatives mentioned so far could be analyzed by adopting a resonance integral of $\beta = -1.9$ eV[79, 80, 82, 84, 85, 86]. This value is identical to the parameter found for the interaction between a cyclopropane ring and a double bond. The interaction integrals, $H_{\pi w}$, between a double bond and a cyclopropane or a cyclobutane ring, however, are quite different due to the wide numerical divergence

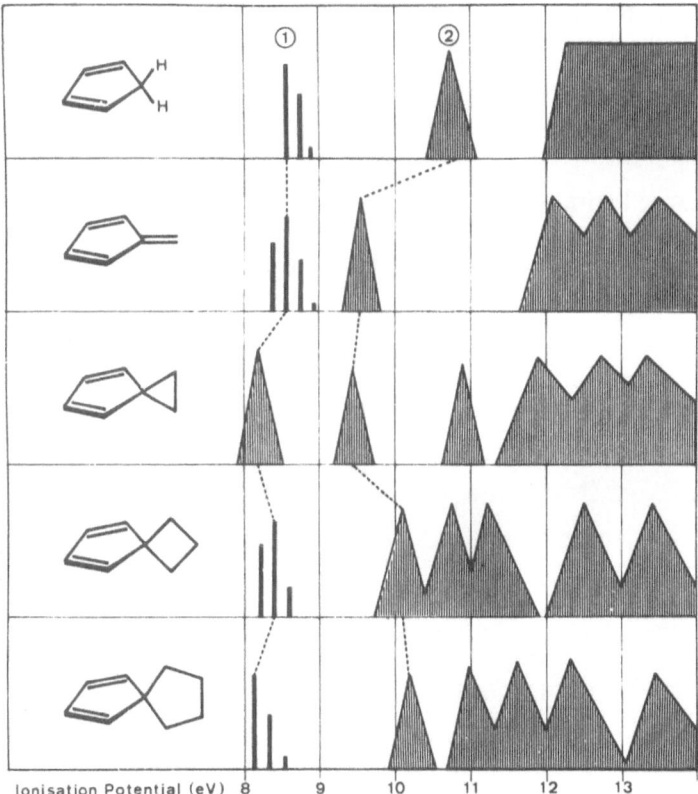

Fig. 21. Comparison between the PE spectra cyclopentadiene, fulvene, homofulvene and spiro[3.4]octa-5,7-diene and spiro[4,4]nona-5,7-diene

227

in the coefficients of the atomic orbitals in the corresponding wave functions. The difference in the interaction integrals should be a measure of the difference in the ability of the particular ring to stabilize a π system.

$$H_{\pi w}(\text{cyclopropane}) = \beta \cdot c_\pi \cdot c_w = -(1.9)\left(\frac{2}{\sqrt{6}}\right)(0.707) = -1.1 \text{ eV}$$

$$H_{\pi w}(\text{cyclobutane}) = \beta \cdot c_\pi \cdot c_w = (-1.9)(0.5)(0.707) = -0.67 \text{ eV}$$

To demonstrate the difference between the interaction integrals, we have compared the PE spectra of fulvene, homofulvene, spiro[3.4]octa-5,7-diene and spiro[4.4]-nonane-5,7-diene in Fig. 21.

Our assignment shows that the ionization potential of each second band in the spectra of Fig. 21 is due to an ionization out of the orbital

$$\psi = N(\pi - \lambda W)$$

where W stands for the exocyclic π bond in fulvene or a Walsh or σ orbital of appropriate symmetry.

A comparison between the ionization potentials of the second bands clearly shows that the effect of the cyclobutane ring is comparable to the effect of a five-membered ring while the shift caused by a cyclopropane ring is close to the one found for a double bond.

4 Bicyclo[2.2.0]hexane Derivatives

In bicyclo[2.2.0]hexane (88) two four-membered rings are joined. The highest occupied Walsh orbitals of 88 are shown below. In Fig. 22 the highest occupied orbitals

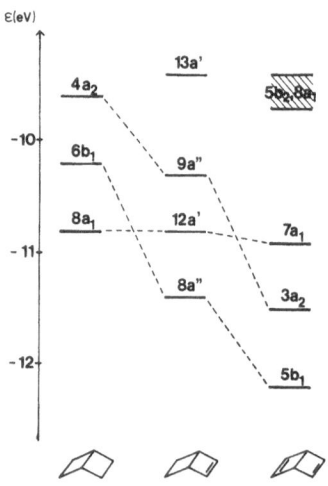

$\varepsilon(eV)$

Fig. 22. Orbital correlation diagram for bicyclo[2.2.0]-hexane, bicyclo[3.3.0]hexene and Dewar-benzene

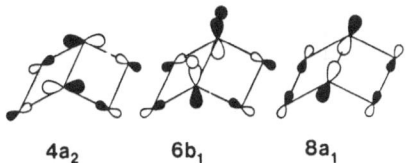

4a₂ · 6b₁ · 8a₁

are correlated with those of bicyclo[2.2.0]hex-2-ene (*89*) and of bicyclo[2.2.0]-hexa-2,5-diene (*90*) (Dewar-benzene).

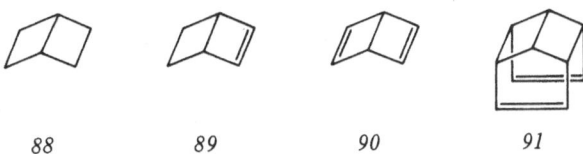

88 · *89* · *90* · *91*

This orbital correlation diagram is in line with the corresponding PE spectra[88]. Introduction of unsaturation into *88* shifts those Walsh-type orbitals with large coefficients on centers 2,3 and 5,6 towards higher energy. In hypostrophene (*91*) the appropriate σ-molecular orbitals of *88* interact strongly with a_1 (π_+) combination of the two π bonds, thus destabilizing π_+ strongly. Therefore the first band of the PE spectrum of *91*[89] at 8.44 eV has been assigned to a_1 (π_+). The assignment of the next two bands at 9.3 and 9.6 eV is not unequivocal.

5 Tricyclo[4.2.0.0^{2,5}]octane Derivatives

In syn- and anti-tricyclo[4.2.0.0^{2,5}]octane (*92*) and (*93*) two cyclobutane rings are connected to each other over two adjacent centers. This will cause a splitting of the

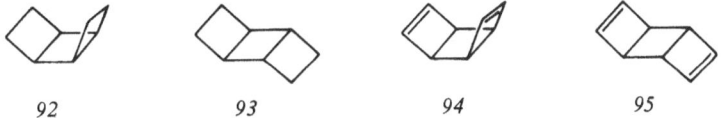

92 · *93* · *94* · *95*

two highest occupied MO's of cyclobutane, resulting in a $b_2(\sigma)$ and $a_2(\sigma)$ linear combination for *92* and a $a_u(\sigma)$ and $b_u(\sigma)$ linear combination for *93*. All four linear combinations are shown below.

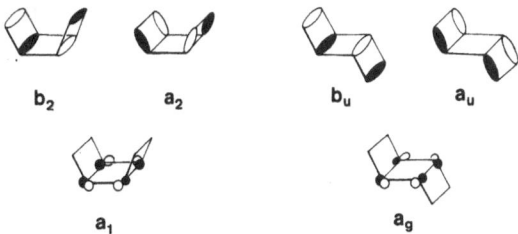

b₂ · a₂ · bᵤ · aᵤ

a₁ · a_g

Semiempirical[90, 91] and *ab initio*[92] calculations predict for *92* b_2 (σ) on top of a_2, for *93* a_u on top of b_u. The assignment of the first two bands in the PE spectrum of *92* and *93* is based on this result[90, 91].

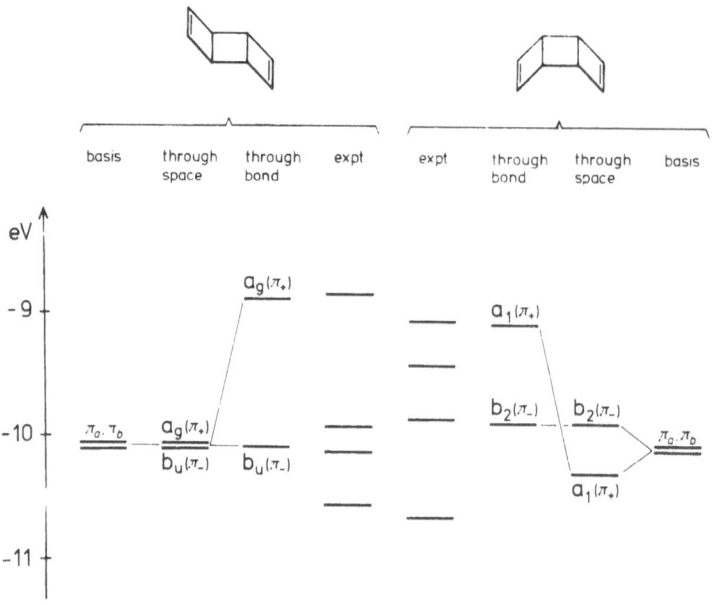

Fig. 23. Comparison between the calculated and measured values of syn- and anti-tricyclo-[4.2.0.0^{2,5}]octadiene (*94* and *95*)

The ordering of the highest occupied molecular orbitals of anti- and syn-tricyclo-[4.2.0.0^{2,5}]octadiene *94* and *95* is still debated. Different MO calculations give widely different results; in fact different orderings of "plus" and "minus" π type combinations for *94* have been realized[90–95]. This has been explained[94] as mainly due to the extreme sensitivity of the competition between through space and through bond interactions with respect to small changes in MO approximation and molecular geometry.

In Fig. 23 we compare the highest molecular orbital energies of *94* and *95* with the experimental results. This correlation illustrates the derivation of the energies of the π$_+$ and π$_-$ orbitals and the dominance of the through bond contribution for both compounds. The assignment suggested in this Fig.[95] has been supported by correlating the PE spectra of *94* and *95* with those of the corresponding homoderivatives *96* to *99*[95]. The orbital sequence a$_1$(π) on top of b$_2$(π) for *94* together with a$_2$(π*) as the LUMO may explain why the irradiation of *94* does not lead to cubane[90].

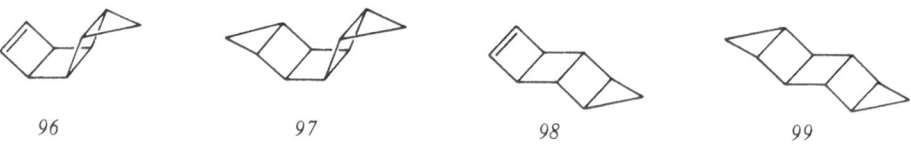

96   97   98   99

The PE spectrum of anti-1,2,5,6-tetramethyl, 3,4,7,8-tetramethylenetricyclo-[4.2.0.0^{2,5}]octane (*100*) has been reported[96] and the first three bands have been

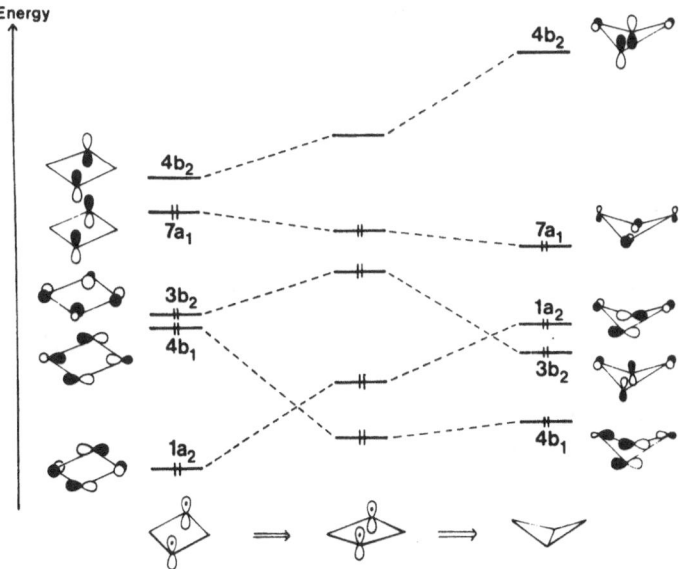

100

assigned. A considerable through bond interaction (0.82 eV) of the two butadiene moieties in the radical cation has been encountered.

IV Systems Containing a Bicyclobutane Moiety

1 Bonding Models of Bicyclobutane

The valence orbitals of bicyclobutane are derived in Fig. 24 from those of cyclo-butane-1,3-diyl in two steps. On the left side of this Fig. the highest occupied and the lowest unoccupied molecular orbitals of a planar cyclobutane-1,3-diyl are shown schematically. The HOMO and LUMO are the linear combinations arising from the $2p_z$ orbitals in positions 1 and 3 of the diradical. The other three occupied molecular orbitals drawn are those of cyclobutane (see Fig. 17). By squeezing the diradical in the 1,3 position and leaving the ring planar, those molecular orbitals which are 1,3-bonding ($4b_1$, $7a_1$) will be stabilized, and the ones which are 1,3-antibonding ($1a_2$, $3b_2$ and $4b_2$) will be destabilized. Bending the squeezed diradical and allowing

Fig. 24. Derivation of the highest molecular orbitals of bicyclobutane from those of a planar cyclobutane-1,3-diyl

231

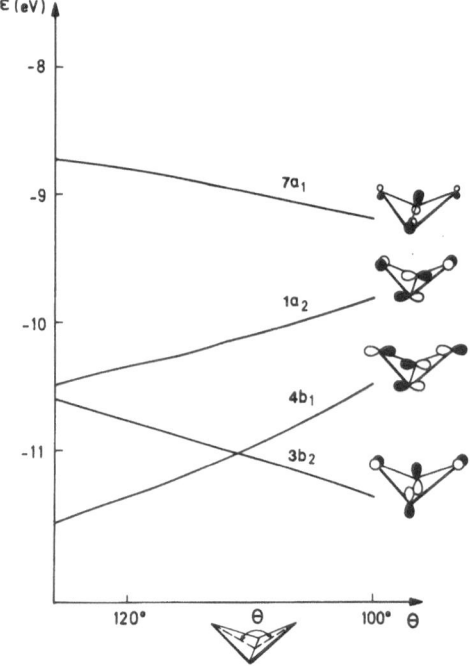

ϵ (eV)

-8

$7a_1$

-9

$1a_2$

-10

$4b_1$

-11

$3b_2$

120° θ 100° θ

Fig. 25. Orbital energies of bicyclobutane as a function of dihedral angle θ according to MINDO/3

the orbital of B_2 symmetry to interact with each other, the orbital scheme on the right of Fig. 24 results.

From this Fig. it is seen that the highest occupied orbital of bicyclobutane is $7a_1$. It is essentially localized in the transannular σ bond. The $1a_2$, $3b_2$ and $4b_1$ orbitals of bicyclobutane are localized in the peripheral bonds and their relation to the cyclobutane orbitals is obvious from Fig. 24.

From this diagram it is evident that the valence orbitals of bicyclobutane[97, 98] depend strongly upon the dihedral angle, θ, between the cyclopropyl rings[99]. This is illustrated in Fig. 25 where the orbital energies of bicyclobutane calculated by the MINDO/3 method are shown as a function of θ[99]. From this diagram it is apparent that the orbitals $7a_1$ and $3b_2$ are shifted towards lower energies while $1a_2$ and $4b_1$ are shifted towards higher energies as the angle θ becomes smaller.

2 PE Spectrum of Bicyclobutane

The PE spectrum of 101[100, 101], shown in Fig. 26, exhibits five well separated peaks between 8 and 18 eV. A comparison with the predicted molecular orbital energies[98, 100] suggests that the second and fourth peak is due to two strongly overlapping bands. Also the approximate ratio below the envelopes is in accordance with this assignment.

102   *103*

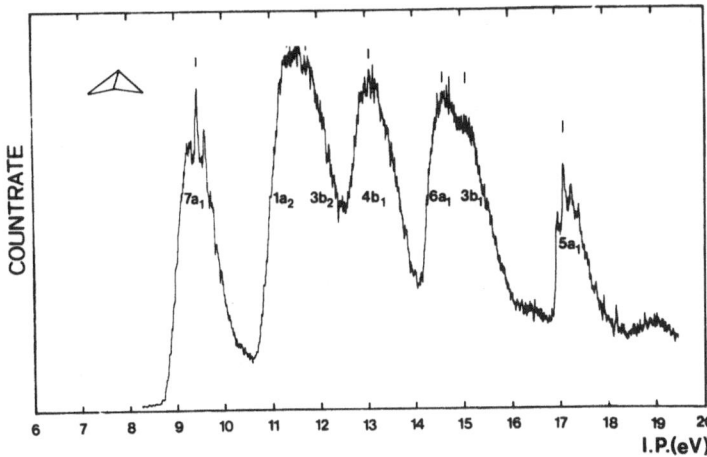

Fig. 26. PE spectrum of bicyclobutane. The assignment based on the validity of Koopmans' theorem is indicated

In agreement with the correlation diagram of Fig. 25 are the PE spectra of tricyclo[4.1.0.0^{2,7}]heptane (*102*) and tricyclo[3.1.0.0^{2,6}]hexane (*103*). Compound *102* with a dihedral angle of 119° shows a split between the first two ionization potentials of 1.73 eV, in compound *103*, however, the angle is reduced to 108° and the split found is 0.80 eV. A number of other alkyl substituted bicyclobutanes listed in the Table confirm the orbital sequence given for *101*.

3 Vinylbicyclobutane Derivatives

3.1 Benzvalene

The He(I) spectrum of benzvalene (*104*) has been reported recently[102]. The first four bands together with an interpretation based on molecular orbital calculations is given in Fig. 27. This interpretation suggests a strong interaction between the

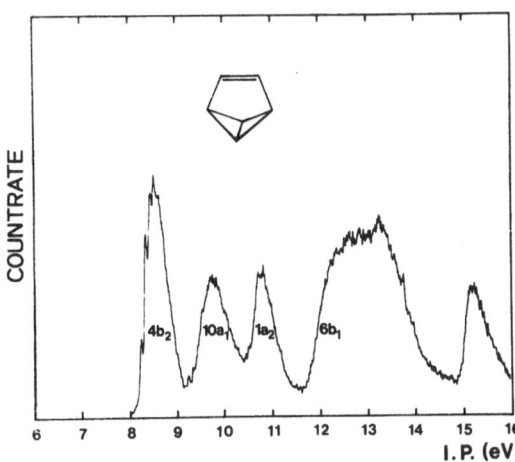

Fig. 27. PE-spectrum of benzvalene. The assignment based on the validity of Koopmans' theorem is indicated

233

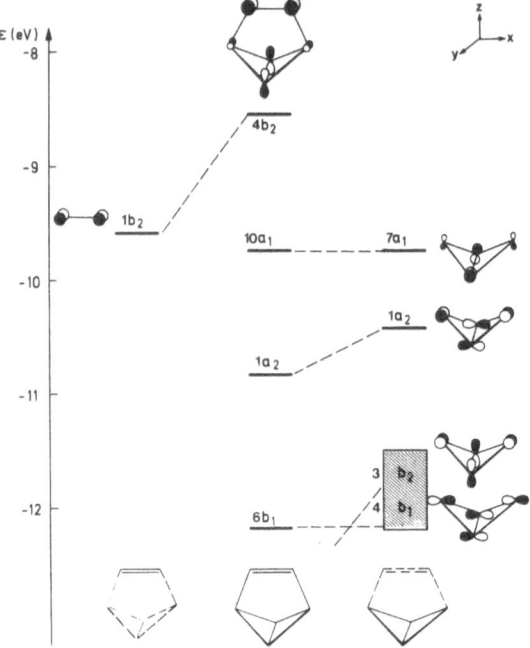

Fig. 28. Interaction diagram between the basis orbitals of a double bond in benzvalene and those of bicyclobutane to yield the orbital energies of benzvalene

b_2 (π) orbital of the ethylenic moiety and the corresponding orbital in the bicyclo-butane unit. This is demonstrated in Fig. 28 by means of a qualitative interaction diagram between an ethylene moiety (left) and a bicyclobutane fragment (right). A similar strong interaction has been found in 1,8-naphthotricyclo[4.1.0.0²,⁷]hep-tene (105)[103].

104   105

3.2 Tricyclo[4.1.0.0²,⁷]hept-3-enes

The PE spectra of tricyclo[4.1.0.0²,⁷]hept-3-ene (106) as well as of several alkyl substitution products (107 to 110) have been reported[104]. In Fig. 29 the recorded

R	R'	R''	R'''	
H	H	H	H	106
CH_3	H	H	H	107
H	CH_3	H	H	108
CH_3	H	CH_3	H	109
CH_3	H	H	CH_3	110

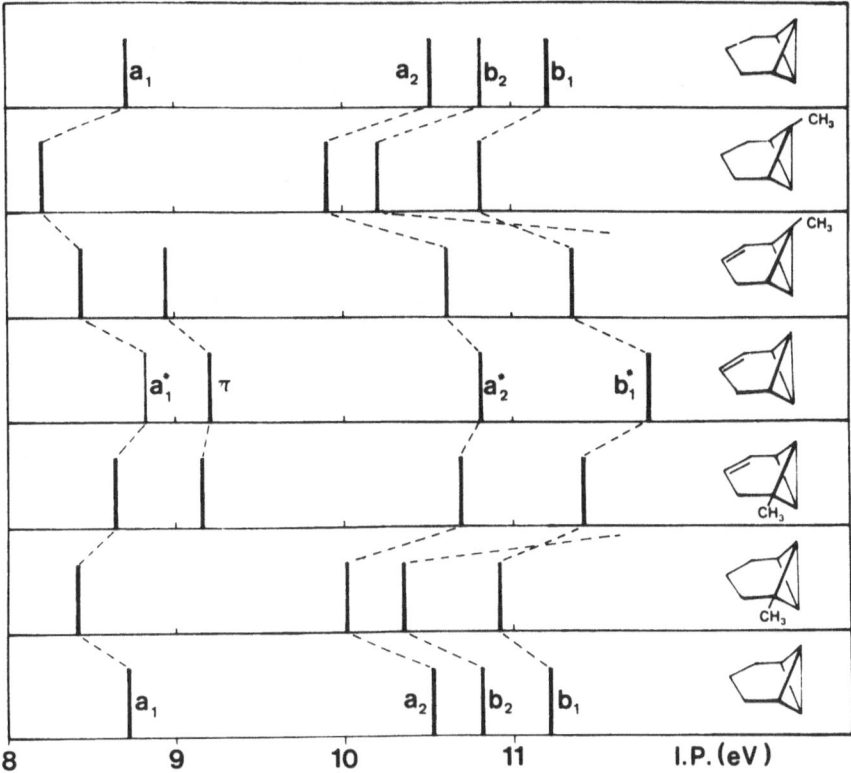

Fig. 29. Comparison between the first bands in the PE spectra of *106* to *108* with the bands of the corresponding saturated species

PE bands are compared with those of tricyclo[4.1.0.0^{2,7}]heptane and the corresponding alkyl products. A comparison between the results in Figs. 27 and 29 shows one significant difference: in benzvalene the HOMO is π type, in tricyclo[4.1.0.0^{2,7}]hept-3-ene it is of a_1 (σ) type. This interpretation for *106* is nicely corroborated by comparison of the PE spectra given by *107* to *110* as illustrated in Fig. 29. For *109* and *110*, the methyl group attached to the double bond shifts the band which is due to ionization out of the π-orbital more strongly toward lower energy than the other bands.

The observation that the π level of *104* is the HOMO is in line with the observation that electrophiles react with the double bond. Thus, reaction of *104* with such reagents as bromine[105], chlorosulfonylisocyanate[106] or benzenesulfenyl chlorid[106] occurs by initial π bond attack. In the case of *106* and its congeners, a marked preference for electrophilic attack by D$^+$ at the edge bicyclobutane bonds has been found[104].

3.3 1,2,5,6-Tetramethyl-3,4-dimethylenetricyclo[3.1.0.0^{2,6}]hexane

One additional compound in which double bonds interact with a bicyclobutane moiety is 1,2,5,6-tetramethyl-3,4-dimethylenetricyclo[3.1.0.0^{2,6}]hexane (*111*). Its

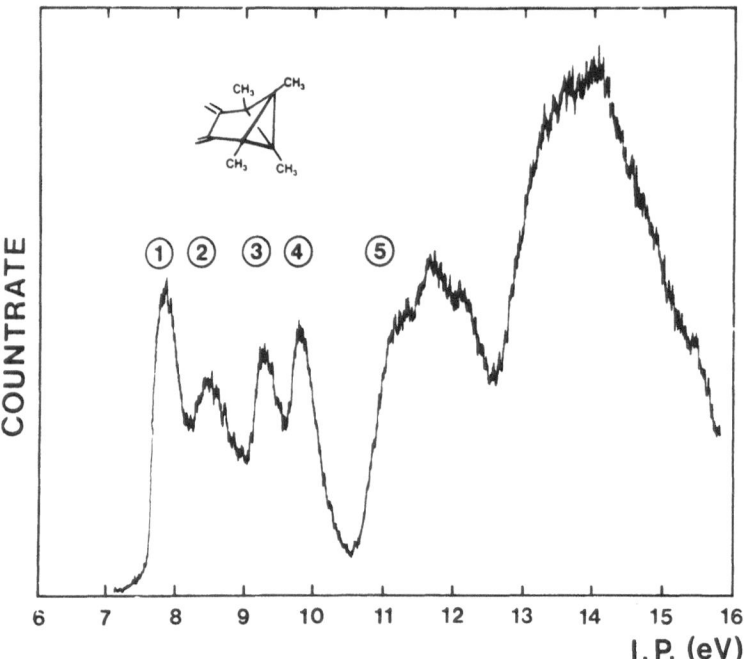

Fig. 30. PE spectrum of 1,2,5,6-tetramethyl-3,4-dimethylenetricyclo [3.1.0.0^{2,6}]hexane (*111*)

PE spectrum has been mentioned in connnection with its reactivity towards dieno-philes[107]. The spectrum is shown in Fig. 30 [108]. An interpretation of the first bands, based on MINDO/3 calculations[108] is given in the Table 1 (see Appendix).

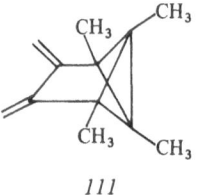

111

V Cubane and Derivatives

1 Bonding Model of Cubane

Cubane *112* (pentacyclo[4.2.0^{2,5}0.^{3,8}0.^{4,7}]octane) belongs to point group O_h which has been confirmed by X-ray analyses[109]. The 28 occupied molecular orbitals of *111* span the following irreducible representations of the group O_h:

$$\Gamma = 3A_{1g} + 2A_{1u} + 1E_g + 3T_{1u} + 3T_{2g} + 1T_{2u}$$

236

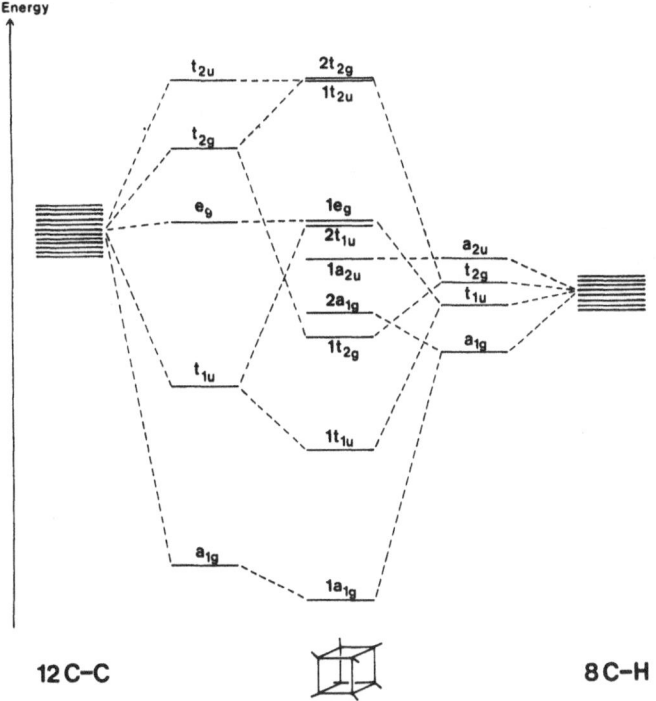

Fig. 31. Valence orbitals of cubane derived from 12 degenerate C–C localized molecular orbitals and 8 degenerate C–H localized molecular orbitals

Several molecular orbital calculations of *111* have been reported[110–112] most of them, however, have not tabulated the molecular orbital energies. Other theoretical treatments[113] have been concerned with the computation of the strain energy of cubane.

In Fig. 31 we derive the valence MO's of cubane in a qualitative interaction diagram[111]. On the left, twelve degenerate C-C localized MO's interact to yield the five CC energy levels of t_{2u}, t_{2g}, e_g, t_{1u} and a_{1g} symmetry. On the right of Fig. 30, eight

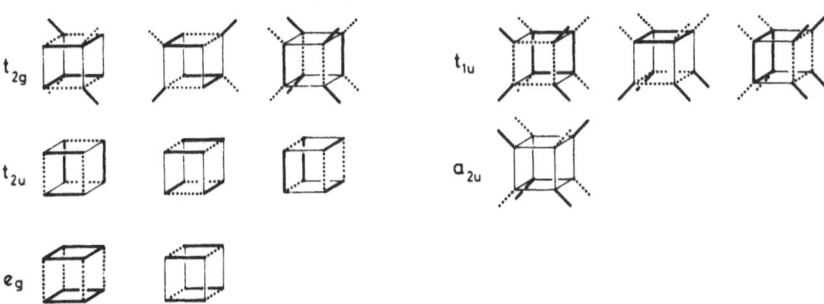

Fig. 32. Schematic drawing of the canonical orbitals of the highest occupied levels. Full and broken lines indicate different phases

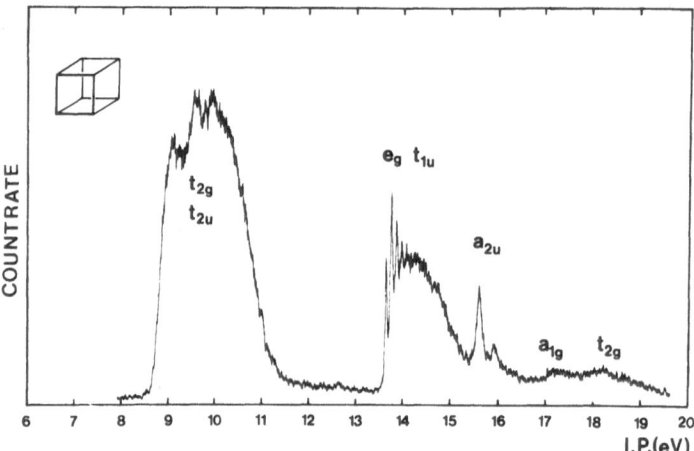

Fig. 33. PE spectrum of cubane

degenerate C-H localized MO's interact and yield four energy levels of a_{2u}, t_{2g}, t_{1u} and a_{1g} symmetry. The energy levels of cubane, shown in the center of Fig. 30, result from the interaction between CC and CH orbitals of the two sets of like symmetry. It is interesting to note that the $1e_g$ and the $1t_{2u}$ levels are built exclusively from CC basis orbitals and that $2a_{2u}$ is a pure linear combination of C-H localized orbitals.

In Fig. 32 the canonical valence orbitals of cubane are drawn schematically. Full and broken lines indicate different phases.

2 PE Spectrum of Cubane

In Fig. 33 the PE spectrum of cubane[50, 112)] is shown. The most striking features are two broad peaks around 9.6 and 14.0 eV. According to the results of ab initio and MINDO/3 calculations[111, 112)], the first broad peak is due to the overlap of the two bands corresponding to the states $^2T_{2g}$ and $^2T_{2u}$ of cubane radical cation. Both states are Jahn-Teller unstable and lead to complicated band envelopes. A detailed assignment of the three maximas observed is thus not possible.

The second band system is again associated with two states namely ^2E_g and $^2T_{1u}$. Here MINDO/3 predicts the reverse order. The fine structure (800 cm^{-1}) associated with the first band (^2E_g) favors the sequence given by the ab initio result. Also the fifth state shows vibrational fine structure (3000 cm^{-1}) which confirms the assignment given ($^2A_{2u}$).

3 Cubane Derivatives

The PE spectra of homocubane (113)[114)], basketane (114)[114)] and secocubane (115)[114)] are compared in Fig. 34. Analogously to the PE spectrum of cubane, one notices broad strongly overlapping bands. The broadening of bands ① – ⑥ ob-

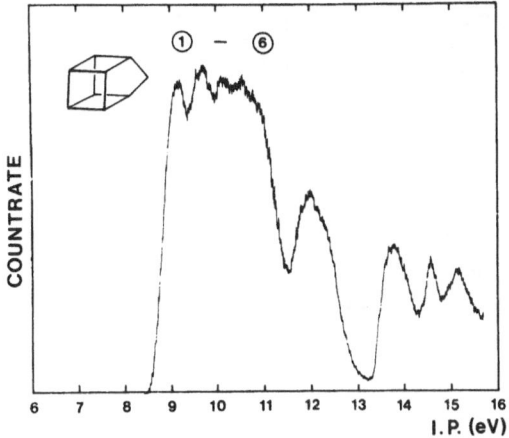

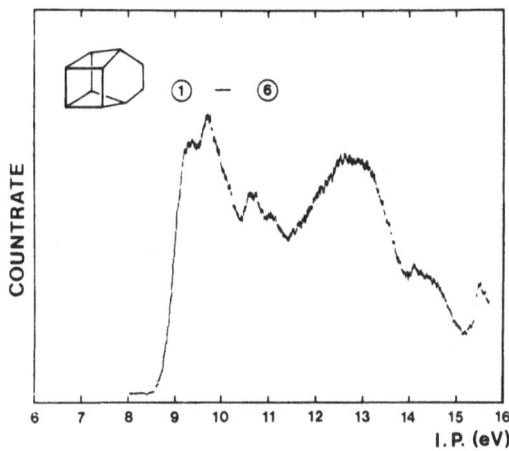

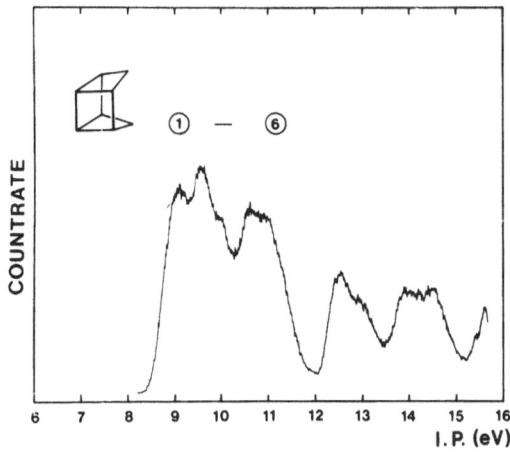

Fig. 34. PE spectra of homocubane, basketane and secocubane

239

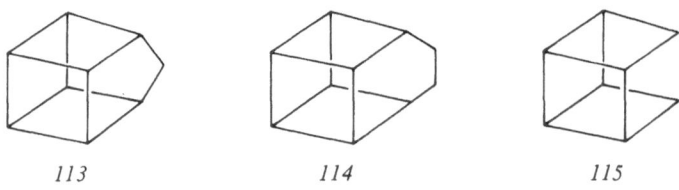

113   114   115

served in the PE spectra of *113* to *115* is substantiated by MINDO/3 calculations[114] as shown in Fig. 34. Due to the strong overlap of the bands and the small energy differences of the orbital energies a definite assignment seems difficult. The PE spectra

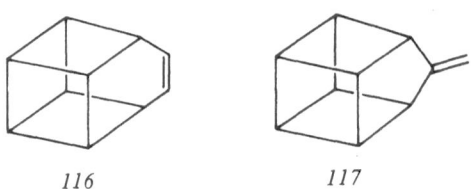

116   117

of basketene (*116*)[115] and exomethylene-homocubane (*117*)[52] also show strongly overlapping bands and are of no help in analyzing the PE spectra of *113 – 115*.

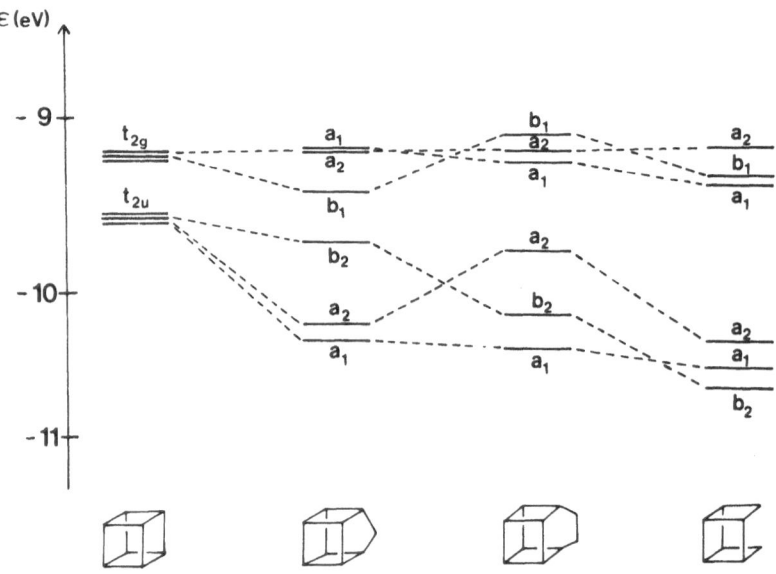

Fig. 35. MINDO/3 molecular orbitals of homocubane, basketane and secocubane

VI Miscellaneous Cyclobutane Derivatives

In Table 1 (see Appendix) several compounds are listed containing an exomethylene cyclobutane ring. Among them are exomethylene cyclobutane (*118*)[68], exomethylene cyclobutene (*119*), the two isomeric exo-dimethylenecyclobutanes *121*[116] and *122*[116] and 3,4-dimethylenecyclobutene (*122*)[35]. A comparison between ex-

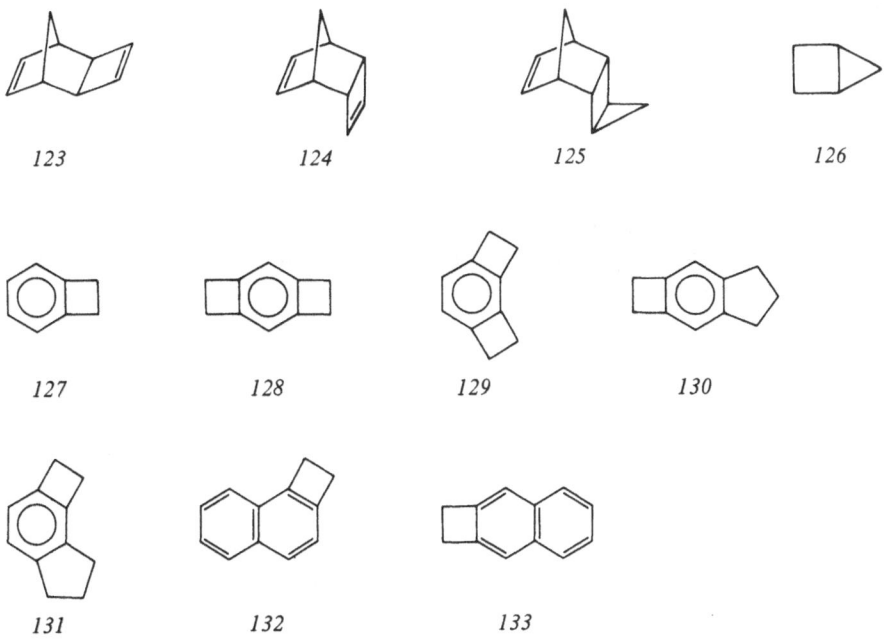

118   *119*   *120*   *121*   *122*

periment and various MO calculations on the ground state led to the assignment of the bands up to 20 eV. It is interesting to note that the split between the first two bands in *120* (2.05 eV) is considerably smaller than for cyclobutadiene (2.4 eV) due to the interaction of the $b_2(\pi)$ orbital with the b_2 orbital localized on the CH_2 groups. The strong interaction between $b_2(\pi)$ and $b_2(CH_2)$ leads to an inverted orbital sequence for *121* [$b_{1u}(\pi_+)$ on top of $b_{3g}(\pi_-)$].

In the literature the PE spectra of a few bicyclic and polycyclic cyclobutane and cyclobutane derivatives have been studied in order to elucidate the through space and through bond interaction of π-orbitals. Examples are *123* to *125* and the corresponding hydrogenated products. In case of *123*[117] the observed split between the first two peaks (1.2 eV) is due to a through bond interaction. The reduced split

123   *124*   *125*   *126*

127   *128*   *129*   *130*

131   *132*   *133*

R. Gleiter

(0.3 eV) found for *124*[118] is interpreted as due to an almost cancellation of through bond and through space effects. For *125* the split of the first two bands is reported to be 0.5 eV. Further PE spectroscopic investigations which should be mentioned here are dealing with bicyclo[2.1.0]pentane (*126*) and the benzo- and naphthocyclo-butene derivatives *127*[71, 119] to *133*[119]. As anticipated from the foregoing chapters the perturbation of the aromatic ring by one C_2H_4 bridge is very similar to that by two CH_3 groups.

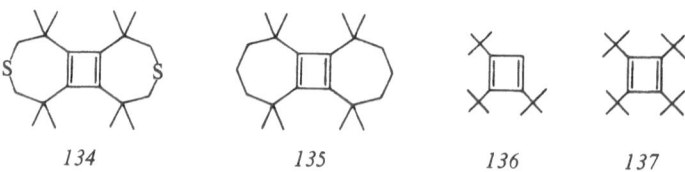

| *134* | *135* | *136* | *137* |

The PE spectra of four cyclobutadiene derivatives *134* – *137*[120, 121] have been recorded. Due to the large alkyl groups only the first two bands could be discriminated and assigned to the two π orbitals.

VII Tetrahedrane

The highest occupied molecular orbitals of tetrahedrane (*138*) can be derived from those of bicyclobutane in two steps as indicated below: From bicyclobutane we generate first bicyclobutanediyl-2,3 (*139*) which we then react to tetrahedrane.

| *101* | *139* | *138* |

Formally we can derive the molecular orbitals of the diradical *139* from those of bicyclobutane by removing two bonding and two antibonding molecular orbitals mainly localized on the C-H bonds (irreducible representations A_1 and B_1, not shown in Fig. 25) and by adding the two linear combinations ($7a_1$ and $4b_1$) obtained from the two sp^n lobes of the diradical *139*. The highest molecular orbitals of *139* are shown on the left of Fig. 36 and can easily be derived from those of bicyclobutane shown in Fig. 25.

The energy difference between the HOMO and LUMO of *139* depends very strongly upon the distance between the centers 2 and 4[122]. At large distances $4b_1$ emerges below $7a_1$. In this case the ring closure to tetrahedrane is symmetry forbidden. Only if $7a_1$ turns out to be below $4b_1$ this reaction is symmetry allowed[122].

If we assume the orbital sequence given at the left of Fig. 36 ($7a_1$ below $4b_1$) and if we reduce the distance between centers 2 and 4, those molecular orbitals

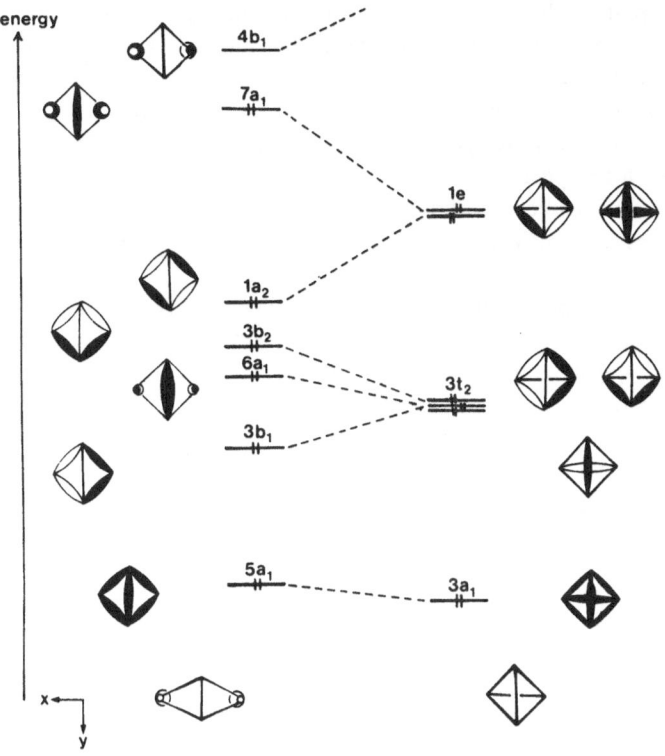

Fig. 36. Derivation of the highest occupied molecular orbitals of tetrahedrane from those of bicyclobutanediyl-1,4

which are bonding between centers 2 and 4 of *139* ($7a_1$, $3b_2$, $6a_1$ and $5a_1$ of Fig. 36) will be stabilized, those which are antibonding ($4b_1$ and $3b_1$) will be destabilized. Finally the molecular orbitals of tetrahedrane (T_d point group) emerge (see Fig. 36, right).

This simple correlation diagram is in accord with more sophisticated molecular orbital calculations carried out for tetrahedrane[123]. The PE spectrum of tetra t-butyl-tetrahedrane has been recorded recently[121]. It confirms the orbital model insofar as the first band has been assigned to 1e. The recorded Jahn-Teller split of 0.7 eV corroborates this assignment.

VIII Conclusions

This review surveys PE spectroscopic work on small hydrocarbon rings carried out in the last ten years. From this research the following conclusions emerge:
1) Koopman's theorem is a useful approximation to interpret the first bands of the PE spectra of small ring hydrocarbons.

R. Gleiter

2) Despite strong overlap of bands in the PE spectra of small ring hydrocarbons it is possible to interpret the PE spectra by studying a large number of similar compounds using perturbation theory.
3) The models for the bonding in small ring hydrocarbons based on single determinant molecular orbital theory are suitable in a qualitative and quantitative sense to interpret their PE spectra.
4) The PE studies on small ring hydrocarbons reveal a considerable interaction between their highest occupied molecular orbitals and those of a π system or another strained ring system.
5) The interaction between a three-membered ring fragment and a π part is found to be comparable to that between two π-moieties. The interaction between a four-membered ring fragment and a π-system is found to resemble that of an unstrained ring and a π-moiety.

Acknowledgment. It is a pleasure to acknowledge the contribution of excellent co-workers whose names appear in the references, prime among them Peter Bischof and Jens Spanget-Larsen. Financial support has been obtained by the Deutsche Forschungsgemeinschaft and the Fonds der Chemischen Industrie. I am grateful to Prof. E. Heilbronner for the communication of results prior to publication and to Prof. W. Lüttke for a sample of bicyclopropane.

IX Appendix

This section contains a compilation of PE data for small ring hydrocarbons. The ordering of the compounds corresponds approximately to the one given in the text. The following informations is given in the Table:
a) Structural formula.
b) Sum formula.
c) Molecular symmetry point group. An asterix indicates that the symmetry is idealized.
d) Vertical ionization energies I_v and their (sometimes tentative) orbital assignment. The vertical ionization energies were obtained by He ($I\alpha$) PE spectroscopy in the gas phase, and were generally estimated from the maximum of the PE band.
e) Leading references, providing a key to the literature.

Table

(a)	(b)	(c)	(d)					(e)
	C_3H_6	D_{3h}	10.6 11.3 ⎫⎬⎭ $3e'$	13.0 $1e''$	15.7 $3a_1'$	16.6 $1a_2$		1, 26, 28, 70)
	C_4H_8	C_s	10.1 e_A	10.9 e_S				62)
	C_7H_{10}	C_{3v}	9.40 10.06 ⎫⎬⎭ $8e$	10.89	12.46	14.16	15.17	31, 50)
	C_7H_{12}	C_s	9.46 e_S	10.01 e_A				57)
	C_8H_{12}	C_s	9.40 e_S	10.20 e_A	10.70	11.40	12.00	62)

Table (continued)

(a)	(b)	(c)	(d)								(e)
	C_8H_{12}	C_s	9.4 e_S	10.00 e_A	10.80	11.50	11.80				62)
	C_9H_{10}	C_{2v}	8.37 b_2	9.26 a_1	10.05 b_2	10.8	11.1	12.5			126)
	C_8H_{14}	C_{2v}	9.46 e_A	10.20 e_S	10.50 σ	11.50 σ					48)
	C_9H_{12}	D_{3h}	8.67 — 9.16 $7e'$	10.23 $4e''$	10.79 $3e''$	12.65 $1a_2'$	·13.05	14.30	15.65		31)
	C_9H_{14}	C_1	9.45 e_A	10.20 e_S	10.86 σ						49)
	C_9H_{14}	C_{2v}	9.6 b_2	10.9 a_1							126)

Compound	Symmetry	Ionization energies (eV) and orbital assignments	Ref.
C_9H_{14}	C_s	9.3 a'(e_S); 10.0 a''(e_A); 10.45 σ	124)
C_9H_{16}	C_s^*	9.4 a'(e_S); 10.0 a''(e_A)	62)
$C_{10}H_{14}$	C_s	9.05 e_S; 9.62 e_A	45)
$C_{10}H_{16}$	C_{3v}	9.05, 9.62 e; 10.25 σ; 10.7; 11.1; 11.9	32)
$C_{10}H_{16}$	C_{2h}	9.17 a_g; 9.91 b_u; 10.18 a_g; 10.58 b_u; 11.61; 11.92	39)
$C_{11}H_{16}$	C_1	8.93 e_S; 9.40 e_A	45)
C_5H_8	C_s	9.2 5a''(π); 10.7 14a'(σ); 11.7 4a''(π)	35, 40)

Table (continued)

(a)	(b)	(c)	(d)					(e)
(structure, CH_3)	C_6H_{10}	C_s	9.12 $a''(\pi)$	10.60 $a'(\sigma)$	11.31 $a''(\pi)$	12.02	12.27	63)
(structure)	C_7H_8	C_{2v}	8.14 $2a_2(\pi)$	9.46 $4b_1(w)$	10.9 $12a_1(\sigma)$	11.89 $7b_2(\sigma)$	12.7 $3b_1(\pi)$	50, 36)
(structure)	C_7H_{10}	C_s	8.48 $\pi - e_A$	10.3 e_S	11.05 $e_A + \pi$	11.6 σ		40)
(structure)	C_7H_{10}	C_1	8.69 $\pi - e_A$	9.96 e_S	10.92 $e_A + \pi$			57)
(structure)	C_8H_{12}	C_{2v}	8.87 π	10.22	10.61	11.08	(11.74) 12.04	63)
(structure, C_2H_5, C_2H_5)	$C_{11}H_{16}$	C_{2v}^*	8.20 $a_2(\pi)$	9.10 $b_1(\pi)$	10.0 $a_1(w)$			37)

Formula	Sym.							Ref.
$C_{11}H_{14}$	C_{2v}	7.87 $a_2(\pi)$	8.63 $b_2(\pi)$	9.08 $b_1(\pi-w)$	10.85 $a_1(w)$			37)
C_8H_{12}	C_s	8.44 $\pi-w$	10.04 w	10.54 $\pi+w$	10.86 σ	11.80		48)
C_8H_{10}	C_s	7.89 $\pi-w$	10.02 w	10.46 $\pi+w$	11.04 π	11.42		48)
C_8H_8	C_s	8.5 a'	8.8 a''	10.4 a''	11.4 a'			38)
C_8H_{10}	C_1	8.5 $\pi-w$	9.7 w	10.8 w				38)
C_8H_{10}	C_s	8.79 a''	9.46 a'	10.5 a''	10.9 a'	12.0	12.4	32)
C_9H_{10}	C_s	8.4 a'	8.8 a''	10.2 a''	11.0 a'			38)

249

R. Gleiter

Table (continued)

(a)	(b)	(c)	(d)							(e)
	$C_{10}H_{10}$	D_{3h}	8.34 (e)	8.7 (e)	9.2 (a_1)	11.4 (e,σ)	11.7 (e,σ)			32, 50)
	$C_{10}H_{12}$	D_{2h}	7.82 (b_{1u})	9.44 (b_{3g})	10.29 (a_{1g})	10.93 (b_{3u})	12.61	12.94		39)
	$C_{10}H_{12}$	C_s	8.32 (a')	8.73 (a'')	10.4 (a'')	10.9 (a')	11.4 (σ)	12.3		32)
	$C_{10}H_{14}$	C_s	8.26 (a'')	9.50 (a')	10.5 (a'')	10.9 (σ)	11.9	13.2		32)
	C_7H_6	C_s	9.83 (14a')	9.98 (10a')	10.43 (9a'')	10.49 (13a')	12.05 (8a'')	12.28 (12a')	13.22 (11a')	41)
			13.83 (7a'')	15.63 (10a')	16.25 (9a')	16.28 (6a'')	17.40 (8a')			

Compound	Sym.	IP 1	IP 2	IP 3	IP 4	IP 5	IP 6	Ref.
C$_7$H$_6$	C$_2$	9.84 / 13a	10.03 / 11b	10.54 / 12a	10.59 / 10b	12.04 / 9b	12.24 / 11a	41)
		13.28 / 8b	13.75 / 10a	15.66 / 9a	16.09 / 7b	16.67 / 8a	17.32 / 6b	
C$_9$H$_{10}$	C$_s$	8.66 / a'(π)	9.21 / a"(π)	10.6 / a'(w)	11.0 / a"(w)	11.7 / σ	12.1	40, 42, 43)
C$_{10}$H$_{12}$	C$_s$	8.73 / a'(π)	9.17 / a"(π)	10.09 / a'(w)	10.59 / a"(π)			42)
C$_{11}$H$_{14}$	C$_s^*$	8.70 / a'(π)	9.17 / a"(π)	9.95 / a'(w)	10.50 / a"(w)			42)
C$_{12}$H$_{16}$	C$_s^*$	8.63 / a'(π)	9.12 / a"(π)	9.74 / a'(w)	10.38 / a"(w)			42)
C$_{13}$H$_{18}$	C$_s$	8.63 / a'(π)	9.15 / a"(π)	9.63 / a'(w)	10.33 / a"(w)			42)
C$_{15}$H$_{12}$	C$_s$	7.77 / a"(π)	8.53 / a"(π)	8.70 / a'(π)	9.76 / a'(π)	10.36 / a'(σ)		43)

Substituent notes: C$_9$H$_{10}$ (cyclopropyl-phenyl); C$_{10}$H$_{12}$ (—CH$_3$); C$_{11}$H$_{14}$ (—C$_2$H$_5$); C$_{12}$H$_{16}$ (—HC(CH$_3$)$_2$); C$_{13}$H$_{18}$ (—C(CH$_3$)$_3$).

Table (continued)

(a)	(b)	(c)	(d)								(e)
	C_{10}H_{12}O	C_s	8.70	8.97	10.57						45)
			n	e_S	e_A						
	C_{11}H_{13}O	C_1	8.75	9.80	10.28						45)
			n	e_S	e_A						
	C_6H_{10}	C_{2h}, C_2	9.6–9.75	10.1–10.4	11.0	11.75–11.9					50, 51, 54)
			b_g, a	a_g, b	b_u, a	a_u, b					
	C_7H_8	C_{2v}	8.33	9.78	10.13	11.00	11.86	12.16	13.33	14.61	17.44 47)
			b_2	b_1	a_2	a_1	b_1				
	C_8H_8	C_{2v}	8.48	8.90	9.86	11.25	11.45	11.90			47)
			b_2	b_1	a_2	a_1	b_1				
	C_8H_{10}	C_{2v}	8.75	9.75	10.2	10.5					66)
			b_1(π)	a_2(w)							

252

Structure	Formula	Symmetry							Ref
	C_8H_{10}	C_{2v}	9.12 (a_2)	9.60 (a_1)	10.25 (b_2)	10.67 (b_1)	11.25	12.25	57)
	C_8H_{12}	C_2	9.39 (14b)	9.82 (16a)	10.14 (13b)	11.24 (15a)			57)
	C_8H_{12}	C_s^*	8.95 (14a″)	9.66 (16a′)	10.14 (13a″)	11.34 (15a′)			58)
	C_9H_8	C_1	8.5 (π)	9.8 (w)	10.2 (w)	10.5 (w)	11.3 (w)		58)
	C_9H_8	C_s	9.0 (π)	9.56	10.0	10.3 } w′, w″, w″	11.4 (w′)		58)
	C_9H_8	C_1	8.96 (π)	9.37 (w_b)	9.88 (w_a)	10.9 (w_b)	11.4 (w_a)		58)

R. Gleiter

Table (continued)

(a)	(b)	(c)	(d)						(e)
	C_9H_8	C_1	8.90 π	9.34 w_a	10.1 w_b	10.4 w_a	11.1 w_b		58)
	C_9H_{10}	C_s	8.43 a'	9.23 a''	10.21 a''	10.67 a'	11.27 a''	11.61	46)
	$C_{10}H_{10}$	C_{2v}	8.80 b_2	(9.19) a_2	9.78 } a_1, b_2	10.20	10.53	11.28	46)
	$C_{10}H_{10}$	C_{2v}	9.00 $7b_2$	9.40 $3a_2$	10.00 $10a_1$	10.15 $6b_2$	10.65 $6b_1$		52, 53)
	$C_{10}H_{12}$	C_{2v}	7.74 a_2	(9.87) (b_2)	9.68 a_2	(10.20) (a_1)	10.46 b_1	11.46 b_1 12.36	48)

Structure	Compound	Point group	IE₁	IE₂	IE₃	IE₄	IE₅	IE₆	Ref.
	$C_{10}H_{12}$	C_{2v}	9.08 (a_2)	9.53 (a_1)	10.06 (b_2)	(10.34) (b_1)	(10.54)		46)
	$C_{10}H_{14}$	C_1	8.48 ($\pi - w$)	9.43	10.16 (w)	10.46 ($\pi + w$)	11.29		48)
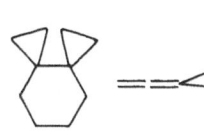	$C_{10}H_{16}$	C_{2v}^*	9.22 (a_2)	9.68 (b_2)	10.20 (a_1)	10.74 (b_1)	11.21	12.18	48)
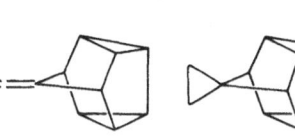	$C_{11}H_{10}$	C_{2v}	8.75 ($7b_2$)	9.22 ($6b_1$)	9.48 ($3a_2$)	9.9 ($11a_1$)	10.2 ($6b_2$)	11.1 ($5b_1$)	53)
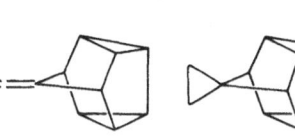	$C_{11}H_{12}$	C_{2v}	9.05 ($6b_2$)	9.35 ($4a_2$)	10.2 ($10a_1$)	10.5 ($5b_1$)	10.7 ($6b_1$)	11.2 ($9a_1$)	53)
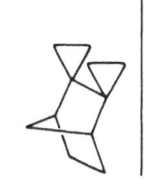	$C_{11}H_{14}$	C_s	8.76 (a')	9.82 (a')	10.30 (a'')	10.60 (a'')	11.02		49)

R. Gleiter

Table (continued)

(a)	(b)	(c)	(d)						(e)
	$C_{12}H_{14}$	C_{2v}	8.35 $7b_2$	9.14 $4a_2$	9.65 $12a_1$	9.95 $6b_2$	10.3 $7b_1$		53)
	$C_{12}H_{18}$	C_{2v}	8.67 a_2	9.58 a_1	9.99 b_2	10.61 b_1	10.86		49)
	C_9H_{12}	C_s	9.0 a''	9.4 a'	9.9 a''	10.5 a'	11.3 a''		59)
	$C_{10}H_{10}$	D_{3h}	8.50 8.97 $11e$	9.42 $3a_2$	10.61 11.13 $10e$	11.42 $8a_1$			58)
	$C_{10}H_{12}$	C_s	8.78 a''	9.20 a''	9.56 a'	10.38 a'	10.90 a''	11.20 a'	58, 59)

Compound	Point group	Structure	Ionization energies (eV) and assignments	Ref.
$C_{10}H_{14}$	C_s		8.8 (w); 9.5 (w''); 10.0, 10.4 (w', w''); 10.75 (w''); 11.4 (w')	54)
$C_{10}H_{14}$	C_s		9.0 (w'); 9.3 (w''); 10.0, 10.5 (w', w''); 10.9 (w''); 11.3 (w')	54)
$C_{10}H_{14}$	C_1		8.88 (w); 9.45 (w); 10.0 (w); 10.4 (w); 10.85 (w); 11.55 (w)	54)
$C_{15}H_{24}$	C_s		8.5 (a''); 8.8 (a'); 9.6 (a''); 9.8 (a''); 10.3 (a')	59)
C_6H_8	D_{2h}		8.93 ($2b_{1u}$); 9.63 ($2b_{1g}$); 10.58 ($4a_g$); 11.48 ($3b_{2u}$); 12.48 ($2b_{3u}$)	60)
C_9H_{12}	D_{3h}		9.12, 9.32 ($3e''$); 9.52, 10.02 ($5e'$); 11.02 ($4a_1'$); 12.22 ($3a_1'$); 12.68 ($2e''$); 13.03 ($4e'$); 13.32 ($1a_2'$)	60)

257

Table (continued)

(a)	(b)	(c)	(d)							(e)
	$C_{12}H_{16}$	D_{4h}	8.22 (2b$_{2u}$)	9.21 9.42 (5e$_u$)		10.12 10.32 (3e$_g$)		10.91 (2b$_{1g}$)	11.21 (4a$_{1g}$)	60)
	C_6H_8	C_{2h}	9.66 9.8 10.05 (bg, bg, ag, ag)			10.4				61)
	C_8H_{10}	C_s	9.05 (π)	9.50 (e$_S$)	10.30 (e$_A$)	11.30	12.20			62)
	C_8H_{10}	C_s	8.90 (π)	9.60 (e$_S$)	10.20 (e$_A$)	11.55	11.90	12.55		62)
	C_9H_{10}	C_s	8.65 (a'(π))	9.5 (a'(π))	9.8 (a'(e$_S$))	10.15 (a''(e$_A$))	11.3 (σ)			124)
	C_9H_{12}	C_s	8.8 (a'(π))	9.5 (a'(e$_S$))	10.0 (a''(e$_A$))	10.5 (σ)				124)

Compound	Sym.									Ref.
C$_6$H$_8$	C$_s$	8.6 a"(π)	9.5 a'(π)	10.3 a'(σ)	11.25 a"(π)					63)
C$_7$H$_{10}$	C$_s$	8.83 a"(π)	9.75 a'(π)	10.54	11.21	12.49	13.60	14.16		63)
C$_8$H$_{12}$	C$_s$	8.81 a"(π)	9.67 a'(π)	10.25	10.81	12.17	12.65	13.80		63)
C$_8$H$_{12}$	C$_s$	8.78 a"(π)	8.98 a'(π)	10.63	11.47	12.70	13.05	14.49		63)
C$_8$H$_{12}$	C$_1$	8.96 a"(π)	9.29 a'(π)	10.25	11.05	12.68	13.76			63)
C$_9$H$_{14}$	C$_s^*$	8.60 a"(π)	9.15 a'(π)	10.41	11.04	12.05	12.64	13.68	14.61	63)
C$_9$H$_{12}$	C$_{2v}$	8.62 a"(π)	9.68 a'(π)	10.19	10.65	11.11	11.74	12.62	15.12	63)

259

Table (continued)

(a)	(b)	(c)	(d)					(e)
(cyclopropylidene, CH_3, CH_3, CH_3, CH_3)	C_9H_{14}	C_s	8.16 $a''(\pi)$	9.00 $a'(\pi)$	10.05 $a'(w)$	11.09 a''	11.75	65)
(cyclopropylidene, CH_3, CH_3, CH_3, CH_3)	C_9H_{14}	C_2	8.13 a''	8.98 a'	10.10 a'	11.08 a''	~11.9	65)
(CH_3, CH_3, H_3C, CH_3)	C_9H_{14}	C_s	8.08 a''	8.96 a'	10.07 a'	11.24 a''	11.95	65)
(CH_3, CH_3, CH_3, H_3C, CH_3)	$C_{10}H_{16}$	C_1	8.00 π	8.90 π	9.72	10.96	11.63	65)
(CH_3, CH_3)	$C_{11}H_{14}$	C_{2v}	8.01 b_1	8.88 b_2	9.79 a_1	10.84 b_1	11.72	65)

Structure	Formula	Symmetry							Ref.
	$C_{11}H_{18}$	C_{2v}	7.87 b_1	8.78 b_2	9.42 a_1	10.78 b_1	11.21		65)
	C_5H_8	D_{2d}	9.45	11.89	15.50	18.04	(18.69)	(19.17)	50)
	$C_{11}H_{14}$	C_3	8.80	9.70	10.1	10.4	11.9	12.4	66)
	$C_{13}H_{16}$	C_{3v}	9.3	9.65	10.0				66)
	$C_{13}H_{16}$	C_3	8.5	9.1	9.4	9.8	10.1	11.0	66)
	C_3H_4	C_{2v}	9.82 $2b_1$	10.95 $3b_2$	12.59 $6a_1$	14.95 $1b_1$	16.68 $5a_1$	18.28 $2b_2$ 19.51 $4a_1$	67, 68, 70)
	C_7H_{12}	C_{2v}	8.52 b_1	9.20 b_2	10.77 a_1				37)

261

Table (continued)

(a)	(b)	(c)	(d)						(e)
	C_4H_6	C_{2v}	9.57	10.47	11.35	13.08	14.52	15.75	68, 70)
			$2b_1$	$4b_2$	$8a_1$	$1a_2$	$3b_2$	$7a_1$	
			17.45	19.72	22.20				
			$6a_1$	$2b_2$	$5a_1$				
	C_6H_6	D_{3h}	8.94	13.0	10.7	11.6	14.1		69)
			$1e''$	$1a_2''$	$5e'$		$(1a_2')$		
	$C_{12}H_{18}$	D_{3h}	7.49	9.71	10.45	11.50			69)
			e''	e'		a_2''			
	C_7H_6	C_{2v}	8.82	9.48	10.17				71)
			$b_1(\pi)$	$a_2(\pi)$	σ				
	C_4H_8	D_{2d}	10.7	11.3	11.7	12.5	13.4	13.6	15.9	16.2	18.2	75)
			$4e$		$4a_1$	$1b_1$	$3e$		$3a_1$	

Formula	Symmetry	Ionization energies (eV) and assignments	Ref.
C_5H_{10}	C_s	10.37 (a′), 10.94 (a″)	80)
C_6H_{12}	C_s	10.12 (a′), 10.80 (a″), 11.15 (a′), 11.55 (a″), 12.35 (a′), 12.75 (a″)	83)
C_7H_{12}	C_{2v}	9.90, 10.44	84)
C_8H_{12}	D_{2d}	9.78 ⎱ 10.45 (e), 10.73 (a_1), 10.60 (b_1)	80)
C_8H_{12}	C_{2v}	8.90 ($6b_1$), 10.15 ($9b_2$), 10.75 ($5b_1$), 11.15 ($13a_1$)	82)
C_8H_{12}	C_s	9.40, 10.05, 15.00, 16.75	50)
C_8H_{14}	C_{2v}	10.05 (a′), 10.62 (a″), 11.03 (a′), 11.60 (a″), 12.35 (a′)	79)

Table (continued)

(a)	(b)	(c)	(d)						(e)
bicyclic structure	C_8H_{14}	C_2	10.0 — 16b	10.5 — 15b	11.0 — 15a				82)
cyclobutane, CH₃, CH₃ / CH₃, CH₃	C_8H_{16}	D_{2h}	9.68 — b₃u	10.69 — b₂u	11.1 — b₁u	11.5 — b₁g	12.4 — b₂g	12.7 — b₁u	83)
cyclobutane, C₂H₅ / C₂H₅	C_8H_{16}	C_2	9.88 — b	10.39 — b	10.98				84)
tricyclic structure	C_9H_{12}	C_s	8.92 — a′(π)	10.00 — a′(e_S)	10.46 — a″(e_A)				118)
tricyclic structure	C_9H_{14}	C_s	9.65 — a′(e_S)	10.13 — a″(e_A)					118)

Compound	Symmetry							Ref.
$C_{10}H_{12}$	C_s	8.8 a′(π)	9.45 a′(π)	9.8 a′(e_S)	10.3 a″(e_A)	10.7 w	11.5 σ	124)
$C_{10}H_{14}$	C_s	9.0 a′(π)	9.5₅ a′(e_S)	9.9₅ a″(e_A)	10.3 a′(w)	11.3 σ		124)
$C_{10}H_{16}$	C_s	9.45 a′(e_S)	9.9 a″(e_A)	10.2 a′(w)	10.8 σ			124)
$C_{12}H_{16}$	C_s	8.39 π	8.94 π	9.84 a′(e_S)	10.36 a″(e_A)			118)
$C_{12}H_{18}$	C_s	8.30 π						118)
C_6H_{10}	C_s	9.44 a″(π)	10.75 a′	11.15 a″	11.5 a′	12.45 a″	12.75 a′	40, 83)

265

R. Gleiter

Table (continued)

(a)	(b)	(c)	(d)						(e)
	C_7H_{10}	C_1	7.98 π	10.22 w	10.82 w	11.16			84)
	C_8H_{10}	C_{2v}	8.38 a_2	10.12 b_1	10.80 a_1	11.22 b_1	12.58 b_1		79)
	C_8H_{10}	C_{2v}	8.11 $2a_2$	10.15 $5b_1$	10.72 $5b_2$	11.13 $4b_1$	12.1 $13a_1$	12.5 $1a_2$	82)
	C_8H_{12}	C_s	9.24 $a'(\pi)$	9.67 $a''(\pi)$	10.69	11.14			84)
	C_8H_{12}	C_{2v}	9.46 b_1	9.78 a_2	10.88	11.3			84)
	C_8H_{12}	C_2	9.3 a	9.98 b	10.87				84)
	C_8H_{12}	C_s	8.97 a''	10.38 a'	10.83 a''	11.75 a''	12.60		79)

Compound	Point group							Ref.
C_8H_{14}	C_s	9.40 (a″)	10.30 (a′)	10.95 (a″)	11.2 (a′)	11.9 (a″)	12.7 (a′)	40, 83)
C_9H_{14}	C_1	8.88 (π)	10.00 (w)	10.45 (w)	10.75			84)
$C_{10}H_{14}$	C_s	8.85, 9.11 (π)		10.6	10.9			81)
$C_{10}H_{14}$	C_s	7.81 (a″(π))	9.6 (a′(π))	10.3	10.56			85)
$C_{10}H_{18}$	C_s	9.10 (a″)	10.05 (a″)	10.60 (a′)	11.0 (a′)	11.4 (a″)	12.1 (a″)	83)
$C_{16}H_{20}$	D_{2d}	7.97 (b₁)	8.80 (a₂)	9.26 (e)				85)

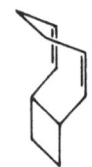

267

Table (continued)

(a)	(b)	(c)	(d)							(e)
	C_6H_8	C_s	10.02 a'', a'(π)	11.4 a'	11.85 a'	12.3 a''	13.25 a'	13.9 a''		83)
	C_8H_8	C_s	9.76 a'	9.9 a''	10.28 a'	10.37 a''				84)
	C_8H_8	C_2	9.86 a	10.00 b	10.37 a	10.74 b				84)
	C_8H_{12}	C_s	9.78 a'	9.92 a''	10.7 a'	10.9 a''	11.25 a'	11.65 a''	12.5 a''	83)
	$C_{10}H_{16}$	C_s	9.33 a''(π)	9.55 a'(π)	10.3 a''	11.1 a'	11.35 a'	11.7 a''	12.6 a''	83)

Compound	Sym.								Ref.
$C_{10}H_{12}$	C_{2v}^{*}	8.77 $b_1(\pi\text{-}w)$	9.21 $a_2(\pi)$	10.6 $a_1(w)$	11.0 $b_1(\pi+w)$	11.7 σ	12.1 $b_1(\pi)$		40)
C_6H_6	C_{2v}	9.4 $5b_2$ / 14.2 $6a_1$	9.7 $8a_1$ / 15.9 $3b_2$	10.9 $7a_1$ / (16.6) $5a_1$	11.5 $3a_2$	12.2 $5b_1$	13.2 $4b_2$		70, 88)
C_6H_8	C_s	9.4 $13a'$ / 15.1	10.3 $9a''$ / 15.5	10.8 $12a'$ / 16.7 $7a'$	11.4 $8a'$	12.4 $11a'$	12.9		70, 88)
C_6H_{10}	C_{2v}	9.6 $4a_2$ / 14.0	10.2 $6b_1$ / 15.1	10.8 $8a_1$	11.5 $3a_2$	12.3 $5b_2$	13.2		70, 88)
$C_{12}H_{18}$	C_{2v}	7.8 $8a_1$	8.3 $5b_1$	9.5 σ	9.9	10.8	11.1	11.8	88)

269

R. Gleiter

Table (continued)

(a)	(b)	(c)	(d)					(e)
	$C_{12}H_{20}$	C_s	8.1 $13a_1$	9.1 σ				88)
	$C_{12}H_{24}$	C_{2v}	8.8 σ					88)
	$C_{10}H_{10}$	C_{2v}	8.44 $a_1(\pi^+)$	9.3 $\overline{b_1(\pi^-), a_1(\sigma)}$	9.6	10.4		89)
	C_8H_8	C_{2h}	8.96 $7a_g(\pi^+)$	9.93 $6b_u$	10.13 $5b_u(\pi^-)$	10.57 $4a_u$	11.6 $3a_u$	90, 91, 95)
	C_8H_8	C_{2v}	9.08 $7a_1(\pi+)$	9.44 $5b_2$	9.87 $6b_2(\pi^-)$	10.67 $3a_2$	10.9 $4b_1$	90, 91, 95)

Compound	Symmetry								Ref.
C_9H_{10}	C_s	9.3 $15a'$	9.3 $14a'(\pi)$	9.86 $8a''$	10.46 $13a'$	11.06 $7a''$			95)
C_9H_{10}	C_s	8.91 $15a'$	9.3 $14a'(\pi)$	9.9 $13a'$	10.3 $8a''$	10.6 $7a''$			95)
$C_{10}H_{12}$	C_{2h}	9.1 $8b_u$	9.5 $9a_g$	9.8 $5a_u$	10.6 $7b_u$	10.9 $4a_u$			95)
$C_{10}H_{12}$	C_{2v}	8.5 $8b_2$	9.4 $9a_1$	~10.0 $(5b_1)$	~10.3 $(7b_2)$	10.6 $(4a_2)$			95)
$C_{16}H_{20}$	C_{2h}	8.10 a_u	8.92 b_g	10.89 b_u					96)
C_4H_6	C_{2v}	9.39 $7a_1$	11.30 $1a_2$	11.70 $3b_1$	12.99 $4b_2$	14.51 $6a_1$	14.99 $3b_2$	17.0 $5a_1$	100, 101)
C_6H_8	C_{2v}	9.2 $10a_1$	10.0 $2a_2$	11.53 $4b_2$					102)

271

Table (continued)

(a)	(b)	(c)	(d)				(e)
	C_7H_{10}	C_{2v}	8.75 $12a_1$	10.5 $2a_2$	10.8 $5b_2$		102)
	C_8H_{12}	C_s	8.53 "a_1"	10.2 "a_2"	10.5 "b_2"	11.1 "b_1"	104)
	C_8H_{12}	C_s	8.69 "a_1"	10.3 "a_2"	10.6 "b_2"	11.2 "b_1"	104)
	C_8H_{14}	C_{2v}	8.6 a_1	10.1 a_2	10.6 b_2	10.75 b_1	101)
	C_6H_6	C_{2v}	8.55 $4b_2$	9.75 $10a_1$	10.83 $1a_2$	12.18 $6b_1$	102)

Structure	Formula	Symmetry								Ref.
	C_7H_8	C_s	8.82 "a1"	9.20 π	10.80 "a2"	11.80 "b1"				104)
	C_8H_8	C_s	8.8 "a1"	9.70 π	9.95 "a2"	10.75				125)
	C_8H_{10}	C_s	9.1 "a1"	9.7 "a2"	10.35					125)
	C_8H_{10}	C_1	8.45 "a1"	8.95 π	10.60 "a2"	11.35 "b1"				104)
	C_8H_{10}	C_s	8.64 "a1"	9.16 "π"	10.67 "a2"	11.4 "b1"				104)
	C_9H_{12}	C_1	8.26 "a1"	8.66 π	10.39 "a2"	11.3 "b1"				104)

273

Table (continued)

(a)	(b)	(c)	(d)					(e)
	C_9H_{12}	C_1	8.30 "a_1"	8.70 π	10.40 "a_2"	11.25 "b_1"		104)
	$C_{14}H_{10}$	C_{2v}	7.55 $3a_2$	8.30 $14a_1$	8.85 $5b_1$	9.14 $4b_1$		103)
	$C_{12}H_{16}$	C_{2v}	7.84 $a_2(\pi)$	8.40 $a_1(\sigma)$	9.28 $b_1(\pi)$	9.78 $a_1(\sigma)$		107, 108)
	C_8H_8	O_h	9.0–9.1 $\}$ $3t_{2g}$, $1t_{2u}$ 9.5–9.6 9.85–9.9	13.75 $1e_g$	14.25–14.35 $3t_{1u}$	15.6–15.65 $2a_{2u}$	17.6 $3a_{1g}$	50, 112)

Compound	Symmetry	Values						Ref.
C_8H_{10}	C_{2v}	9.12	9.63	10.0	10.65	10.97	11.2	114)
			$a_2, b_1, a_1, a_2, a_1, b_2$					
C_9H_{10}	C_{2v}	9.14	9.64	10.12	10.4	10.7	11.1	114)
			$a_1, a_2, b_1, b_2, a_2, a_1$					
$C_{10}H_{10}$	C_{2v}	8.97	9.18	9.50	9.8	10.80		115)
		11.06	12.7	13.0	14.2	14.5		
$C_{10}H_{12}$	C_{2v}	9.15	9.53	9.7	9.95	10.5	10.85	114)
			$b_1, a_2, a_1, a_2, b_2, a_1$					
$C_{12}H_{12}$	C_2	8.30	9.2	9.35	9.6			127)
		π	π					
C_4H_6	C_{2v}	9.59	11.04	11.80	12.84	13.40		67, 68, 70)
		$2b_1$	$5b_2$	$7a_1$	$1a_2$	$6a_1$		
		15.54	16.44	17.19	20.72			
		$1b_1$	$5a_1$	$4b_2$	$3b_2$			

Table (continued)

(a)	(b)	(c)	(d)	(e)
(structure)	C_5H_8	C_{2v}	9.35 ($3b_1$), 10.93 ($5b_2$), 11.68 ($10a_1$), ~12.2 ($4b_2$), 12.72 ($2b_1$), 13.53 ($1a_2$); 14.53 ($3b_2$), 15.72 $\{9a_1\ 1b_1\}$, 17.05 ($8a_1$), 19.54 ($7a_1$), 21.73 ($2b_1$)	67, 68, 70)
(structure)	C_6H_6	C_{2v}	8.80 ($2b_1$), 9.44 ($1a_2$), 11.50, 12.3 ($1b_1$), 13.3, 14.1	35)
(structure)	C_6H_8	D_{2h}	9.08 ($2b_{1u}$), 9.94 ($1b_{3g}$), 11.0 ($3b_{3u}$), 11.45 ($2b_{1g}$), 12.45 ($3b_{2u}$), 13.5 ($1b_{2g}$); 14.2 ($2b_{3u}$), 15.1 ($1b_{1u}$), 15.5 ($1b_{1g}$), 17.1 ($3a_g$), 17.4 ($2b_{2u}$)	116)
(structure)	C_6H_8	C_{2v}	8.66 ($2a_2$), 10.62 ($2b_2$), 11.2 ($7a_1$), 11.7 ($5b_1$), 12.1 ($6a_1$), 12.9 ($1a_2$); 13.9 ($4b_1$), ~15.0 ($5a_1$), ~15.3 ($1b_2$), ~15.6 ($4a_1$), ~16.1 ($3b_1$), ~19.1 ($3a_1$), ~19.7 ($2b_1$)	116)

276

							Ref.
C_7H_8	C_1	10.30 π	11.26 π	12.41	14.18		50)
C_9H_{10}	C_s	9.03 π+	9.36 π−	10.46	10.90		118)
C_9H_{10}	C_s	8.65 π+	9.80 π−	10.4			118)
C_9H_{12}	C_s	9.0 π	10.0				118)
$C_{12}H_{14}$	C_s	8.33 π	8.94 π	9.52 π	10.15	(10.71)	118)
$C_{14}H_{10}$	C_s	7.72 14a″	8.71 19a′	9.19 18a′	9.73 17a′		103)

277

Table (continued)

(a)	(b)	(c)	(d)				(e)
	C_5H_8	C_s	9.55	10.38	11.50	11.90	108)
	$C_{10}H_{12}$	C_s	8.83	9.35	10.15		118)
			$\pi - e_S$	$e_S + \pi$	e_A		
	$C_{10}H_{14}$	C_s	9.20	9.92			118)
			e_S	e_A			
	$C_{10}H_{10}$	C_{2v}	8.22	8.91			119)
			$a_2(\pi)$	$b_1(\pi)$			
	$C_{10}H_{10}$	C_{2v}	8.35	8.69			119)
			$a_2(\pi)$	$b_1(\pi)$			

Structure	Formula	Symmetry					Ref.
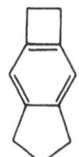	$C_{11}H_{12}$	C_{2v}	8.15 $a_2(\pi)$	8.77 $b_1(\pi)$			119)
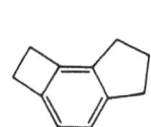	$C_{11}H_{12}$	C_{2v}^*	8.38 $a_2(\pi)$	8.71 $b_1(\pi)$			119)
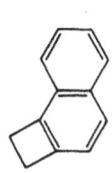	$C_{12}H_{20}$	C_s	7.84 "a_{1u}"	8.65 "b_{1u}"	9.69 "b_{3g}"	10.7 "b_{2g}"	119)
	$C_{12}H_{20}$	C_{2v}	7.92 "a_{1u}"	8.54 "b_{1u}"	9.83 "b_{3g}"	10.56 "b_{2g}"	119)
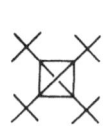	$C_{20}H_{12}$	T_d	$\left.\begin{array}{c}7.5 \\ 8.2\end{array}\right\}$ 2e	10.2 σ			121)
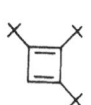	$C_{16}H_{13}$	C_s	6.83 π				120)

279

Table (continued)

(a)	(b)	(c)	(d)				(e)
	$C_{20}H_{12}$	D_{2h}	6.35 b_{3g}	9.20 b_{3u}	9.5 b_{2u}		121)
	$C_{22}H_{36}$	D_{2h}^{*}	6.5 b_{3g}	9.2 b_{3u}			121)
	$C_{20}S_2H_{32}$	D_{2h}^{*}	6.89 π	8.20 n	10.20 σ	11.10 π	120)

X References

1. Turner, D. W., Baker, A. D., Baker, C., Brundle, C. R.: Molecular photoelectron spectroscopy. New York: Wiley-Interscience 1970
2. Brundle, C. R., Robin, M. B., In: Determination of organic structures by physical methods, Vol. III. Nachod, F. C., Zuckermann, J. J. (eds.). New York: Academic Press 1971
3. Eland, J. H. D.: Photoelectron spectroscopy. London: Butterworths 1974
4. Rabalais, J. W.: Principles of ultraviolet photoelectron spectroscopy. New York: Wiley 1977
5. Heilbronner, E., Maier, J. P., In: Electron spectroscopy, theory, techniques and application. Brundle, C. R., Baker, A. D. (eds.). New York: Academic Press 1977
6. Worley, S. D.: Chem. Rev. *71*, 295 (1971)
7. Bock, H., Ramsay, B. G.: Angew. Chem. *85*, 773 (1973); Angew. Chem. Int. Ed. Engl. *12*, 743 (1973)
8. Heilbronner, E. In: The world of quantum chemistry. p. 211. Daudel, R., Pullman, B. (eds.). Dordrecht, Holland: Reidel 1974
9. Bock, H., Mollère, P. D.: J. Chem. Ed. *51*, 506 (1974)
10. Heilbronner, E. In: Molecular Spectroscopy, Proc. of Sixth Conference on Molecular Spectroscopy. Durham 1976, London: Heyden 1977
11. Bock, H.: Angew. Chem. *89*, 631 (1977), Angew. Chem. Int. Ed. Engl. *16*, (1977); Wittel, K., McGlynn, S. P.: Chem. Rev. *77*, 745 (1977)
12. Turner, D. W.: Ann. Rev. Phys. Chem. *21*, 107 (1970)
13. Kutzelnigg, W.: Einführung in die Theoretische Chemie, Vol. 2. Weinheim – New York: Verlag Chemie 1978
14. Koopmans, T.: Physica *1*, 104 (1934)
15. Cederbaum, L. S., von Niessen, W.: J. Chem. Phys. *62*, 3824 (1975)
16. Cederbaum, L. S., Domcke, W., von Niessen, W.: Chem. Phys. *10*, 459 (1975)
17. Domcke, W., Cederbaum, L. S., von Niessen, W., Kraemer, W. P.: Chem. Phys. Lett. *43*, 258 (1976)
18. Von Niessen, W., Cederbaum, L. S., Kraemer, W. P.: Theoret. Chim. Acta *44*, 85 (1977)
19. Franck, J.: Trans. Faraday Soc. *21*, 536 (1925)
    Condon, E. U.: Phys. Rev. *32*, 858 (1928)
20. Coulson, C. A., Moffitt, W. E.: J. Chem. Phys. *15*, 151 (1947); Philos. Mag. *40*, 1 (1949)
21. Förster, Th.: Z. Physik. Chem. B. *43*, 58 (1939)
22. Walsh, A. D.: Nature (Lond.) *159*, 167, 712 (1947); Trans. Faraday Soc. *45*, 179 (1949)
    Sugden, T. M.: Nature (Lond.) *160*, 367 (1947)
23. Coulson, C. A., Goodwin, T. H.: J. Chem. Soc. 2851 (1962); 3161 (1963)
    Peters, D.: Tetrahedron *19*, 1539 (1963)
    Veillard, A., Del Re G.: Theore. Chim. Acta *2*, 55 (1964)
    Klasinc, L., Maksic, Z., Randić, M.: J. Chem. Soc. A, 755 (1966)
24. Pauling, L.: J. Am. Chem. Soc. *53*, 1367 (1931)
    Slater, J. C.: Phys. Rev. *37*, 481 (1931)
25. Mulliken, R. S.: Phys. Rev. *41*, 751 (1932)
26. Basch, H., Robin, M. B., Kuebler, N. A., Baker, C., Turner, D. W.: J. Chem. Phys. *51*, 52 (1969)
27. Buenker, R. S., Peyerimhoff, S. D.: J. Phys. Chem. *73*, 1299 (1969)
    Radom, L., Lathan, W. A., Hehre, W. J., Pople, J. A.: J. Am. Chem. Soc. *93*, 2603 (1971)
    Jorgensen, W. L., Salem, L.: The Organic Chemist's Book of Orbitals. New York: Academic Press 1973
    Lathan, W. A., Radom, L., Hariharan, P. C., Hehre, W. J., Pople, J. A.: Topics in Current Chemistry, Vol. 41, 1. Berlin–Heidelberg–New York: Springer 1973
    Deakyne, C. A., Allen, L. C., Laurie, V. W.: J. Am. Chem. Soc. *99*, 1343 (1977)
28. Lindholm, E., Fridh, C., Åsbrink, L., J.: Chem. Soc. Faraday Disc. *54*, 127 (1972)
29. Haselbach, E.: Chem. Phys. Lett. *7*, 428 (1970)
30. Bischof, P.: J. Am. Chem. Soc. *99*, 8145 (1977)

R. Gleiter

31. Haselbach, E., Heilbronner, E., Musso, H., Schmelzer, A.: Helv. Chim. Acta 55, 302 (1972)
32. Bischof, P., Gleiter, R., Heilbronner, E., Hornung, V., Schröder, G.: Helv. Chim. Acta 53, 1645 (1970)
33. Schweig, A., Thiel, W.: Chem. Phys. Lett. 21, 541 (1973)
34. Lüttke, W., de Meijere, A.: Angew. Chem. 78, 544 (1966); Angew. Chem. Int. Ed. Engl. 5, 512 (1966)
    Günther, H., Wendisch, D.: Angew. Chem. 78, 266 (1966); Angew. Chem. Int. Ed. Engl. 5, 251 (1966)
    de Mare G. R., Martin, J. S.: J. Am. Chem. Soc. 88, 5033 (1966)
    de Meijere, A., Lüttke, W.: Tetrahedron 25, 2047 (1969)
    Günther, H., Klose, H., Cremer, D., Chem. Ber. 104, 3884 (1971)
    Carreira, L. A., Towns, T. G., Malloy Jr., T. B.: J. Am. Chem. Soc. 100, 385 (1978)
35. Heilbronner, E., Gleiter, R., Hopf, H., Hornung, V., de Meijere, A.: Helv. Chim. Acta 54, 783 (1971)
36. Gleiter, R., Heilbronner, E., de Meijere, A.: Helv. Chim. Acta 54, 1029 (1971)
37. Bischof, P., Gleiter, R., Dürr, H., Ruge, B., Herbst, P.: Chem. Ber. 109, 1412 (1976)
38. Askani, R., Gleiter, R., Heilbronner, E., Hornung, V., Musso, H.: Tetrahedron Lett. 4461 (1971)
39. Asmus, P., Klessinger, M., Meyer, L. W., de Meijere, A.: Tetrahedron Lett. 381 (1975)
40. Bruckman, P., Klessinger, M.: Chem. Ber. 107, 1108 (1974)
41. Brogli, F., Heilbronner, E., Wirz, J., Kloster-Jensen, E., Bergman, R. G., Vollhardt, K. P. C., Aske III, A. J.: Helv. Chim. Acta 58, 2620 (1975)
42. Prins, I., Verhoeven, J. W., de Boer, T. J., Worrell, C.: Tetrahedron 33, 127 (1977)
43. Shudo, K., Kobayashi, T., Utsunomiya, C.: Tetrahedron 33, 1721 (1977)
44. Closs, G. L., Klinger, H. B.: J. Am. Chem. Soc. 86, 908 (1964),
    Fischer, P., Kurtz, W., Effenberger, F.: Chem. Ber. 106, 549 (1973)
45. Gleiter, R., Kobayashi, M., Babiak, K. A., Ford, T. M., Goff, D. L., Morgan, T. K., Munay, Jr., R. K.: unpublished results
46. Martin, H. D., Heller, C., Werp, J.: Chem. Ber. 107, 1393 (1974)
47. Martin, H. D., Heller, C., Haselbach, E., Lanyjova, Z.: Helv. Chim. Acta 57, 465 (1974)
48. De Meijere, A.: Chem. Ber. 107, 1684 (1974)
49. Asmus, P., Klessinger, P.: Liebigs Ann. Chem. 1975, 2169
50. Bodor, N., Dewar, M. J. S., Worley, S. D.: J. Am. Chem. Soc. 92, 19 (1970)
51. Asmus, P., Klessinger, M.: Angew. Chem. 88, 343 (1977); Angew. Chem. Int. Ed. Engl. 15, 310 (1977)
52. Martin, H. D., Heller, C., Haider, R., Hoffmann, R. W., Becherer, J., Kurz, H. R.: Chem. Ber. 110, 3010 (1977)
53. Bischof, P., Böhm, M., Gleiter, R., Snow, R. A., Doecke, C. W., Paquette, L. A.: J. Org. Chem. 43, 2387 (1978)
54. Spanget-Larsen, J., Gleiter, R., Detty, M. R., Paquette, L. A.: J. Am. Chem. Soc. 100, 3005 (1978)
55. Bastiansen, O., de Meijere, A.: Angew. Chem. 78, 142 (1966); Angew. Chem. Int. Ed. Engl. 5, 125 (1966); Acta Chem. Scand. 20, 516 (1966)
    Hagen, K., Hagen, G., Traetteberg, M.: Acta Chem. Scand. 26, 36449 (1972);
    de Meijere, A., Lüttke, W., Heinrich, F.: Liebigs Ann. Chem. 1974, 306
56. Braun, H., Lüttke, W.: J. Mol. Struct. 28, 391 (1975)
57. Braun, S., Traetteberg, M.: J. Molecular Struct. 39, 101 (1977)
58. Heilbronner, E., Gleiter, R., Hoshi, T., de Meijere, A.: Helv. Chim. Acta 56, 1594 (1973)
59. Spanget-Larsen, J., Gleiter, R., de Meijere, A., Binger, P.: Tetrahedron 35, 1385 (1979)
60. Gleiter, R., Haider, R., Conia, H. M., Barnier, J. P., de Meijere, A., Weber, W.: J. C. S. Chem. Comm. 1979, 130
61. Gleiter, R., Martin, H. D., Szeimies, G.: to be published
62. Bischof, P., Heilbronner, E., Prinzbach, H., Martin, H. D.: Helv. Chim. Acta 54, 1072 (1971)
63. Gleiter, R., Hopf, H.: unpublished results
64. Dewar, M. J. S., Fouken, G. J., Jones, T. B., Minter, D. E.: J. C. C. Perkin II, 1976, 764

65. Pasto, D. J., Fehlner, T. P., Schwartz, M. E., Baney, H. F.: J. Am. Chem. Soc. *98*, 530 (1976)
66. Gleiter, R., de Meijere, A.: unpublished results
67. Bischof, P., Heilbronner, E.: Helv. Chim. Acta *53*, 1677 (1970)
    Clary, D. C., Lewis, A. A., Morland, D., Murrell, J. N., Heilbronner, E.: J. Chem. Soc. Faraday Trans. II *70*, 1889 (1974)
68. Wiberg, K. B., Ellison, G. B., Wendoloski, J. J., Brundle, C. R., Kuebler, N. A.: J. Am. Chem. Soc. *98*, 7179 (1976)
69. Bally, Th., Haselbach, E.: Helv. Chim. Acta *58*, 321 (1973); unpublished results, as references in 70
70. Bieri, G., Burger, F., Heilbronner, E., Maier, J. P.: Helv. Chim. Acta *60*, 2213 (1977)
71. Brogli, F., Giovannini, E., Heilbronner, E., Schuster, R.: Chem. Ber. *106*, 961 (1973)
72. Maksić, Z., Klasinc, Z., Randić, M.: Theoret. Chim. Acta *4*, 273 (1966)
    Yonezawa, T., Shimizu, K., Kato, H.: Bull. Chem. Soc. Jap. *40*, 456 (1967)
73. Salem, L., Wright, J. S.: J. Am. Chem. Soc. *91*, 5947 (1969)
    Salem, L.: Chem. Brit. *5*, 449 (1969) Wright, J. S., Salem, L.: Chem. Commun. 1370 (1969)
74. Hoffmann, R., Davidson, R. B.: J. Am. Chem. Soc. *93*, 5699 (1971)
75. Bischof, P., Haselbach, E., Heilbronner, E.: Angew. Chem. *82*, 952 (1970); Angew. Chem. Int. Ed. Engl. *9*, 953 (1970)
76. Mollère, P. D., Houk, K. N.: J. Am. Chem. Soc. *99*, 3226 (1977)
77. Gleiter, R., Kobayashi, T.: Helv. Chim. Acta *54*, 1081 (1971)
    Jorgensen, W. L.: J. Am. Chem. Soc. *97*, 3082 (1975)
    Jorgensen, W. L., Borden, W. T.: J. Am. Chem. Soc. *95*, 6649 (1973)
    Gleiter, R., Bischof, P., Volz, W. E., Paquette, L. A.: J. Am. Chem. Soc. *99*, 8 (1977)
    Bischof, P., Gleiter, R., Haider, R.: Angew. Chem. *89*, 122 (1977); Angew. Chem. Int. Ed. Engl. *16*, 110 (1977); J. Am. Chem. Soc. *100*, 1036 (1978)
    Borden, W. T., Gold, A., Jorgensen, W. L.: J. Org. Chem. *43*, 491 (1978)
78. Hehre, W. J.: J. Am. Chem. Soc. *94*, 6592 (1972)
79. Bischof, P., Gleiter, R., de Meijere, A., Meyer, L. U.: Helv. Chim. Acta *57*, 1519 (1974)
80. Bischof, P., Gleiter, R., Kukla, M. J., Paquette, L. A.: J. Electron Spectrosc. Relat. Phenom. *4*, 177 (1974)
81. Bischof, P., Gleiter, R., Grimme, W.: unpublished results
82. Gleiter, R., Bischof, P., Volz, W. E., Paquette, L. A.: J. Am. Chem. Soc. *99*, 8 (1977)
83. Bruckmann, P., Klessinger, M.: Chem. Ber. *111*, 944 (1978)
84. Bischof, P., Gleiter, R., Hopf, H., Musso, H.: unpublished results
85. Borden, W. T., Young, S. D., Frost, D. C., Westwood, N. P. C., Jorgensen, W. L.: J. Org. Chem. *44*, 737 (1979)
86. Gleiter, R.: unpublished results
87. Hoffmann, R.: Acc. Chem. Res. *4*, 1 (1971)
    Gleiter, R.: Angew. Chem. *86*, 770 (1974); Angew. Chem. Int. Ed. Engl. *13*, 696 (1974)
88. Bieri, G., Heilbronner, E., Kobayashi, T., Schmelzer, A., Goldstein, M. J., Leight, R. S., Lipton, M. S.: Helv. Chim. Acta *59*, 2657 (1976)
    Bieri, G., Heilbronner, E.: Tetrahedron Lett. 581 (1975)
    Delwiche, J. P., Praet, M. T.: J. Electron Spectros. *7*, 317 (1975)
89. Schmidt, W., Wilkins, B. T.: Tetrahedron *28*, 5649 (1972); Spanget-Larsen, J., Gleiter, R., Klein, G., Doecke, C. W., Paquette, L. A.: Chem. Ber. (in press)
90. Gleiter, R., Heilbronner, E., Hekman, M., Martin, H. D.: Chem. Ber. *106*, 28 (1973)
91. Bodor, N., Chen, B. H., Worley, S. D.: J. Electron Spectros. *4*, 65 (1974)
92. Wipff, G.: Thèse, Université Louis Pasteur de Strasbourg (1971)
93. Jaffe, H. H.: (1974) as quoted in ref. 67; Lehn, J. M., Wipff, G.: (1974) as quoted in ref. 67 and 70
94. Heilbronner, E., Schmelzer, A.: Helv. Chim. Acta *58*, 936 (1975)
95. Spanget-Larsen, J., Gleiter, R., Paquette, L. A., Carmody, M. J., Degenhardt, C. R.: Theore. Chim. Acta (Berl.) *50*, 145 (1978)
96. Borden, W. T., Young, S. D., Frost, D. C., Westwood, N. P. C., Jorgensen, W. L.: to be published

97. Pomerantz, M., Abrahamson, W.: J. Am. Chem. Soc. *88*, 3970 (1966)
98. Newton, M. D., Schulman, J. M.: J. Am. Chem. Soc. *94*, 767 (1972)
99. Bischof, P., Gleiter, R., Müller, E.: Tetrahedron *32*, 2769 (1976)
100. Newton, M. D., Schulman, J. M., Manus, M. M.: J. Am. Chem. Soc. *96*, 17 (1974)
101. Gleiter, R., Bischof, P., Müller, E.: unpublished results
102. Bischof, P., Gleiter, R., Müller, E.: Tetrahedron *32*, 2769 (1976)
    Harman, P. J., Kent, J. E., Gan, T. H., Peel, J. B., Willett, G. D.: J. Am. Chem. Soc. *99*, 943 (1977)
103. Gleiter, R., Haider, R., Murata, I., Pagni, R. M.: J. Chem. Research: *1979*, 72
104. Bischof, P., Gleiter, R., Taylor, R. T., Browne, A. R., Paquette, L. A.: J. Org. Chem. *43*, 2391 (1978)
105. Roth, R. J., Katz, T. J.: J. Am. Chem. Soc. *94*, 4770 (1972)
106. Katz, T. J., Nicolaou, K. C.: J. Am. Chem. Soc. *96*, 1948 (1974)
107. Hogeveen, H., Huurdeman, W. F. J., Kok, D. M.: J. Am. Chem. Soc. *100*, 871 (1978)
108. Bischof, P., Gleiter, R., Martin, H. D., Kunze, M.: unpublished results
109. Fleischer, E. B.: J. Am. Chem. Soc. *86*, 3889 (1964)
110. Veillard, A., Del Re, G.: Theore. Chim. Acta *2*, 55 (1964)
    Preuss, H., Janoschek, R.: J. Mol. Struct. *3*, 423 (1969)
    Bodor, N., Dewar, M. J. S., Worley, S. D.: J. Am. Chem. Soc. *92*, 19 (1970)
    Martensson, O.: Acta Chem. Scand. *24*, 1495 (1970)
    Iwamura, H., Morio, K., Oki, M., Kunii, T.: Tetrahedron Lett. *1970*, 4575
    Iwamura, H., Morio, K., Kunii, T. L.: J. Chem. Soc. (D) *1971*, 1408
111. Schulman, J. M., Fischer, C. R., Solomon, P., Venanzi, T. J.: J. Am. Chem. Soc. *100*, 2949 (1978)
112. Bischof, P., Eaton, P. E., Gleiter, R., Heilbronner, E., Jones, T. B., Musso, H., Schmelzer, A., Stober, R.: Helv. Chim. Acta *61*, 547 (1978)
113. Miller, I. J.: Tetrahedron *25*, 1349 (1969); Aust. J. Chem. *24*, 2013 (1971)
    Schleyer, V. R. P., Williams, J. E., Blanchard, K. R., J. Am. Chem. Soc. *92*, 2377 (1970)
    Baird, N. C.: Tetrahedron *26*, 2185 (1970)
    Shimizu, K., Morimoto, H., Kato, H., Yonezawa, T.: Nippon Kagaku Zasshi *90*, 865 (1969)
    Kovačević, K., Maksić, Z. B.: J. Org. Chem. *39*, 539 (1974); Kovačević, K., Eckert-Maksić, M., Maksić, Z. B.: Tetrahedron Lett. 101 (1975)
    Allinger, N. L., Tribble, M. T., Millter, M. A., Wertz, D. H.: J. Am. Chem. Soc. *93*, 1637 (1971)
    George, P., Trachtman, M., Bock, C. W., Brett, A. M.: Tetrahedron *32*, 317 (1976)
114. Bischof, P., Gleiter, R., Musso, H.: unpublished results
115. Boyd, R. J., Bünzli, J. C., G., Snyder, J. P.: J. Am. Chem. Soc. *98*, 2398 (1976)
116. Hemmersbach, P., Klessinger, M., Bruckmann, P.: J. Am. Chem. Soc. *100*, 6344 (1978)
117. Brogli, F., Eberbach, W., Haselbach, E., Heilbronner, E., Hornung, V., Lemal, D. M.: Helv. Chim. Acta *56*, 1933 (1973)
118. Martin, H. D., Kagabu, S., Schwesinger, R.: Chem. Ber. *107*, 3130 (1974)
119. Santiago, C., Gandour, R. W., Houk, K. N., Nutakul, W., Cravey, W. E., Thummel, R. P.: J. Am. Chem. Soc. *100*, 3730 (1978)
120. Lauer, G., Müller, C., Schulte, K. W., Schweig, A., Krebs, A.: Angew. Chem. *86*, 597 (1974); Angew. Chem. Int. Ed. *13*, 544 (1974)
    Lauer, G., Müller, C., Schulte, K. W., Schweig, A., Maier, G., Alzérreca, A.: Angew. Chem. *87*, 194 (1975); Angew. Chem. Int. Ed. *14*, 172 (1975)
121. Heilbronner, E., et al.: J. Am. chem. Soc., to be published
122. Böhm, M. C., Gleiter, R.: Tetrahedron Lett. 1179 (1978)
123. Baird, N. C., Dewar, M. J. S.: J. Chem. Phys. *50*, 1262 (1969)
    Buenker, R. J., Peyerimhoff, S. D.: J. Am. Chem. Soc. *91*, 4342 (1969)
    Schulman, J. M., Venanzi, T. J.: J. Am. Chem. Soc. *96*, 4739 (1974)
    Hehre, W. J., Pople, J. A.: J. Am. Chem. Soc. *97*, 6941 (1975)
124. Bruckmann, P., Klessinger, M.: Angew. Chem. *84*, 543 (1972); Angew. Chem. Int. Ed. Engl. *11*, 524 (1972)

125. Gleiter, R., Meinwald, J., Smith, L. R.: unpublished results
126. Brogli, F., Heilbronner, E., Ipaktschi, J.: Helv. Chim. Acta *55*, 2447 (1972)
127. Martin, H. D., Pföhler, P.: Angew. Chem. *90*, 901 (1978); Angew. Chem. Int. Ed. Engl. *17*, 847 (1978)

Received March 2, 1979

Author Index Volumes 26–86

The volume numbers are printed in italics

Dürr, H.: Reactivity of Cycloalkene-carbenes. *40*, 103–142 (1973).

Dürr, H.: Triplet-Intermediates from Diazo-Compounds (Carbenes). *55*, 87–135 (1975).

Dürr, H., and Kober, H.: Triplet States from Azides. *66*, 89–114 (1976).

Dürr, H., and Ruge, B.: Triplet States from Azo Compounds. *66*, 53–87 (1976).

Dugundji, J., and Ugi, I.: An Algebraic Model of Constitutional Chemistry as a Basis for Chemical Computer Programs. *39*, 19–64 (1973).

Dugundji, J., Kopp, R., Marquarding, D., and Ugi, I J.: *75*, 165–180 (1978).

Eglinton, G., Maxwell, J. R., and Pillinger, C. T.: Carbon Chemistry of the Apollo Lunar Samples. *44*, 83–113 (1974).

Eicher, T., and Weber, J. L.: Structure and Reactivity of Cyclopropenones and Triafulvenes. *57*, 1–109 (1975).

Epiotis, N. D., Cherry, W. R., Shaik, S., Yates, R. L., and Bernardi, F.: Structural Theory of Organic Chemistry. *70*, 1–242 (1977).

Erni, F., see Clerc, T.: *39*, 139–167 (1973).

Eujen, R., see Bürger, H.: *50*, 1–41 (1974).

Faber, D. H., see Altona, C.: *45*, 1–38 (1974).

Fietzek, P. P., and Kühn, K.: Automation of the Sequence Analysis by Edman Degradation of Proteins and Peptides. *29*, 1–28 (1972).

Finocchiaro, P., see Mislow, K.: *47*, 1–22 (1974).

Fischer, G.: Spectroscopic Implications of Line Broadening in Large Molecules. *66*, 115–147 (1976).

Fluck, E.: The Chemistry of Phosphine. *35*, 1–64 (1973).

Flygare, W. H., see Sutter, D. H.: *63*, 89–196 (1976).

Fowler, F. W., see Gelernter, H.: *41*, 113–150 (1973).

Freed, K. F.: The Theory of Raditionless Processes in Polyatomic Molecules. *31*, 105–139 (1972).

Frei, H., Bauder, A., and Günthard, H.: The Isometric Group of Nonrigid Molecules. *81*, 1–98 (1979).

Fritz, G.: Organometallic Synthesis of Carbosilanes. *50*, 43–127 (1974).

Fry, A. J.: Stereochmistry of Electrochemical Reductions. *34*, 1–46 (1972).

Gandolfi, M. T., see Balzani, V.: *75*, 1–64 (1978).

Ganter, C.: Dihetero-tricycloadecanes. *67*, 15–106 (1976).

Gasteiger, J., and Jochum, C.: EROS – A Computer Program for Generating Sequences of Reactions. *74*, 93–126 (1978).

Gasteiger, J., Gillespie, P., Marquarding, D., and Ugi, I.: From van't Hoff to Unified Perspectives in Molecular Structure and Computer-Oriented Representation. *48*, 1–37 (1974).

Geick, R.: IR Fourier Transform Spectroscopy. *58*, 73–186 (1975).

Geist, W., and Ripota, P.: Computer-Assisted Instruction in Chemistry. *39*, 169–195 (1973).

Gelernter, H., Sridharan, N. S., Hart, A. J., Yen, S. C., Fowler, F. W., and Shue, H.-J.: The Discovery of Organic Synthetic Routes by Computer. *41*, 113–150 (1973).

Gerischer, H., and Willig, F.: Reaction of Excited Dye Molecules at Electrodes. *61*, 31–84 (1976).

Gillespie, P., see Gasteiger, J.: *48*, 1–37 (1974).

Gleiter, R., and Gygax, R.: No-Bond-Resonance Compounds, Structure, Bonding and Properties. *63*, 49–88 (1976).

Gleiter, R. and Spanget-Larsen, J.: Some Aspects of the Photoelectron Spectroscopy of Organic Sulfur Compounds. *86*, X–X (1979).

Gleiter, R.: Photoelectron Spectra and Bonding in Small Ring Hydrocarbons. *86*, X–X (1979).

Günthard, H., see Frei H.: *81*, 1–98 (1979).

Guibé, L.: Nitrogen Quadrupole Resonance Spectroscopy. *30*, 77–102 (1972).

Gundermann, K.-D.: Recent Advances in Research on the Chemiluminescence of Organic Compounds. *46*, 61–139 (1974).

Lecture Notes in Chemistry

Edited by G. Berthier, M. J. S. Dewar, H. Fischer,
K. Fukui, H. Hartmann, H. H. Jaffé, J. Jortner,
W. Kutzelnigg, K. Ruedenberg, E. Scrocco, W. Zeil

A Selection

Springer-Verlag
Berlin Heidelberg New York

Inorganic Chemistry Concepts

Editors: M. Becke, C. K. Jørgensen,
M. F. Lappert, S. J. Lippard, J. L. Margrave,
K. Niedenzu, R. W. Parry, H. Yamatera

Volume 1
R. Reisfeld, C. K. Jørgensen

Lasers and Excited States of Rare Earths

1977. 9 figures, 26 tables. VIII, 226 pages
ISBN 3-540-08324-3

Contents:
Analogies and Differences Between Mon-
atomic Entities and Condensed Matter. –
Rare-Earth Lasers. – Chemical Bonding
and Lanthanide Spectra. – Energy Transfer. –
Applications and Suggestions.

Volume 2
R. L. Carlin, A. J. van Duyneveldt

Magnetic Properties of Transition Metal Compounds

1977. 149 figures, 7 tables. XV, 264 pages
ISBN 3-540-08584-X

Contents:
Paramagnetism: The Curie Law. – Thermo-
dynamics and Relaxation. – Paramagnetism:
Zero-Field Splittings. – Dimers and Clus-
ters. – Long-Range Order. – Short-Range
Order. – Special Topics: Spin-Flop, Meta-
magnetism, Ferrimagnetism and Canting. –
Selected Examples.

Volume 3
P. Gütlich, R. Link, A. Trautwein

Mössbauer Spectroscopy and Transition Metal Chemistry

1978. 160 figures, 19 tables, 1 folding plate.
X, 280 pages
ISBN 3-540-08671-4

Contents:
Basic Physical Concepts. – Hyperfine Inter-
actions. – Experimental. – Mathematical
Evaluation of Mössbauer Spectra. – Inter-
pretation of Mössbauer Parameters of Iron
Compounds. – Mössbauer-Active Transition
Metals Other than Iron. – Some Special
Applications.

Volume 4
Y. Saito

Inorganic Molecular Dissymmetry

1979. 107 figures, 28 tables. IX, 167 pages
ISBN 3-540-09176-9

Contents:
Introduction. – X-Ray Diffraction. – Conformational Analysis. – Structure and Isomerism
of Optically Active Complexes. – Electron-
Density Distribution in Transition Metal
Complexes. – Circular Dichroism. – References. – Subject Index.

Springer-Verlag
Berlin
Heidelberg
New York